PULMONARIAS

AND THE BORAGE FAMILY

MASHA BENNETT

TIMBER PRESS
Portland, Oregon

To my daughter Anya.

Acknowledgements

This book would not have been possible without the generous help and support of many people, especially Joe Sharman of Monksilver Nursery. Many others have helped in numerous ways – with invaluable information and advice, moral and practical support, childcare, lodgings, and companionship on plant-hunting trips. Among these wonderful people are Crinan Alexander (Royal Botanic Gardens Edinburgh), Ann Bennett, Daniel Bennett, Chris Brickell, David Bunting, Valerie Finnis (The Merlin Trust), Kirill and Yana Forer (Israel), Dan Heims (Terra Nova Nurseries, USA), Jennifer Hewitt, Harry Jans (Germany), Andrew Jordan, Bodil Larsen (Canada), Alan Leslie (RHS Garden Wisley), Bjørn Malkmus (Rareplants, Germany), George McKay (Cruickshank Botanic Garden), Ian and Margaret Nimmo-Smith, Sylvia Norton, Chris Oldershaw, Erich Pasche (Germany), Howard Rice, Gordon Smith (University of Aberdeen), Anne Staines, Rosie Steel, Doreen Townley, Chris Wilcock (University of Aberdeen) and many others – thank you for making this project come to life!

Masha Bennett
August 2002

Published in North America in 2003
by Timber Press, Inc.

Illustrations in Introduction
and Chapter 1 by Joe Sharman.

ISBN 0-88192-589-6

A CIP record for this book is available
from the Library of Congress.

Printed in China

Timber Press, Inc.
The Haseltine Building
133 S.W. Second Avenue, Suite 450
Portland, Oregon 97204, U.S.A.

www.timberpress.com

CONTENTS

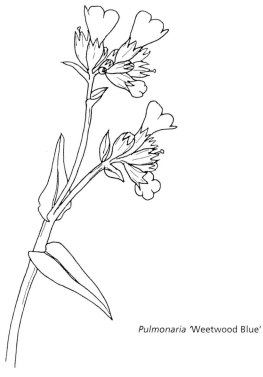

Pulmonaria 'Weetwood Blue'

FOREWORD

An enthusiasm for a single genus of plants is a common enough occurrence among gardeners and botanists. Dedication to a complete family of plants, however, as exemplified by the author in this fascinating horticultural and botanical account of the Boraginaceae, is much rarer.

Coverage of a family containing some 150 genera and 2500 species, as well as a considerable number of cultivars, is a daunting task for anyone to undertake, particularly as the Borage family is not generally considered to be a source of that many desirable garden plants.

Ask even a well-versed gardener about which plants of horticultural merit he or she grows and which belong to the Borage family, and they will almost certainly say forget-me-not (*Myosotis*), common borage (*Borago*) and lungworts (*Pulmonaria*), but may find it difficult to suggest very many other genera known to them. Rock gardeners, perhaps, would also come up with *Lithodora*, *Eritrichium* and the recently popularized *Cerinthe*.

As readers will find by delving into the different sections of Masha Bennett's treatise, there is a host of different borage relatives that are attractive garden plants. These include many annuals and perennials that are deserving of wider cultivation, as well as a number of woody plants, such as *Cordia* and *Ehretia*, which are suitable mainly for Mediterranean and subtropical climes.

Others will interest growers of herbs. Several genera are rich in herbal lore and have been valued for their actual or reputed medicinal qualities as well as for use as dye plants, for their oils, as bee-fodder and even, in the case of the genus *Cordia*, for valuable timber. Plus, of course, for the indispensable inclusion of the brilliant blue flowers of common borage in claret cup and Pimm's No. 1!

The author's enthusiasm for the Borage family is infectious. This became very apparent to me on a plant-hunting visit to Kyrgyzstan in 1999, where she persuaded many of the participants to search willingly for as many borage relatives as possible for her to examine. Even some of the minute Boraginaceous 'finds', whose beauty was apparent only under a lens, were greeted with great delight by Masha.

Such was, and is, her commitment to the Borage family. Had any comfrey or knitbone, *Symphytum officinale*, been found during our visit, I have no doubt that a broken arm I sustained towards the end of the trip would have been strapped by her with a plaster made from its bruised foliage and roots so that she could see if the plant deserved its reputation for healing broken bones!

Reginald Farrer says of *Eritrichium nanum* in his book *The English Rock Garden* that 'In a race of dowdy and impermanent weeds lies lurking, like the precious jewel in the head of the toad, the Crowned King of the Alps, the Herald of Heaven, Woolly-hair the Dwarf.' Be assured that when you read this book you will find many other 'jewels' that can be enjoyed in your garden, most of which will be much less difficult to please than the 'King of the Alps'.

Chris Brickell, CBE VMH
May 2002

INTRODUCTION TO THE FAMILY BORAGINACEAE

Lungwort, forget-me-nots, comfrey, borage, alkanet – these common names are familiar to most gardeners, and you may well wonder: what else is there to the Borage family?

Boraginaceae is, in fact, a vast family containing about 150 genera and 2500 species, just some of which are included in this book, and the variety of shapes and habits, colours and features that this group of plants offers to the gardener is truly tremendous, enough to satisfy the needs of the most discerning plantsman.

Some gardening books refer to the members of the borage family disparagingly, describing them as 'coarse'. This label may be appropriate for the foliage of some species, but surely the charm of the little forget-me-not flower, the elegance of Virginia bluebell, the spicy fragrance of heliotrope, or bumblebee acrobatics on the borage flower, more than make up for the less-than-neat appearance of some of the family members.

The intention of this book is to dispel the notion that the borage family are rather unexciting plants and I hope that, after reading it, many gardeners will consider even the most lowly member of Boraginaceae with a sense of curiosity and wonder.

This Introduction gives an overview of the family Boraginaceae, its classification, biology and cultivation. Chapter 1 is devoted solely to the genus *Pulmonaria*, the humbly named 'lungwort', whose popularity has deservedly soared in recent years. Chapter 2 details all other genera commonly encountered in cultivation, and the final chapter deals with those genera that are rarely grown but are of interest to gardeners.

Burnet moth (*Zygaena* species) on *Echium vulgare*. (Masha Bennett)

Geography and Ecology of Boraginaceae

DISTRIBUTION

Boraginaceae are found mainly in temperate regions of the northern hemisphere, but also extend into South America, South Africa and Australasia, and a number of genera are found mostly or exclusively in the tropics. The flora of the Mediterranean is especially rich in representatives of the borage family.

HABITATS

Members of the borage family are to be found in a wide variety of habitats in the wild, which is also reflected in the versatility of this family in cultivation.

Temperate woodland accommodates shade-loving species including *Pulmonaria*, *Omphalodes verna*, *O. cappadocica* and *O. nitida*, *Brunnera macrophylla*, *Trachystemon orientale* and *Symphytum*. These are mostly deciduous woods, but sometimes forests of pine, fir or spruce.

Tropical and sub-tropical forests are home to *Cordia* and *Ehretia* species.

Numerous Boraginaceae are found in the **mountains**, and those that grow in high alpine areas covered in snow in winter include *Eritrichium*, *Chionocharis hookeri* and some *Mertensia* species. Those that are found growing in rock crevices include *Omphalodes luciliae* and many *Onosma* species.

Arable land and **waste places** are home to a number of annual *Anchusa* and *Myosotis* species and some *Nonea* and *Cynoglossum* species.

Few plants are adapted to the harsh, salt-laden conditions of **maritime habitats**, but *Mertensia maritima* and *M. simplicissima* are the prime examples of coastal plants, growing in sand or shingle on northern sea-shores in Asia and Europe. *Myosotidium hortensia* is another member of the family, found in similar habitats in New Zealand.

Worldwide distribution of the family Boraginaceae.

Only very few Boraginaceae can be found in **wetlands**, among them *Myosotis scorpioides* and a couple of other forget-me-not species. Stream-sides are popular with a number of mertensias, including *M. ciliata*, *M. franciscana*, *M. sibirica*, and some comfreys, *Symphytum*.

LONGEVITY

Annuals, biennials, perennials – these are terms familiar to gardeners, who may, however, use them somewhat loosely. In Boraginaceae we can see examples of all these.

- Annuals complete their life cycle within one season, commonly germinating in spring and dying in late summer or autumn after flowering and setting seed; some germinate in autumn and overwinter. Annuals include *Borago officinalis*, *Omphalodes linifolia* and *Cerinthe major*. Some species frequently grown as annuals are normally biennials in the wild, for example *Anchusa capensis*.
- Biennials complete their life cycle within two seasons, dying after flowering and fruiting once. Among these are many *Echium*, *Cynoglossum* and some *Anchusa* species. Some plants grown as biennials in gardens, such as bedding strains of *Myosotis sylvatica*, are in reality short-lived perennials.
- Perennials can be monocarpic (dying after flowering only once), or polycarpic (flowering more than once). Strictly speaking, biennials are monocarpic perennials, as they do not necessarily flower in their second year. This is true of many Canary Islands *Echium* species.
- Perennials can be further subdivided into herbaceous and woody. If we exclude trees and shrubs belonging to Ehretioideae and Cordioideae, there are few woody members remaining in the borage family – these will include sub-shrubby *Lithodora*, *Moltkia*, *Lobostemon* and the shrubby Canary Islands *Echium* species.

RAUNKIAER LIFE FORMS

This classification system, developed by the Danish botanist Christen C. Raunkiaer in the 1920s, is still considered the most valuable among other types of classification of plant growth forms. Raunkiaer life forms are used to define the position of the resting buds or persistent stem apices in relation to soil level, which is a convenient method for indicating how a plant passes the unfavourable season.

I think it is useful for gardeners to be aware of this classification, if only in terms of considering the type of winter protection that a plant may need if grown in colder areas.

ABOVE *Onosma sericea* on the wall of Old Jerusalem. (Masha Bennett)

ABOVE *Borago officinalis* and *Cerinthe major* growing in an olive grove in southern Italy. (Masha Bennett)

Echium vulgare growing with *Pilosella aurantiacum* and *Codonopsis clematidea* in the Tian Shan Mountains. (Masha Bennett)

- Phanerophytes (literally, 'visible plants') – woody plants with buds more than 25cm (10in) above soil level; e.g. *Ehretia*, *Cordia*, shrubby *Echium*, *Lobostemon*, *Heliotropium arborescens*.
- Chamaephytes ('small plants') – woody or herbaceous plants with buds above the soil surface but below 25cm (10in); e.g. *Lithodora diffusa*, *Buglossoides purpurocaerulea*.
- Hemicryptophytes ('half-hidden plants') – herbaceous or very rarely woody plants with buds at soil level; e.g. *Pulmonaria*, *Cynoglossum*, *Anchusa azurea*.
- Geophytes ('earth plants') – herbaceous plants with buds below the soil surface; these include all bulbous plants and, among Boraginaceae, those plants which have deeper-lying rhizomes, e.g. *Mertensia pulmonarioides*. (Sometimes geophytes, helophytes and hydrophytes are grouped together as cryptophytes, 'hidden plants'.)
- Helophytes ('marsh plants') – plants growing in permanently wet, marshy places; there are very few Boraginaceae belonging to this life form, but a well-known example is the water forget-me-not, *Myosotis scorpioides*, and a few closely related species.
- Hydrophytes ('water plants') – plants that grow in water, with the buds passing the unfavourable season under water. Apart from the unusual *Rotula aquatica* (related to *Ehretia*) from the tropics of the Old World and Brazil, there are no true hydrophytes among the Boraginaceae family, although *Myosotis scorpioides* occasionally grows in quite deep water and, if this is the case, can be classified as such.

- Therophytes ('summer plants') – plants that pass the unfavourable season as seeds (annuals), e.g. *Omphalodes linifolia*, *Borago officinalis*, *Cerinthe major*, but this refers only to plants germinating in spring – if an annual germinates in autumn and overwinters as a rosette, it belongs to hemicryptophytes.

POLLINATION

FLOWER VISITORS AND POLLINATORS

Many Boraginaceae are absolute favourites with bees. It has been noted that bees show preference to blue flowers, which abound among this family. The flowers are visited by honeybees (*Apis mellifera*), bumblebees (*Bombus* species) and various solitary bees of genera such as *Osmia*, *Andrena* and *Halictus*.

Butterflies and moths (Lepidoptera) are attracted to the common heliotrope (*Heliotropium arborescens*), and the white flowers of bindweed heliotrope (*Heliotropium convolvulaceaum*) are most likely to be pollinated by night-flying moths.

Hummingbirds in the USA are known to visit *Mertensia, Macromeria* and *Pulmonaria* species (this is curious as the latter genus is not native in America). Some Canary Island *Echium* species have characteristics of bird-pollinated flowers, such as dilute nectar and often red-coloured blooms, although insects have been observed to pollinate the flowers.

TRICKS

Long tubular flowers of many Boraginaceae, such as comfrey (*Symphytum*) or honeywort (*Cerinthe*), and languid-ladies (*Mertensia*), allow access to nectar to long-tongued insect species only, such as bumblebees. However, the bumblebee will often bite a hole in the corolla and steal the nectar without pollinating the flower – this is called 'nectar robbing'. Short-tongued bees and even flies, whose mouthparts are not powerful enough to pierce a hole in the corolla, will often use the bumblebee-made holes 'second-hand'.

Some plants also have a means of tricking the insects – certain populations of *Cerinthe major* are known to have only a proportion of their flowers producing nectar, others remaining barren but nevertheless visited by insects who anticipate finding a reward in all flowers.

FLOWER COLOUR CHANGE

In many Boraginaceae, including certain species of lungwort (*Pulmonaria*), comfrey (*Symphytum*), forget-me-nots (*Myosotis*), languid-ladies (*Mertensia*) and viper's-bugloss (*Echium*), flowers go through a series of colour changes, usually from pink in bud to a variety of shades of blue. This is due to the change in acidity of the cell sap. When the flower is still in bud or just beginning to open, the cell sap is acidic, with a low pH, and the corresponding colour of corolla is pink or red. The acidity decreases (with the pH rising) in older flowers, and the colour changes to blue.

This phenomenon is a signal to pollinators, indicating when the flower is brimming with nectar and ready for visiting, and informing them when it has been fertilized, as the nectar levels drop dramatically. This is advantageous both for the insect, which can tell at a distance where they can expect a generous reward, and for the plant as the potential pollinators do not waste time in visiting old flowers but are attracted to young, fertile blooms. As the opening of the flowers is staggered on the same plant in most Boraginaceae, so is the colour change, so you often get different shades of pink and blue on the same plant. This also encourages insects to visit a number of different plants and cross-pollinate them.

FRUIT DISPERSAL

Once the seed is ripe, the fruits must be dispersed from the parent plant into new niches that the seedlings may be able to occupy. To avoid competition with parent plants, and colonize new areas, many plants have developed sophisticated mechanisms for dispersing far away from their parents. Few Boraginaceae possess any specialized mechanisms for fruit dispersal.

Anemochory, or dispersal by wind, in its simplest form involves the seed being shaken out of the seedhead onto the ground beneath the parent plant. In some cases, the seeds possess wings or other appendages that allow them to travel a long distance aided by wind. In Boraginaceae, a few genera have nutlets equipped with a small wing that helps their dispersal, such as in *Omphalodes*. In some *Cordia* species, for example *C. alliodora* and *C. gerascanthus*, the fruit is a dry nut, which is surrounded by persisting corolla lobes that act as a parachute when the fruit is ripe and falls from the tree.

Species with nutlets that are covered in either hooks or bristles, such as *Cynoglossum* and *Solenanthus*, show adaptations to **zoochory** (dispersal by animals), where the well-armed nutlet attaches itself to the fur of any mammals brushing past or, indeed, the clothing of humans.

Fleshy fruits of trees and shrubs belonging to *Cordia*, *Ehretia* and *Carmona* are eaten by birds and can be dispersed for large distances, even some thousands of miles, in their digestive tract; *Carmona retusa*, native in East Asia, is believed to have been introduced on Hawaii by birds.

Semi-desert habitat of *Arnebia guttata*. (Masha Bennett)

Dispersal by water, or **hydrochory**, is quite rare in the plant kingdom, as most seeds rapidly deteriorate when immersed in water. However, there is evidence that at least in *Mertensia maritima* nutlets are dispersed, at least to some extent, by the sea and remain afloat for several days without being adversely affected, and the same is probably true for *Myosotidium hortensia*. Though not a true case of hydrochory, water from melting snow probably plays part in the dispersal of some alpines at high altitudes, such as *Eritrichium nanum*.

PLANT NAMING AND CLASSIFICATION

BINOMINAL NOMENCLATURE

The system of plant naming as we know it was first introduced by the Swedish botanist Carl Linnaeus (Linné) in the eighteenth century. He proposed what we now call **binominal nomenclature**, whereby the name of each species consists of two words (genus and species), as opposed to rather lengthy descriptions in Latin that had been used to denote a particular plant – for example 'The yellow Iris of watersides', referring to *Iris pseudacorus*.

Plant structure in Boraginaceae: *Pulmonaria*

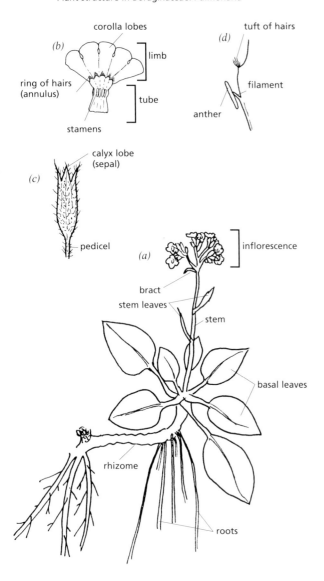

(a) whole plant; (b) dissected corolla; (c) calyx; (d) stamen attached inside corolla tube.

TAXONOMIC RANKS

All flowering plant species are grouped under a number of ranks, including **Class**, **Subclass**, **Order**, **Family**, **Genus** and **Species** (sometimes intermediate ranks, such as Subfamily and Tribe, are also distinguished). An example of classification for three members of Boraginaceae is shown below. For horticultural purposes, it is usually sufficient to know the generic (genus) name and the species name, although it is very useful to be aware of the family to which the plant belongs.

There may also be ranks below the level of species, such as **subspecies** (abbreviated as 'subsp.' or 'ssp.'), **varieta** (abbreviated as 'var.' or 'v.') and **forma** (abbreviated as 'f.'). The subspecies may be isolated from the typical population, and will have some features that distinguish it from the species. Naturally occurring varieties and forms are normally only slightly different from the typical species, perhaps in flower colour or degree or hairiness.

The names of the genus and species (plus the ranks below them) are written in italics. The genus name always begins with a capital letter, and all the ranks below it with a lower-case initial, even though in the past the species names commemorating people were written with an upper-case initial.

If the plant is a hybrid, this is usually indicated by a cross 'x' between the genus and species names, thus *Symphytum* x *uplandicum*.

AUTHORITY

After each name of the species, or subspecies, in this book follows a name of authority – that is, the person who has originally described the plant under this name. If the plant has since been moved to a different genus, the original authority name will be given in brackets, followed by the name of the person who assigned it to the new genus. The authority name is frequently abbreviated – for example, Linnaeus is commonly shortened to L., Rydberg to Rydb., Lehmann to Lehm., and de Candolle to DC.

	Pulmonaria officinalis (lungwort)	*Myosotis alpestris* (alpine forget-me-not)	*Heliotropium arborescens* (heliotrope)
Class	Dicotyledonae	Dicotyledonae	Dicotyledonae
Subclass	Asteridae	Asteridae	Asteridae
Order	Lamiales	Lamiales	Lamiales
Family	Boraginaceae	Boraginaceae	Boraginaceae
Subfamily	Boraginoideae	Boraginoideae	Heliotropioideae
Tribe	Boragineae	Lithospermeae	–
Genus	*Pulmonaria*	*Myosotis*	*Heliotropium*
Species	*P. officinalis*	*M. alpestris*	*H. arborescens*

ETYMOLOGY

The meaning of the scientific plant name is not only fascinating in itself, but to a curious gardener is often useful in indicating the plant's preferred habitat, its country of origin, colour of flowers or particular botanical characteristic. Here are a few examples of specific name meanings:

HABITAT

alpestris – alpine	(*Myosotis alpestris*)
arvensis – of cultivated fields	(*Anchusa arvensis*)
maritima – of sea-shores	(*Mertensia maritima*)
sylvatica – of woods	(*Myosotis sylvatica*)

COUNTRY OF ORIGIN

cappadocica – from Cappadocia (Turkey)	(*Omphalodes cappadocica*)
capensis – of Cape (South Africa)	(*Anchusa capensis*)
caucasicum – of the Caucasus	(*Symphytum caucasicum*)
russicum – of Russia	(*Echium russicum*)

FLOWER COLOUR

azurea – azure-blue	(*Anchusa azurea*)
lutea – yellow	(*Nonea lutea*)
purpurocaerulea – purple-blue	(*Buglossoides purpurocaeruleum*)
rubra – red	(*Pulmonaria rubra*)

RESEMBLANCE TO ANOTHER PLANT

anchusoides – *Anchusa*-like	(*Lindelofia anchusoides*)
linifolia – with leaves like *Linum* (flax)	(*Omphalodes linifolia*)
moltkioides – *Moltkia*-like	(*Mertensia moltkioides*)
pulmonarioides – *Pulmonaria*-like	(*Mertensia pulmonarioides*)

Specific epithets that are derived from people's names, such as *Omphalodes luciliae* or *Cynoglossum wallichii*, do not give us any clues about the plant, but are interesting historically, as they often commemorate great explorers, botanists and horticulturists.

CULTIVARS

Plant cultivars, or cultivated varieties, are the selections made by man for their desirable characteristics, such as flower size or colour and foliage variegation. Cultivars may be hybrids created specially or by chance, or selections of a particular species, and may originate in gardens or in the wild. The cultivar name follows the species name, for example *Pulmonaria rubra* 'David Ward' or, if it is a hybrid of two or more species, it follows the genus name, e.g. *Symphytum* 'Hidcote Blue'. Cultivar names must always be written in single quotes. Since 1959, it has not been permitted to 'Latinize' cultivar names, as in 'Alba', 'Variegata' and 'Grandiflora', though such names that existed prior to that date can still be used.

The great majority of plant cultivars can only be propagated vegetatively, as they do not come true from seed. With some cultivated varieties, however, seedlings arise that are reasonably close to the true form – this is the case, for example, with *Pulmonaria saccharata* 'Argentea'. In fact, the original clone is probably lost to cultivation, and the plants currently grown in gardens are named *Pulmonaria saccharata* Argentea Group, indicating that there is variation between the plants included under that name.

The detailed rules for naming garden plants are set out in Trehane *et al.*'s *International Code of Nomenclature for Cultivated Plants* (see Bibliography).

SYNONYMS

Owing to complexities of plant classification, many plant genera and species have been described and named more than once, so, apart from the currently accepted name, in this book I list synonyms under which the plants may sometimes be found in other publications. Some cultivars also have more than one name and these are also given here.

COMMON NAMES

Vernacular plant names in a number of languages are included in this book. Though a potential source of confusion when used without reference to scientific names, common names can be a source of amusement and fascination in their own right, and often even the most staunchly professional horticulturists cannot refrain from using them – after all, what is more acceptable to our ears, *Myosotis scorpioides* or Water forget-me-not, *Buglossoides purpurocaerulea* or Purple Gromwell?

Some common names are straightforward translations of their Latin counterparts, but some are imaginative and fascinating: some of my favourites include Languid-ladies (for various *Mertensia* species), Golden-drop (*Onosma* species), Prophet flower (*Arnebia pulchra*), Tower of Jewels (*Echium wildpretii*), King-of-the-Alps (*Eritrichium nanum*) and Chocolate-creams (*Trichodesma physaloides*).

Though common names may be fun and easy to use, it is essential for every modern gardener to be well versed in scientific plant names, or at least to be aware where to look them up. Vernacular names are of limited practical use and can cause much confusion. An example of a common name that is applied to a number of vastly different plants is 'bluebell' – this may refer to *Mertensia* species (Boraginaceae) in the United States, *Hyacinthoides non-scripta* (Liliaceae) in England, *Campanula rotundifolia* (Campanulaceae) in Scotland, and *Sollya heterophylla* (Pittosporaceae) in Australia.

CLASSIFICATION OF FAMILY BORAGINACEAE

There are differing opinions on the relationships between Boraginaceae and other families of flowering plants. Some include it in the order Lamiales, with a number of other families such as Lamiaceae, or Labiatae (including mints, thymes, sages) and Verbenaceae (verbenas). Others consider it more closely related to Solanaceae (potato, tomato, nightshade, tobacco), Hydrophyllaceae (nemophila, phacelia) and Convolvulaceae (bindweeds, morning-glory), grouped under the order Solanales or, alternatively, Polemoniales. Thankfully, for a gardener it is quite sufficient to know the family name.

According to Mabberley's *Plant-Book*, the family Boraginaceae is subdivided into five subfamilies that are distinguished primarily by flower structure. This classification is shown below (with some examples of genera belonging to each group).

Subfamily I   Cordioideae: *Cordia*
    Mainly trees, with a terminal, 4-branched style.

Subfamily II   Ehretioideae: *Ehretia, Carmona*
    Mainly trees, with a terminal, 2-branched style.

Subfamily III   Heliotropioideae: *Heliotropium, Tournefortia*
    Mainly herbs and some shrubs, with a terminal, undivided style.

Subfamily IV   Boraginoideae:
    Mainly herbs, with a gynobasic style (i.e. style arising near the base of a deeply lobed ovary). This subfamily is further subdivided into 5 tribes:
    Tribe 1   Cynoglosseae: *Cynoglossum, Omphalodes, Rindera*
    Flowers regular, base of style more or less conical, tips of nutlets not projecting above points of attachment.
    Tribe 2   Eritricheae: *Cryptantha, Eritrichium, Trigonotis*
    Flowers regular, base of style more or less conical; tips of nutlets projecting above points of attachment.
    Tribe 3   Boragineae: *Anchusa, Borago, Alkanna, Symphytum, Pulmonaria*
    Flowers regular, base of style flat or slightly convex; nutlets with concave attachment surface.

Tribe 4   Lithospermeae: *Lithospermum, Lithodora, Moltkia, Myosotis, Arnebia, Cerinthe*
Base of style flat or slightly convex; nutlets with flat attachment surface.
Tribe 5   Echieae: *Echium*
Flowers zygomorphic (i.e. bilaterally symmetrical).

Subfamily V   Wellstedioideae: *Wellstedia*
    Woody herbs, flowers 4-parted, fruit a capsule that splits longitudinally.

Cordioideae and Ehretioideae are often separated into a distinct family, Ehretiaceae, and the genus *Wellstedia* is frequently allocated to its own family, Wellstediaceae. Representatives of all subfamilies apart from the last one can be found in cultivation.

PLANT STRUCTURE

The borage family includes mostly perennial, biennial or annual herbs, some small shrubs and, rarely, larger shrubs or trees. A typical characteristic of many Boraginaceae is that they are rough to the touch. Plants are frequently generously covered in bristly hairs (the bristles often having a swollen base) or, more rarely, may be softly hairy, or have a combination of both. There are nearly-hairless exceptions, but these are few.

ROOTS

Roots play the role of an anchor and allow plants to take up the essential nutrients and water from the soil. In Boraginaceae, roots may be either fibrous or fleshy, and some species, like *Echium* and *Cynoglossum*, have a long taproot. Oyster plant (*Mertensia maritima*) and the closely related *M. simplicissima* have a unique, cable-like root structure that allows the plants to anchor themselves firmly in the unstable sand and gravel of sea-shores. This happens through internal thickening in the taproot, which subsequently divides into separate strands.

Roots of certain species, especially alkanets (*Alkanna* and *Anchusa*) contain a red dye that can be extracted and used in cosmetics and economically; roots are also a useful source of propagation material for certain Boraginaceae, such as *Anchusa* and some *Pulmonaria* species, which can be propagated by root cuttings.

RHIZOMES

Many Boraginaceae possess rhizomes – modified underground stems – that, similarly to roots, perform a function of anchorage, and also vegetative reproduction. Rhizomes may be thin or thick and fleshy, and vertical or horizontal. In some species, such as tuberous comfrey (*Symphytum tuberosum*), the creeping rhizome has alternating thick tuberous and thin sections.

Division of rhizomes offers a convenient method of propagation of many species in the garden (some *Mertensia* species, *Symphytum* and *Trachystemon*).

STEMS

Stems in the borage family are usually rounded and very often covered in stiff bristles or hairs. They become woody in trees and shrubs, such as *Ehretia* and *Lobostemon*, and are often woody at the base in smaller, herbaceous plants like *Moltkia* and *Onosma*.

Some members of this family, including a number of *Mertensia* species, have the base of the stem modified into the so-called caudex – a persistent swollen organ, probably with a function of storage.

Stems are a source of propagation material for plants with woody stems or stem bases, such as *Lithodora* and *Onosma*, but can also be also used for herbaceous perennials, including some *Pulmonaria* species and forms.

LEAVES

As with most flowering plants, the leaf of Boraginaceae consists of a blade, or **lamina**, and a stalk, or **petiole**. Plants of this family lack stipules – smallish leaflike structures found in leaf axils of many other plants. Their leaves are arranged alternately on the stem, with some very rare exceptions, such as in some *Cordia* species that have opposite leaves, and in *Cryptantha* species that have opposite leaves at the base and alternate ones above. The leaves are simple (not divided) and normally have entire margins, though in some *Pulmonaria* forms the margin can be rather wavy and, in some *Ehretia* and *Cordia* species, may be toothed.

Leaves are frequently covered in coarse hairs, although sometimes they may be hairless or nearly so, as in the oyster plant (*Mertensia maritima*), or be softly downy. The white tubercles, or 'warts', at the base of hairs in some Boraginaceae are, in fact, deposits of calcium carbonate.

INFLORESCENCE

The basic type of inflorescence in Boraginaceae is usually a **cincinnus**, or **scorpioid cyme**, which is often one-sided, coiled like a snail shell or scorpion's tail while in bud, and uncoiling as flowers open so that newly opened flowers face in the same direction. The cymes may be simple or forked. Occasionally the flowers may be solitary, as in *Myosotis pulvinaris*. **Bracts** – modified leaves that subtend a flower or

inflorescence – may be present or absent in the inflorescence.

FLOWERS

Flower shape in Boraginaceae is usually regular, or **actinomorphic**, radially symmetrical (with more than one plane of symmetry), though in *Echium* the flowers are somewhat bilaterally symmetrical – so-called **zygomorphic**.

The **calyx** is five-parted, often deeply divided, sometimes almost to the base, usually hairy and persistent. Calyx lobes are commonly called **sepals**. Calyx protects the corolla and the reproductive organs within it, especially while still in bud.

The **corolla** is five-parted, saucer-, funnel-, salver- or bell-shaped, or constricted at the mouth. It protects the reproductive organs of the plant and consists of a **tube**, and a **limb** that is often more or less expanding, and usually 5-lobed. There are often appendages in the throat – for example, a ring of hairs in *Pulmonaria*, *Nonea*, *Onosma*; scales (fornices) in *Mertensia*, *Omphalodes*, *Symphytum*; or folds in *Solenanthus*, which may partly or completely obscure the throat.

The flowers are normally bisexual (carry both male and female reproductive organs). There are some exceptions – in *Echium* female-only and bisexual flowers can be found on different plants, while in *Cordia* flowers are unisexual, borne on separate plants.

There are five **stamens** – male reproductive organs – each consisting of a **filament** supporting an **anther**, the latter containing the pollen. Stamens are attached to the corolla tube and alternate with the corolla lobes, and may be protruding from the corolla. Filaments often have a nectar-secreting disc at their base.

Some genera, including *Pulmonaria*, *Mertensia* and *Amsinckia*, show a phenomenon called **distyly**, or **heterostyly** – when there are distinct populations present with long and short styles. Due to the specifics of the positioning of pollen on the pollinating insect's body, long-styled flowers can be fertilized only with pollen from short-styled ones, and vice-versa, which encourages cross-pollination between populations.

Female reproductive parts consist of a 4-celled or sometimes 2-celled **ovary** situated at the base of the corolla. The solitary **style** is normally inserted in the depression between the lobes of the ovary, or more rarely is terminal, and has a capitate (headlike) or two-lobed **stigma** (receptive tip on the style on which pollen adheres and germinates).

FRUIT

Boraginaceae fruit is usually composed of four, occasionally two, **nutlets**. A nutlet is a small, dry fruit that contains one seed and does not dehisce (split), but to gardeners it looks very much like a seed, and in this book the terms nutlet/seed will be used interchangeably. Nutlets in Boraginaceae are often ornamented with hooks, bristles, warts or wrinkles, but they can be smooth and shiny, as in *Lithospermum*. They are attached to the **receptacle**, which may be flat or conical. In Ehretioideae and Cordioideae the fruit is usually a fleshy **drupe** with 2 to 4 seeds. Very rarely (in Wellstedioideae), fruit is a capsule splitting longitudinally.

ALKALOIDS

It is important to be aware that many members of this family contain pyrrholizidine alkaloids, which are highly toxic if ingested in sufficient quantities, and may be potentially carcinogenic or cause liver failure. Though it is sometimes argued by herb enthusiasts that you would have to consume a very large quantity of leaves, roots and so on, for any harm to be done, it is likely that some individuals may be susceptible even to small amounts. For this reason, NEVER take any plant parts or preparations internally without prior consultation with a doctor. The genera, members of which are known to contain alkaloids, include *Symphytum*, *Lindelofia*, *Heliotropium*, *Amsinckia*, *Trichodesma*, *Solenanthus*, *Rindera* and others.

CULTIVATION

OBTAINING PLANTS

NURSERIES
Specialist herbaceous and alpine nurseries are the best bet. Mail order is expensive, but this is often the only way to obtain a particular plant that may be stocked by only one nursery in the country, unless you want to travel far. Most nurseries do mail order, but some do not and you may need to travel to obtain that special plant. A list of good nurseries that stock a good variety of pulmonarias and other Boraginaceae is given in Appendix II, but I would urge every gardener in the UK to use *The RHS Plant Finder*, the invaluable reference for finding plants.

When requesting a catalogue, bear in mind that most nurseries require a number of stamps or a payment. A well-produced, descriptive catalogue can be of more value than some of the lavishly illustrated gardening books by popular authors.

GARDEN CENTRES
Large garden centres may stock a selection of *Pulmonaria* species and varieties, cultivars of *Anchusa azurea*, *Omphalodes*, *Brunnera*, *Symphytum* and *Mertensia pulmonarioides*. The frequent problem with garden centres is that naming may be inaccurate, and not all staff are knowledgeable about plants.

SPECIALIST SOCIETIES
Many specialist plant societies hold plant sales, where the plants not available to the general public may be sold. These are excellent places to meet enthusiastic gardeners and to learn a lot more about plants. A list of specialist societies of interest is given in Appendix VI.

FRIENDS AND NEIGHBOURS
An invaluable source for obtaining plants. Do not be afraid to ask for seeds or cuttings; offer to give plants away, swap, and so on – this is how some of the more unusual and interesting plants keep going in cultivation!

WHAT TO LOOK FOR AND WHAT TO AVOID
When obtaining your plants, look for even, balanced growth, a healthy appearance and clear labelling. Avoid a plant if you can see evidence of pests and diseases, the plant is pot-bound with roots visible on the surface of the compost, the compost is too dry and the plant wilting, and finally if there are weeds, moss or liverwort in the compost.

PLANTING OUT
If you cannot transfer to the garden immediately after purchase, keep the plant in a cool, lightly shaded spot, protecting it from frost, if necessary. Most plants offered for sale now are pot-grown and can comfortably wait for weeks or even months in a pot, even though ideally they should be planted straight away, if the weather conditions allow. Planting should be avoided in very cold or wet conditions, certainly if the soil is frozen or waterlogged.

PREPARING THE PLANTING SITE
Unless the soil is compacted, it is unnecessary to dig over the area, especially as it would bring a lot of dormant weed seeds to the surface. The planting hole should be made somewhat larger than the size of rootball of the plant. Most Boraginaceae do not require any special additions to the soil prior to planting, though a little bonemeal is always useful to help good root establishment. If the soil is poor in organic matter, many plants will appreciate some leaf mould or other well-rotted organic material mixed into the planting hole – this is especially true of plants native to woodlands.

PREPARING THE PLANT

If the compost in the pot is dry, it should be stood in a tray of water for several hours and then allowed to drain well, prior to planting. After taking the plant out of its pot, tease out congested roots and remove the top layer of compost, plus any damaged or diseased leaves. Place the plant into the planting hole and add or remove some soil as necessary to make sure that the soil level is correct. One of the most important things is to plant firmly so that the roots are in close contact with the soil. Water thoroughly after planting, trying not to get water on the leaves, and then at regular intervals throughout the first growing season.

LABELLING

If you are seriously interested in plants, you will want to know what they are. White plastic labels are adequate, and can be written on with a permanent marker or lead pencil. Unless you keep separate records of each plant, it may be a good idea to write on the label the source of the plant and the date of planting, as well as its full scientific name.

CARE AND MAINTENANCE

WATERING

*Pulmonaria*s, some comfreys (*Symphytum*), woodland species of navelworts (*Omphalodes*) and languid-ladies (*Mertensia*), plus the regal Chatham Island forget-me-not *Myosotidium hortensia*, appreciate plenty of moisture, and will quickly succumb to drought. In such conditions, it is generally preferable to water in the evening, to avoid excessive water loss through evaporation, and leaf scorch, which may happen if leaves with droplets of water on them are exposed to strong sunlight.

When growing high alpine Boraginaceae, such as *Eritrichium* species, in the alpine house, it may be a good idea to water the sand in which the pots are plunged, to avoid overwatering.

FEEDING

Most Boraginaceae grown in the open will appreciate a moderate amount of general-purpose fertilizer applied in spring. Care should be taken not to overfeed, as this may result in production of extra foliage at the expense of flowers and, in the extreme, may damage the plants. Even more care with feeding should be taken when growing any plants in containers, as it is too easy to overfeed. In pots and tubs, slow-release fertilizers are preferred.

WEED CONTROL

Weeds compete with our garden plants for water,

nutrients, sunlight and physical space, and they must be controlled to give the cultivated plants the best chance possible (and to have an aesthetically pleasing garden). Incidentally, some of the plants that are grown as ornamentals in Britain may be troublesome weeds elsewhere – this is the case with some *Echium* species that are commonly grown in gardens in Europe but are considered noxious weeds in some American states and in Australia.

Hand-weeding among bristly members of the family, especially *Onosma* and *Echium*, is rather unpleasant, so try to prevent weed growth in the first place, by generous mulching. This not only helps to control the weeds, but also helps to minimize water loss from soil and, in some cases, protects the base of the plant from frosts. Good mulching materials include leaf mould (though this may contain weed seeds), composted bark chippings, cocoa shell and, especially in rock gardens, gravel. Avoid the use of peat for mulching as it is very wasteful of natural resources, and does not do the job any better than other materials.

Chemical 'touch-weeders' are useful when trying to eradicate individual perennial weeds amongst other plants, where digging them out may not be possible without damaging the ornamentals.

SUPPORTING

A few of the taller Boraginaceae may occasionally need support, among them some *Anchusa azurea* cultivars. Twiggy sticks or metal linkstakes are convenient for this purpose.

DIVISION

Some herbaceous perennials, like *Anchusa* and *Pulmonaria*, benefit from being divided and replanted every few years or so. When splitting up the clump, discard the old woody bit in the middle and replant the young, strong portions.

CUTTING BACK

Tidy gardeners may prefer to cut back old flowering stems to stop the plants from looking scruffy, or prevent self-seeding, though the seedheads of some plants, such as *Cynoglossum* and *Solenanthus*, look quite attractive in their own right, and will provide interest in a winter garden. If you require seed for propagation, leave at least a few stems to allow it to ripen.

PROTECTION FROM WINTER RAIN

Many Boraginaceae from alpine areas, including most *Eritrichium*, some *Onosma* and *Alkanna* species, and New Zealand *Myosotis* species, require protection from wet winter weather in mild maritime climates, such as in Britain and Ireland, otherwise they will be susceptible to rot. The plants can be covered with a pane of glass propped up on stones or sticks or, alternatively, should be grown in pots in the alpine house, where watering can be carefully monitored.

PROPAGATION

FROM SEED VERSUS VEGETATIVE PROPAGATION

There are advantages and disadvantages to propagating plants either from seed or vegetatively. (I am using the word 'seed' in broader terms here, as what we as gardeners would call 'seeds' in Boraginaceae are in reality fruits – nutlets – that contain the seed within them.)

Seed is relatively cheap to obtain, can be stored for prolonged periods of time and is the only method for propagating annuals. Starting with a small packet of seed, you can raise a large number of plants, and can expect to see natural variation in the seedlings, some of which may be interesting new mutations – this is how most plant cultivars have originated (such as those of *Pulmonaria*). With hardy annuals, propagation from seed is easy, and many (like *Borago officinalis* and *Omphalodes linifolia*) can be sown directly into their flowering positions.

On the other hand, **vegetative propagation** is the only method of multiplying most cultivars, to ensure that they are true to type (the identical clone of the parent). Growing from cuttings or division is often a lot faster than sowing seed, and the young plants are usually larger and begin flowering earlier. Some plants, especially those with creeping rhizomes, are so easy to divide that there is hardly any point in sowing their seed, unless you are trying to raise new varieties. Moreover, a number do not produce any viable seed (such as some *Symphytum* hybrids).

PROPAGATION FROM SEED

OBTAINING SEED

Boraginaceae are poorly represented in most **seed catalogues**, apart from a basic selection of strains for bedding, like those of *Myosotis sylvatica*, *Anchusa capensis*, *Echium vulgare* and *Heliotropium arborescens*. Some of the bigger catalogues, however, offer a number of perennial and alpine species.

The most exciting way of obtaining seed is by joining one or more of the **specialist plant societies** (see the list in Appendix II). Most of these run a seed exchange scheme, whereby all members are encouraged to collect and donate seed of a variety of plants and, once the list of all species sent in has been compiled (and this typically contains some thousands of items), everyone is entitled to order a number of packets, mostly free of charge, with the donors getting a few extra. Many of the unusual plants, that are almost never offered for sale by commercial nurseries, may crop up here. One disadvantage is that it is not uncommon for the seed to be named incorrectly, and you will often end up with a completely different plant from what you expected – though sometimes this may be a bonus.

COLLECTING SEED

Collecting the seed of Boraginaceae is a daunting task, as each flower often produces just one or two, and a maximum of four nutlets – compare that to many dozens or even hundreds in a capsule of, say, a *Primula*. Moreover, if you are collecting from plants like *Onosma* or *Echium*, you are likely to end up with pricked fingers, as these plants and their calyces that conceal the nutlets, are covered in irritating bristles. So, to gather a reasonable number of seeds, both time and patience will be required.

Once collected, the seed should be left to dry, unless it is to be sown immediately, and then stored in a cool, dry place – the bottom shelf of a refrigerator, in an airtight container with some silica gel, is ideal.

SOWING SEED

Timing Although seed of most species is traditionally sown in spring, we may wish to imitate nature, which means sowing straight away, as soon as the seed is ripe, usually in late summer and early autumn. This is probably the best policy for those seeds that are difficult or erratic in their germination. Many of the seeds of plants described in this book can be sown in either early autumn or spring. If you obtain your seed through the exchange scheme of a specialist society, seed distribution usually takes place in midwinter, and you could sow these straight away, or wait until spring, depending on whether they need stratification (see below – Special treatments).

Containers Plastic trays and pots are perfectly suitable for sowing most seeds. If you want to raise only a few plants for your own use, a pot will be quite adequate, and the half-sized tray is spacious enough for dozens of seedlings. If you re-use old pots, make sure they are perfectly clean and, ideally, sterilized.

Special treatments Many plant species have developed mechanisms to stop their seed from germinating during an unfavourable period. There are a number of techniques that gardeners can use to overcome this seed **dormancy**.

Some seeds require a period of cold before they can germinate. This can range from a couple of weeks to several months, the average being about six weeks. We can satisfy these conditions by placing the containers with the sown seeds in a cold place (the bottom shelf of a refrigerator is ideal), with temperatures between 0°C and 4°C (32–39°F). Leave a day or two after sowing, to allow the seeds to take up water and swell before putting them in a cold place. Seed sown in autumn and left in a cold frame will receive the cold treatment naturally. After this period of **cold stratification**, the containers can brought into the warmth of, say, a glasshouse, and germination should occur soon afterwards. A few plants are trickier and need a couple of weeks of warm stratification *before* the cold treatment.

Scarification involves mechanically damaging or removing the hard seedcoat (the nutlet coat for Boraginaceae) prior to sowing, and this was shown to be useful in germinating some *Mertensia* species.

Some seeds have a preference for **light** or **dark** conditions, which help to trigger germination. *Myosotis sylvatica* and some other forget-me-nots emerge better in the dark, so the pots or trays may need covering with a piece of foil, while many plants with very fine seeds need light in order to germinate. To my knowledge, there are no cultivated members of Boraginaceae that absolutely *must* have light or darkness, but some have preferences which are mentioned in the accounts of genera or species.

Growing medium Proprietary seed composts purchased in garden centres are not always adequate, and the frequent problem is insufficient drainage. This can be remedied by mixing in extra sharp grit. Do not use potting composts for seeds, as they contain too much fertilizer, which may inhibit germination.

The traditional seed compost is a mixture of loam, peat and sharp sand with a little bit of fertilizer added, but loam-free composts have now almost taken over. For an environmentally conscious gardener (something that, I believe, we should all strive to be), there is a selection of peat-free composts on offer, varying in their composition and quality.

Perhaps most satisfying of all is to mix your own compost. A mixture suitable for sowing annuals and many perennials could be 2 parts sterilized loam to 1 part peat or leaf mould (or other substitute) and 1 part sharp sand. All parts are by volume, not weight. For a soil-free compost, you could try 2–3 parts peat or leaf mould (or other substitute) to 1 part sharp sand. For plants that require extra drainage, like many alpines and plants originating from dry areas, add more sand to the compost, perhaps up to half the volume.

Base fertilizers for home-made composts are obtainable from garden centres, and should be added at the recommended rate, especially if the seedlings are likely to remain in the original compost mix for a while. For alpines and plants typically growing on poor soils, this rate can be reduced.

The seeds of lime-hating plants, such as *Lithodora diffusa*, *Lithospermum canescens* and *Eritrichium nanum*, must be sown into lime-free, acidic compost – the proprietary ericaceous compost with plenty of extra non-alkaline grit added would be suitable.

Sowing Always sow as thinly as possible, to allow plenty of air circulation around geminating seedlings. Overcrowded seedlings may rot or result in poor plants.

As a rule of thumb, the seed should be covered to a depth approximately equalling its own size. Sieved compost can be used, but I would give preference to horticultural grit, which has the advantage of maintaining free drainage at the seedling's collar. Very fine seed is better left uncovered.

Outdoors Many hardy annuals, such as *Borago*, *Cerinthe* and *Omphalodes linifolia*, can be sown directly into their permanent positions, either in spring or, in milder areas, in autumn, or both – this will extend the flowering season considerably. Some biennials, such as *Cynoglossum* and *Echium*, can also be sown *in situ*, and this is usually done in summer for flowering the next year. If perennials, like *Pulmonaria* and *Anchusa*, are to be sown outdoors, this should ideally be in a seedbed, from which the young plants can later be transplanted or potted up.

Soil preparation will involve clearing off all weeds, removing large stones, and raking to break up any clods and produce fine tilth. Unless the soil is compacted, it is probably not a good idea to dig it over, as this will just bring more weed seeds to the surface, where they will germinate prolifically.

Under cover For most cultivated Boraginaceae, it is probably best to sow under cover, in a glasshouse, cold frame or on a windowsill. This is essential for plants that are not fully hardy, such as *Heliotropium*, *Myosotidium*, *Ehretia*, and all alpine and rock garden species. The advantage of sowing indoors is that you have much more control over the growing medium, the environment and the watering regime, which are crucial in raising the more challenging species, such as *Arnebia*, *Eritrichium*, *Alkanna* and the smaller *Mertensia*.

Pricking out In most cases, the seedlings should be pricked out into individual small pots as soon as they have a first pair of true leaves, but with small seedlings you will need to wait until they are large enough to handle. On the other hand, do not leave it too late, as many Boraginaceae have a long taproot that develops quite early on, and they are resentful of disturbance. The compost for pricking out can be similar to that for seed sowing, with extra base fertilizer added according to the recommendations on the packet.

VEGETATIVE PROPAGATION
DIVISION
Division of rootstock is often the easiest method of propagation, and can be used for such plants as *Pulmonaria*, *Symphytum*, rhizomatous *Omphalodes*, *Brunnera* and *Trachystemon orientalis*. Some plants, like *Mertensia*, are trickier and dislike disturbance. In general, for most plants that have creeping rhizomes and those specimens that have more than one basal rosette of leaves, division should be quite straightforward. Most commonly it is done in spring or autumn, but hardy plants do not mind being divided in winter if the ground is neither frozen nor too wet, and many can also be divided in mid- to late summer (in the latter case, it helps to reduce the size/number of leaves to minimize transpiration, and to keep them well-watered). The plant should be lifted out of the ground, the excess soil shaken off. Smaller plants can usually be separated into sections by hand, but for large specimens you may need a sharp spade. For those that are especially valuable or difficult, you may like as a matter of precaution to dust the cuts with fungicide powder, to prevent rot.

STEM CUTTINGS
Cuttings are not as important as seed and division in propagating Boraginaceae, but there are a few plants that can be conveniently increased by this method.

Basal stem cuttings can be taken from a number of perennial plants, like *Onosma* and *Anchusa caespitosa*. These are young growths at the base of the plant, and can be taken whenever such new shoots are available, often in spring. The shoot is cut off right at the base.

Semi-ripe stem cuttings are a common method of propagation for shrubs, such as *Lithodora*. This is usually done in summer, when this season's growth begins to harden. You may like to apply some hormone rooting powder to the cut ends of the cutting, to improve the chances of rooting. It usually contains fungicide to discourage rot.

Some perennials can be propagated from **root cuttings**, which is usually done in winter. It is a common method of multiplying perennial *Anchusa*, but can also be done with *Symphytum*, *Pulmonaria* and some *Mertensia* species (see the details in Chapter 1).

Typical **cuttings compost** will consist of equal parts peat/peat substitute and sharp sand, but this varies widely according to gardeners' preferences, vermiculite or perlite sometimes replacing sand, and sometimes pure sand is used on its own. The main prerequisite is that the drainage should be perfect.

Cuttings should be inserted in pots or trays filled with compost, and placed in a lightly shaded position in a cold frame or glasshouse. If you lack these facilities, a clear plastic bag placed over a pot on a wire hoop is usually adequate, if you ventilate regularly, not allowing too much condensation to form. Using a heated propagator with a bottom heat of around 15°C (59°F) will speed up the process of rooting stem cuttings, but make sure not to overheat. Never allow the cuttings to dry out completely, but avoid watering from overhead – the best method of watering is probably by standing the containers with cuttings in a tray of water, and draining well afterwards.

TISSUE CULTURE
In recent years, some nurseries, such as Terra Nova Nurseries in the USA, have started propagating new *Pulmonaria* cultivars and some other plants by tissue culture. This method is out of reach of ordinary gardeners, as it requires special equipment, sterile conditions and laboratory expertise, but it does mean that the new introductions can be multiplied rapidly by this method and thus will become available to us, the general public, much sooner than if they were propagated by conventional methods.

PESTS & DISEASES

This is a very brief account of the possible problems that may be encountered with plants of the Boraginaceae family. For details of control and prevention methods, use a good, up-to-date horticultural reference book, such as some of the Royal Horticultural Society's handbooks and encyclopaedias.

SLUGS AND SNAILS
These molluscs are a menace to almost any seedling, and can also ruin a mature plant. They have a preference for smooth, hairless foliage, so plants like *Mertensia maritima* are especially vulnerable, and can be reduced to next-to-nothing almost overnight in warm, wet weather. But even the hairier types will be attacked if nothing else is available. Use a combination of cultural, mechanical and biological methods to protect your plants from these thugs.

APHIDS
Aphids are a particular problem under glass, and smaller plants are particularly vulnerable. Small numbers may be simply squashed, or the infested shoot tips pinched out. Encouraging ladybirds, lacewings and

blue tits in the garden will go a long way in controlling the numbers of these pests. If the infestation is severe, you may want to resort to chemical sprays, but try using the 'softer' options, such as insecticidal soap.

CATERPILLARS

Caterpillars of a variety of moths may cause damage on a number of Boraginaceae. Like slugs, they are more likely to attack the less hairy species. The best environmentally friendly way to deal with these is to inspect your plants regularly and to remove any eggs or young caterpillars that may be found. If you feel you must spray, there is an array of chemicals available in garden centres.

POWDERY MILDEW

Fungi of family Erysiphaceae are the culprits in causing this common garden disease. Leaves and other green parts of the plants become covered in a powdery white coating. Severely affected plants will be stunted and some of the smaller and weaker ones may be killed. Mildew, like many other fungi, usually affects plants that are already weakened or stressed in some way, so good growing conditions are important in preventing its occurrence.

Never allowing the soil to become bone dry, maintaining good hygiene by removing infected and damaged plant parts, and wider spacing for good air circulation are all helpful here. In general, it is better to use preventive measures, than to wait until the disease attacks, but systemic and sulphur-based fungicides are available to treat the plants when the mildew does strike.

RUST

Rust is caused by fungi of the order Urediniomycetes, producing orange or yellow-brown pustules on leaves and weakening the plants by tapping them for nutrients. Among Boraginaceae, *Symphytum* species are susceptible to the rust fungus *Melampsorella symphyti*, which is very difficult to treat, so the affected plants should be destroyed.

VIRUSES

Viral diseases in the garden are transmitted mainly by aphids, and sometimes nematodes. Most commonly, virus infection demonstrates itself in yellow mottling in the leaves, and sometimes distorted growth. Affected plants should be destroyed immediately, as there is no known cure, and controlling the vectors (aphids and nematodes) will help to reduce the chance of infection.

REFERENCES

BIRD, 1993; BRICKELL, 1996; BUCZACKI & HARRIS, 1991; HEYWOOD, 1996; HICKEY & KING, 1988; HUXLEY, 1992; LORD, 2002; LUSBY & WRIGHT, 1996; MABBERLEY, 1997; MAUSETH, 1988; OBERRATH & ZANKE, 1995; TREHANE *et al*, 1995; WALTERS, 2000.

CHAPTER 1
PULMONARIAS
Written with Joe Sharman

This chapter will deal exclusively with the genus *Pulmonaria* – the lungworts – including details of their naming, structure, biology, cultivation and propagation. Here you will find descriptions of Pulmonaria species, forms and cultivars, and some tips on choosing varieties for your garden.

Pulmonaria rubra 'Ann'. (Howard Rice)

The genus *Pulmonaria* Linnaeus, 1735
Lungworts

Ref: Syst. ed. 1 (1735); Gen. ed. I. 37 (1737)

'They form dense tufts of foliage, generally handsomely blotched and speckled with white, and make pretty groups in the spring garden, or in semi-wild places, but are worthy of the best places in the flower garden.'
(William Robinson)

NAMING
ETYMOLOGY
The scientific name of the genus *Pulmonaria* is derived from Latin *pulmo* (the lung). In the times of sympathetic magic, the appearance of a plant was taken as a sign for the type of ailments it could cure. So the spotted leaves oval leaves of *Pulmonaria officinalis* were thought to symbolize diseased, ulcerated lungs and so were used to treat pulmonary infections. The common name of pulmonarias in many languages also refers to lungs, as in the English 'lungwort' and German 'lungenkraut'. In some East European languages, however, the common name is derived from a word for honey, probably as an indication of the attraction of flowers to bees – Russian 'medunitza' and Polish 'miodunka' – a rather more appealing association.

SYNONYMS
In the past, the generic name *Pulmonaria* has been applied to some plants that now belong to different genera, such as *Mertensia* Roth., but in modern literature there is no confusion between this group of plants and the other genera of Boraginaceae – a happy state of affairs, which contrasts the bewildering complexity of species synonyms within the genus *Pulmonaria* (see below).

COMMON NAMES
English: Lungwort, Soldiers and sailors, Spotted dog, Joseph and Mary, Jerusalem cowslip, Bethlehem sage; *Chinese*: Fei cao shu; *Danish*: Lungeurt; *Dutch*: Longkruid; *Finnish*: mikkä; *French*: Pulmonaire; *German*: Lungenkraut; *Italian*: Polmonaria; *Norwegian*: Lungeurt; *Polish*: miodunka; *Russian*: medunitza; *Slovak*: plúcnik; *Slovenian*: pljucnik; *Swedish*: lungört.

Leaves of various *Pulmonaria* forms and species. (Howard Rice) (a) *P.* 'Oliver Wyatt's White'; (b) *P.* 'Sissinghurst White'; (c) *P. officinalis* Cambridge Blue Group; (d) *P.* 'Blauhimmel'; (e) *P.* 'Mawson's Blue'; (f) *P.* 'Blue Ensign'; (g) *P. officinalis*; (h) *P. obscura*; (i) *P. affinis*.

GENUS CHARACTERISTICS
NUMBER OF SPECIES
Taxonomy of pulmonarias is extremely confusing, as the species are notoriously difficult to distinguish one from another, and some are thought to hybridize freely; and, unfortunately, the currently available Floras do not provide adequate keys for identifying pulmonarias.

'Pin' flower **'Thrum' flower**

(a)

corolla lobe — corolla lobe

corolla throat — corolla throat

(b)

corolla — corolla

stamens — stamens

calyx — calyx

pedicel — pedicel

(c)

limb — limb

position of annulus — tube

position of annulus — tube

(d)

calyx — calyx

style — style

pedicel — pedicel

(e)

stigma — stigma

style — style, style

ovary — ovary

receptacle — receptacle

Comparison of 'pin' and 'thrum' flowers:
(a) view of corolla lobes and throat; (b) cross-section of whole flower, showing position of stamen; (c) corolla (whole); (d) calyx with style, showing relative lengths of style; (e) style with ovary.

According to various estimates, there may be between 10 and 18 *Pulmonaria* species found in the wild. Some plants thought to be species in cultivation could be hybrids, or different species, yet to be described. It is possible that there may be more *Pulmonaria* species in cultivation than we realize. Although we make some comments about the status of some pulmonarias in cultivation, it is not the intention of this book to sort out the confusion in taxonomy of the genus, nor is it about the relationship between the different species – these things we will leave to specialists.

DESCRIPTION
Pulmonarias are perennial herbs with a habit of forming clumps or rosettes. They are covered in hairs, which vary in length and stiffness depending on species, and sometimes also bear glands, which make some plants, such as *P. mollis* and *P. vallarsae*, feel sticky to touch.

Their **underground parts** consist of a relatively slowly creeping, compact rhizome with adventitious roots. In some varieties, such as *Pulmonaria* 'Nürnberg', the creeping habit is more prominent. Flowering **stems** of pulmonarias are unbranched, rough, covered with bristly hairs, mostly not exceeding 25–30cm (10–12in) in height, with a few exceptions

Pulmonaria officinalis-type habit.

such as *P. mollis* and *P. vallarsae*. The stems are normally upright, or may be slightly sprawling.

Basal **leaves** of pulmonarias are arranged in rosettes. The blades are usually large, ranging in shape from narrowly lanceolate to oval, with the base ranging from heart-shaped to very gradually narrowing, and can have a sharply pointed or a blunt tip. Leaf margin is always entire but, in some species and cultivars, may be rather wavy (as in *P. vallarsae*). Basal leaves are carried on stalks, or petioles, that can be short or longer than the leaf blade in various species. Stem leaves are smaller and often relatively narrower, and are unstalked or even clasping the stem. All leaves are rough and covered with hairs that are usually bristly, with some exceptions (*P. mollis* is covered with soft hairs). The leaves are often prominently spotted, sometimes in pale green but more commonly in silvery-white, as in *P. officinalis* and *P. longifolia*. In some of the varieties the silver marking is so strong that it covers most of the leaf, as with *P. saccharata* Argentea Group or *P.* 'Excalibur'. In other pulmonarias, such as *P. rubra*, the leaves are normally unspotted. Variability in the degree of spotting is enormous, and it is known for some individuals of the normally

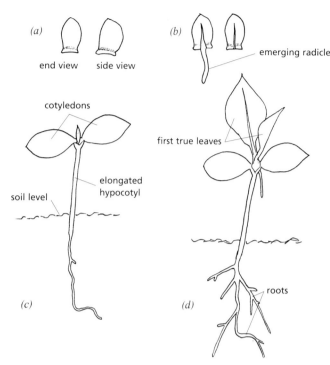

Pulmonaria germination and seedling growth.
(a) nutlets; (b) hard coat splitting; (c) young seedling; (d) growing seedling.

'spotty' species to have plain leaves, and vice versa. In summer, after flowering is finished, basal leaves often enlarge and become the most distinctive feature of the plant.

Inflorescence in pulmonarias is of a typical construction for the Boraginaceae family – it is a terminal scorpioid cyme, with bracts. It is usually quite short, and elongates only a little when in fruit.

Pulmonaria **flowers** are heterostylous – that is, there are two distinct forms of flower within each species: those with short stamens and long styles, and those with long stamens and short styles (the so-called 'pin' and 'thrum' flowers, respectively). This ensures cross-pollination between pin and thrum flowers (see Pollination). 'Pin' flowers are usually larger and more showy.

The **calyx** in pulmonarias is hairy, 5-angled at base, tubular or funnel-shaped in flower, and as the fruit ripens it gradually enlarges and becomes rather bell-shaped. There are 5 triangular calyx lobes up to one-third as long as the corolla tube (see line drawing, left). Calyces of pulmonarias are usually green or brown, but can be purple in some varieties. In *P. mollis* and *P. vallarsae*, calyces feel sticky due to the presence of glands.

Pulmonaria longifolia-type habit.

Pulmonaria 'Mawson's Blue'. (Howard Rice)

The **corolla** is funnel-shaped and consists of a long, cylindrical tube and a limb with five shallow lobes. Within the corolla throat, five tufts of hairs alternating with the stamens form a ring.

The colour of corolla varies from purple, violet or blue to shades of pink and red, or sometimes white. Regardless of the colour of the mature flower, the colour of the flower bud in the majority is almost invariably pink of various intensity. As the flower opens, the colour gradually changes, which probably gives a signal to potential pollinators.

Stamens are included within the corolla, not protruding from it, and are inserted in the throat or middle of the tube, alternating with tufts of hairs. The **style** is also included within the corolla. Its stigma is capitate/headlike or slightly bifid/forked.

The **nutlets** of pulmonarias are smooth, egg-shaped, brownish, up to 4.5mm long and 3mm wide, with a basal annulus (raised ring). Each contains a single seed. The nutlets are usually hairy, at least at first. Up to four nutlets per flower are produced, ripening mostly in summer, typically one to two months after blooming.

Germination of pulmonarias is epigeal: that is, the seed cotyledons (small leaflike structures) are pushed out of the ground, as opposed to hypogeal

germination, where the seed remains in the soil (as, for example, in peas, *Pisum* and *Lathyrus*).

GEOGRAPHY AND ECOLOGY

DISTRIBUTION

Pulmonarias are widely distributed in Europe and western Asia. Their westernmost countries of distribution are England, Spain and Portugal, while in the east they go as far as Siberia and in the south reach Italy and Greece, while the northernmost border is Sweden. They are not native to North America, but a couple of species have been naturalized in the USA.

HABITATS

Pulmonarias are rather consistent in their choice of habitat, the majority being found in shady or partially shaded places, such as woodland and scrub, shady banks and often at rather high altitudes. They tend to inhabit deciduous woods rather than coniferous forests, but some can be found in mixed woodland. Occasionally some species venture out into the open, and are found growing in meadows (such as *P. angustifolia*) or rocky places at higher altitude (*P. filarszkyana*). Sometimes pulmonarias that occur in open places may be relics of former woodland.

Approximate distribution area of the genus *Pulmonaria*.

LONGEVITY

All pulmonarias are **polycarpic perennials** – that is, they live for many years and flower more than once. In cultivation, they often decline and die out with age, but if a plant is divided and replanted every few years, it can in theory last almost indefinitely.

LIFE FORMS

According to the Raunkiaer plant life form classification (see page 9), pulmonarias can be classed as **hemicryptophytes** – that is, their perennating buds overwinter just at the soil level.

POLLINATION

Like the majority of plants in the Boraginaceae family, pulmonarias are pollinated by insects. Because of their long, funnel-shaped flowers, only certain types of visitors are able to reach the nectar. Only bumblebees (*Bombus* species) and possibly solitary bees (*Anthophora* species) have sufficiently long tongues to obtain their reward from *Pulmonaria* flowers, and shorter-tongued bees may steal nectar by piercing holes in the lower part of the corolla tube where nectaries are situated – the so-called 'nectar robbery'.

As already noted in the description of genus characteristics, flowers in pulmonarias may be either 'pin' (long-styled) or 'thrum' (short-styled) within the same species. Pollination can occur only between the two plants with different style length, an adaptation to ensure cross-pollination of pulmonaria populations.

The change in flower colour, most often from pink to blue or purple, in pulmonarias and many other Boraginaceae, is a very distinctive feature of this family. It has been shown that the colours at different stages of flower development are likely to act as signals to pollinating insects – the colour of a pulmonaria flower at exactly the right stage for pollination is usually different from that of an immature bloom, or the one that has already been fertilized. Thus the plant increases the likelihood of pollination by informing the insects exactly when flowers are ready for a visit, and rewarding them with the largest amounts of nectar and pollen if they arrive at the 'right' time.

Interestingly, pollen produced by anthers in 'thrum' *Pulmonaria* flowers is a little larger in size than that produced by 'pin' flowers. Also, there are 45% more 'pin' pollen grains found trapped on 'thrum' stigmas than 'thrum' pollen on 'pin' stigmas. A detailed account of heterostyly can be found in John Richards' magnificent book *Primula* (1993).

FRUIT DISPERSAL

Pulmonarias do not appear to possess any specialized mechanism for fruit dispersal, but there is some evidence that ants of genus *Formica* are attracted to nutlets of certain species, including *P.obscura*, and carry them to their nests. Some of the nutlets are probably lost along the way, and are thus dispersed, even if just a short distance, away from the parent plant. This happens more commonly in plants in which the seeds possess a fleshy appendage – for example, *Cyclamen*. Mice also eat the nutlets and may, perhaps, play a minor role in their dispersal.

ANIMAL FEEDERS

Pulmonarias do play their part in the food chain in their ecosystems by supporting a number of insects, but overall it has to be said that they are not very important as a wildlife food source.

Small moths (Lepidoptera) that feed on pulmonarias include *Coleophora pennella*, *C. pulmonariella*, *Euchalcia modestoides*, *Xestia rhaetica* and *Ethmia quadrillela*. Also, weevils of genus *Ceutorhynchus* have been found living on pulmonarias as have several leaf beetles, including *Longitarsus pulmonariae*, *L. curtus* and *L. lateripunctatus*.

For other animals that you may come across on these plants, see below under Pests.

CULTIVATION

A BRIEF HISTORY

Pulmonaria officinalis has been cultivated in Europe since at least the sixteenth century, for medicinal purposes, and was among the plants grown in the first botanical garden in North America, founded by John Bartram in 1728 near Philadelphia. Both *P. angustifolia* (or the species thought to be this plant) and *P. saccharata* were being grown in England for ornament by the eighteenth century, and *P. mollis* appeared in the early nineteenth century. Selective breeding of pulmonarias did not start until the second part of the twentieth century. Some of the older varieties include 'Mawson's Blue', 'Sissinghurst White', 'Mrs Moon' and the Argentea Group, though it is doubtful whether these cultivars have remained true to the original forms.

Some of the best modern pulmonaria varieties have been introduced in the last one or two decades by such well-known names as Beth Chatto, Monksilver Nursery, Washfield Nurseries (UK) and Terra Nova Nurseries (USA). While the majority of English cultivars have arisen 'by accident', American nurserymen have been using selective breeding in recent years with some very promising results.

HARDINESS/CLIMATE

Most pulmonarias are very hardy, and tolerate freezing temperatures down to -29–34°C/-20–29°F (US Zone 4), and some down to -34–40°C/-29–40°F (Zone 3), especially under snow cover (see the individual species for hardiness zones). Semi-evergreen species and cultivars may lose all their leaves during especially harsh weather but should recover in spring. In areas warmer than Zone 8 (annual minimum temperature -7–12°C/10–19°F) most pulmonarias suffer, as they do in very dry conditions. While they can be grown throughout most of western Europe, cultivation of pulmonarias in the USA is restricted to the states with less extremes of climate, such as Oregon, northern California, and along the east coast.

SUN/SHADE ASPECT

One of the most valuable features of the genus *Pulmonaria* is that virtually all species and forms are very shade-tolerant, perhaps with the exception of *P. longifolia*. Many will happily grow even in heavy shade, although flowering is likely to be reduced in such conditions.

Most pulmonarias fare best in the dappled shade of those trees whose crowns let through a fair amount of light, like *Quercus* (oaks), *Betula* (birches), *Fraxinus* (ash) and *Sorbus* (rowans and whitebeams). They will not mind some direct sun in the morning, but in hot afternoon sun these plants will tend to wilt. In countries where summers are cool, such as England, some pulmonarias will tolerate a sunny position as long as they are kept sufficiently moist.

Pulmonarias that are reasonably tolerant of full sun

P. affinis
P. angustifolia (true species)
P. 'Blue Ensign'
P. 'Glacier'
P. longifolia
P. 'Mary Mottram'
P. 'Milchstrasse'
P. mollis
P. officinalis Cambridge Blue Group

SOIL REQUIREMENTS

Pulmonarias are tolerant of most soils and are especially useful on heavy clay. Their full potential is best achieved on fertile soil that is not too dry, and has a high content of organic matter. These plants will not tolerate continuous waterlogging, but almost any other type of soil can be adapted for growing pulmonarias. The majority will withstand a high degree of alkalinity, with the exception of *P. angustifolia*, *P. angustifolia* hort. and their forms, which suffer from chlorosis on lime-rich soil. Less than ideal soils should be improved to provide optimum conditions (see Planting).

CHOOSING PULMONARIAS FOR YOUR GARDEN

With over 150 forms and cultivars of pulmonarias in existence, it is understandable that a newcomer to this group of plants will be baffled by the wide choice of plants available. In a recent survey (carried out in 2000) by the Pulmonaria Group of the Hardy Plant Society (UK), the following pulmonarias were voted the favourite for their flowers:

Flower colour	Pulmonaria
White	P. 'Sissinghurst White'
Pale/mid-blue	P. Opal ('Ocupol')
Royal/darker blue	P. 'Blue Ensign'
Purple/blue	P. 'Barfield Regalia'
Pink	P. saccharata 'Dora Bielefeld'
Red	P. rubra
Other (incl. striped)	P. rubra 'Barfield Pink'

Four pulmonarias voted in 1999 by the Hardy Plant Society Pulmonaria Group members as favourite for their summer foliage were:

P. 'Cotton Cool' (the winner)
P. 'Mary Mottram'
P. rubra 'David Ward'
P. longifolia 'Ankum'

Although the number of members who took part in these surveys was small, the judgement of these connoisseurs is sure to be sound, and their favourite pulmonarias are likely to become highly prized by most gardeners.

The Royal Horticultural Society's Award of Garden Merit is an excellent indication of whether a plant is a garden-worthy one. The following have so far received this award:

Pulmonaria angustifolia hort. (1993)
Pulmonaria angustifolia hort. 'Blaues Meer'(1998)
Pulmonaria 'Blue Ensign' (1998)
Pulmonaria 'Lewis Palmer' (1993)
Pulmonaria 'Margery Fish' (1993)
Pulmonaria officinalis 'Blue Mist' (1998)
Pulmonaria rubra (1993)
Pulmonaria rubra 'David Ward'(1998)

Pulmonaria saccharata Argentea Group (1993)
Pulmonaria saccharata 'Leopard' (1998)
Pulmonaria 'Sissinghurst White' (1993)
Pulmonaria 'Vera May' (1998) – subject to availability
Pulmonaria 'Weetwood Blue' (1998)

The AGM symbol ♀ appears within the book to highlight these plants.

RHS trials at Wisley in 1996–98, apart from awarding the AGM to selected plants, have noted particular pulmonarias for their good foliage (in alphabetical order):

Pulmonaria 'British Sterling'
Pulmonaria 'Crawshay Chance'
Pulmonaria 'Dawn Star'
Pulmonaria 'Jaak'
Pulmonaria longifolia 'Ankum'
Pulmonaria longifolia subsp. *cevennensis*
Pulmonaria 'Margery Fish'
Pulmonaria 'Mary Mottram'
Pulmonaria 'Merlin'
Pulmonaria officinalis 'Blue Mist'
Pulmonaria 'Oliver Wyatt's White'
Pulmonaria Opal ('Ocupol')
Pulmonaria 'Red Freckles'
Pulmonaria 'Reginald Kaye'
Pulmonaria saccharata Argentea Group

This list is mostly biased towards cultivars originating in the UK, but there have been many forms with good foliage developed in the USA in recent years that were not included in the trials.

It is important to note that a number of pulmonaria cultivars are very similar, and some pulmonaria experts, including the UK National Plant Collection Holder, Vanessa Cook, consider that these should be grouped together, and that no garden needs more than one cultivar from each group, due to their similarity. Such groups are red-flowered, spotted-leaved cultivars 'Elworthy Rubies', 'Cleeton Red', 'Esther' and red-flowered, plain-leaved *P. rubra*, *P. r.* 'Redstart', *P.* 'Barfield Ruby'. Cultivars 'Jill Richardson', 'Paul Aden', 'Reginald Kaye' and 'Silver Mist' are all similar to 'De Vroomen's Pride' and, in Vanessa Cook's opinion, should probably all be called 'Reginald Kaye'.

GARDEN USES OF PULMONARIAS

GROUND COVER

Pulmonarias are perfect ground-cover plants, forming neat clumps of handsome leaves that exclude weeds and conserve moisture in the soil. Unlike many other types of ground cover, they are non-invasive. Many pulmonarias are semi-evergreen, with at least some foliage persisting through winter months. Only *P. angustifolia* (*P. angustifolia* hort) and *P. mollis* and their cultivars are completely or largely deciduous. Use pulmonarias as cover under shrubs – blue-flowered cultivars, such as *P.* 'Blue Ensign' or *P.* 'Mawson's Blue', make a lovely combination with yellow-flowered shrubs, such as *Forsythia* species and cultivars, or winter jasmine, *Jasminum nudiflorum*, or a paler shade like 'Opal' or 'Cambridge Blue' under pale yellow *Corylopsis glabrescens*. Or try *P. rubra* and its forms under white-flowered shrubs, such as *Magnolia stellata* or *Daphne mezereum* 'Album'.

Those that are more tolerant of sun (such as *P. affinis*, *P.* 'Cambridge Blue', or *P.* 'Glacier') would make good cover under roses, as long as the ground is kept moist throughout the summer. Pulmonarias that withstand drier conditions than most include *P. affinis*, *P. mollis*, *P. longifolia* and their cultivars.

Pulmonarias are among the few perennials capable of surviving under black walnut (*Juglans nigra*), which produces toxic allelochemicals that prevent the growth of most plants under the tree's canopy.

Another virtue of this versatile group of perennials is that they will grow underneath shallow-rooted trees, where many other plants are likely to fail.

WOODLAND GARDEN

All pulmonarias and many other spring-flowering species of Boraginaceae are intrinsically woodland plants and are most comfortable in a wooded setting. Most pulmonaria forms look stunning when planted *en masse* in the dappled shade of deciduous trees and shrubs. Good woodland companions for pulmonarias include *Pachyphragma macrophyllum*, *Cardamine pentaphyllos*, *Helleborus*, *Dicentra* (bleeding-heart), *Corydalis*, *Primula*, *Cyclamen*, *Arisaema*, *Asarum*, *Epimedium* (barrenworts), *Hosta*, *Veratrum* (false helleborine), *Fritillaria* (fritillaries), *Hyacinthoides non-scripta* (bluebell), *Brunnera macrophylla*, *Omphalodes verna* (blue-eyed Mary), *Heuchera*, *Luzula* (woodrush) and *Deschampsia flexuosa* (wavy hair-grass).

MIXED BORDERS AND BEDS

With their short, neat habit, pulmonarias are ideal for the front of the border, preferably the one that is lightly to moderately shaded, though certain forms will fare well in the sun if the soil is kept reasonably moist. They will provide interest from early spring into late autumn, and will combine well with many shrubs, perennials and bulbs – try combinations with *Narcissus* (daffodils), *Crocus*, *Geranium* (cranesbills), *Primula*, *Iris*, *Hemerocallis* (daylilies), *Cimicifuga* (bugsbane), *Astrantia* (masterwort), *Astilbe*, *Paeonia* (peonies), ferns, and any of the other woodland plants suggested above.

ROCK GARDEN

Though even the smallest pulmonarias may appear as coarse giants in comparison with some of the choicer alpines, they are well suited for a lightly shaded spot in a larger rock garden. Make sure that the free-draining rock garden mixture does not dry out in hot weather. Best rock plants to combine with pulmonarias are those that are reasonably happy in light shade and enjoy leafy soil, such as certain *Omphalodes* (navelwort), *Fritillaria*, miniature *Narcissus*, small *Scilla* (squills), *Crocus*, *Epimedium* (barrenwort), dwarf *Carex* (sedges) and ferns.

Shorter cultivars (15–20cm/6–8in) especially suitable for the rock garden
Pulmonaria angustifolia hort. 'Munstead Blue'
Pulmonaria 'Blauhimmel'
Pulmonaria Cally hybrid
Pulmonaria 'Little Star'
Pulmonaria longifolia 'Ankum'
Pulmonaria 'Merlin'
Pulmonaria 'Ultramarine'
Pulmonaria 'Weetwood Blue'

WATER GARDEN

Pulmonarias do not tolerate waterlogging, but will be happy in a moist situation on a stream bank or by a pool among waterside planting, in the company of, for example, *Dicentra* (bleeding-heart), *Aconitum* (monkshood), *Primula*, *Filipendula ulmaria* (meadowsweet), *Thalictrum* species (meadow-rue), *Mertensia ciliata* (fringed bluebells), *Peltiphyllum peltatum* and ferns. Although they appreciate a humid, cool microclimate created by the proximity of water, pulmonarias should be positioned so that no water splashes on their leaves, otherwise they may tend to rot, especially the flower buds.

CONTAINERS

Pulmonarias dislike being restricted in containers, and especially drying out (which is often the fate of plants growing in pots) but, with careful maintenance, will provide a good show in large pots. Keep in part shade and do not allow the compost to dry out. The advantage of growing in containers is that in very cold areas (Zones 1–2) they can be brought indoors to protect the plants from the hardest frosts.

ALL-YEAR-ROUND INTEREST
The majority of pulmonarias, apart from *P. angustifolia* and its forms, are at least partly evergreen, and so provide interest throughout much of the year. By selecting a variety of pulmonaria cultivars with different flowering times, you can have continuous blooms from late winter into early summer, and attractive foliage the rest of the year, apart from perhaps in late autumn and early winter, when they may not look their best.

EVENING GARDEN
In her book *Evening Gardens*, Cathy Wilkinson Barash recommends pulmonarias among other plants 'to dazzle the senses after sundown'. Silver-spotted and marbled leaves are of particular interest here, for example *P. saccharata* Argentea Group, *P.* 'Reginald Kaye', *P. longifolia* subsp. *cevennensis*, *P.* 'Crawshay Chance' and others.

White pulmonaria flowers will be especially vivid at dusk – choose *P. rubra* 'Albocorollata', *P.* 'White Wings', *P.* 'Sissinghurst White'. Combine them with evening-scented plants, such as honeysuckle (*Lonicera*) and evening primrose (*Oenothera* species), or perhaps ornamental tobacco (*Nicotiana* species and cultivars) and night-scented stock (*Matthiola longipetala bicornis*).

HERB GARDEN
Pulmonarias have been grown for many centuries as medicinal herbs and edible plants. As they require very different growing conditions from the majority of common garden herbs, which are Mediterranean in their origin, they may not be easy to incorporate into a conventional small herb garden. These plants' medicinal and culinary attributes are of limited significance in the present day, and perhaps they are best assigned to the ornamental garden, especially as there are some concerns about possible toxicity.

GARDENS FOR PEOPLE WITH DISABILITIES
Pulmonarias can be usefully included in the garden for people with physical disabilities. Their short stature allow them to be planted attractively in raised beds, which are more accessible to wheelchair users, for both enjoyment and maintenance. To visually impaired people, pulmonarias offer tactile qualities, with their distinctly textured, rough or softly hairy leaves. The buzzing of bumblebees and other insects attracted to the flowers is another bonus.

WILDLIFE GARDEN
The flowers of pulmonarias are rich in nectar and are very attractive to bumblebees (*Bombus* species), honeybees (*Apis mellifera*) and solitary bees such as *Anthophora*. These plants are a valuable early nectar source for queen bumblebees emerging from hibernation in spring. They are especially popular with the large garden bumblebee (*Bombus hortorum*), early bumblebee (*Bombus pratorum*) and common carder bee (*Bombus pascuorum*). In some parts of the United States, you can expect rather more exotic visitors: hummingbirds are sometimes seen at pulmonaria flowers. Pulmonarias are also attractive to certain butterflies and moths – they are a favourite, for example, with day-flying hawkmoths, *Hemaris tityus* and *H. fuciformis* (the narrow-bordered bee hawkmoth and the broad-bordered bee hawkmoth).

WILDFLOWER GARDEN
Where pulmonarias are native or naturalized species, they deserve a place in a collection of wild plants. Species native to England include *P. longifolia* and *P. obscura*, though the latter is only found in Suffolk, and *P. officinalis* is naturalized.

In European wildflower gardens, they could be grown with native species of *Helleborus* (hellebores), *Anemone nemorosa* (wood anemone), *Ranunculus ficaria* (lesser celandine), *Geranium* (cranesbills), *Arum maculatum* and *A. italicum* (lords-and-ladies), *Galanthus nivalis* (snowdrops), *Narcissus pseudonarcissus* (wild daffodil), *Hyacinthoides non-scripta* (bluebell), *Tulipa sylvestris* (wild tulip), *Luzula* (wood-rushes), *Carex* (sedges), *Melica* (melicks) and *Deschampsia flexuosa* (wavy hair-grass). In the accounts of individual *Pulmonaria* species in this book, examples of plant associations in the wild are given, and these provide some ideas of possible companions for pulmonarias.

Though no pulmonarias are native to America, *P. officinalis* is naturalized in Wisconsin and *P. saccharata* in New York. North American wild flowers that will make attractive combinations with pulmonarias are *Erythronium* (dog's-tooth violets), *Trillium* (wake-robins), *Fritillaria* (fritillaries), *Aquilegia* (columbines), *Mertensia* (languid-ladies), *Sanguinaria canadensis* (bloodroot), *Stylophorum diphyllum* (wood poppy), *Arisaema* (Jack-in-the-pulpit), *Dicentra* (bleeding-heart), *Viola* (violets) and *Asarum* (wild ginger).

CUT FLOWERS
Pulmonarias are not commonly seen in vases, but their flowering stems look charming in fresh flower

arrangements, though they are rather short-lived, lasting not more than about six days in water. The stems should be picked when the blooms are already opened, then the ends dipped in boiling water, and finally soaked for two to four hours in warm water before arranging.

OTHER USES OF PULMONARIAS
CULINARY
The gourmet value of pulmonarias is rather insignificant in comparison with their ornamental garden value but, nevertheless, enthusiasts may be interested in this different facet of their favourite plant. The basal leaves of *Pulmonaria obscura* and *Pulmonaria officinalis* can be eaten raw or cooked when young. These two species are the ones that are more commonly used for culinary purposes, but others may be suitable, too. The flavour is bland, and the hairy texture and mucilage of the leaves make the value of these as a delicacy questionable, but they can be added fresh to mixed salads and soups, or cooked as a vegetable.

Among other pulmonaria species, *P. obscura* is an ingredient of the alcoholic drink Vermouth, and *P. saccharata* has been used as a spice.

A word of warning to those who are keen to try the culinary delights of pulmonarias: the *RHS Encyclopedia of Herbs* states that pulmonaria is a skin irritant and allergen, and may possess toxicity similar to that of *Symphytum* (comfrey).

MEDICINAL
Pulmonaria species have a long history as medicinal plants, and were originally cultivated as such. *Pulmonaria officinalis* and *P. obscura* are the species most commonly used in herbal medicine, although other species are also important locally, such as *P. mollis* in Siberia. These plants contain mucilage, tannins, carotene, ascorbic acid (vitamin C), and microelements including manganese, iron and copper.

Pulmonarias have expectorant, soothing and astringent properties. These herbs, when taken internally, are used for treatment of bronchitis, chronic cough, catarrh, and in the past were even considered effective for pulmonary conditions (hence the name). They can be effective as antiseptic for wounds, and haemorrhoids and diarrhoea can also be treated with the aid of these plants.

Ailments such as asthma, tuberculosis and even hernias have been treated with pulmonaria in the past, though there is little scientific evidence of the effectiveness of such treatments.

The parts of the plant used medicinally are the leaves and flowering stems, harvested from spring to early summer. They are then dried, and made into infusions and extracts.

WARNING: Do not use any herbal remedy without a prior consultation with a doctor.

OTHER
In those regions where wild fires occur, like some parts of the USA, ground cover of pulmonarias can be doubly useful in landscaping (as long as it is not warmer than Zone 8), as they are relatively resistant to fire and may prevent it from spreading.

CULTIVATION
BUYING A PULMONARIA
Pulmonarias have become increasingly popular in the last few years and are now often available in some of the better garden centres, though the best place to obtain a good-quality, correctly named plant must be a specialist nursery.

In garden centres, pulmonarias are most often offered for sale in spring, while in bloom – and it is perhaps the best time to buy one, when you are able to see the flowers for yourself.

Many specialist nurseries offer mail order in autumn and/or spring, and sometimes throughout winter. You can find the details of all nurseries offering pulmonarias in the British Isles in *The RHS Plant Finder*. The number of nurseries currently supplying each particular *Pulmonaria* in the UK is given at the end of the entry for individual species and cultivars as an index of 'availability'. Buying plants by mail order has obvious disadvantages, including not being able to inspect the plants, the cost of postage and packing, and possible damage to the parcel. However, this is the only way to find many of the unusual and new varieties.

Young pulmonarias sold as 'liners', in 7cm (2³⁄₄in) or 9cm (3¹⁄₂in) square pots, are cheaper than those in 1- or 2-litre containers, but may take about a year to achieve the same effect.

When selecting a plant to buy, inspect it thoroughly to ensure that it is in good condition and has been cared for properly. Cracked pots and liverworts and weeds growing in compost are an indication of neglect. A pulmonaria that has been sitting in a small pot for too long is unlikely to make a good garden plant. At the other extreme, young plants may sometimes be potted in large pots and offered for sale almost immediately, at an inflated price. If given a tug, such plants may pull out of the compost quite easily, revealing the true size of the root-ball.

Make sure that the plant you choose is clearly labelled, especially if obtaining more than one variety. Pulmonarias are an especially confusing group of plants and our memory is never as good as we expect. It is clear that on many occasions in the past certain pulmonarias have changed hands with a wrong or inaccurate name attached to them. Although labels do not guarantee correct identification, they can help to keep confusion to a minimum.

If you obtain a bare-root pulmonaria, it is usually best to pot it up in a 1- or 2- litre pot, depending on the plant's size. Keep it in a sheltered, shady spot and plant out as soon as the roots fill the pot nicely.

PLANTING

Try to plant as soon as possible, as pulmonarias do not like sitting in pots. Autumn or spring are probably the best for planting, while the weather is relatively mild and moist. In milder, temperate climates, such as in the UK, it is possible to plant pot-grown plants throughout the year, but avoid dry and hot periods, as well as the times when the soil is frozen.

As with all plants, soil preparation is most important. Pulmonarias are tolerant of many different types of soil, but by preparing soil correctly we can improve their chances of quick establishment and good growth.

Light sandy soils tend to drain fast, with the essential plant nutrients leaching out rapidly. To improve the retention of both moisture and nutrients, you need to add plenty of organic matter to sandy soil. Clay soils, on the other hand, retain too much moisture, sometimes to the point of waterlogging, and lock the nutrients away between their tiny particles. Though the problem here is different, the cure is the same – by mixing in as much organic matter as possible, you will improve the structure of clay, making it more open, and the nutrients will become more freely available to plants.

Leaf mould is probably the best source of organic matter, as this occurs naturally in pulmonarias' habitats, but it takes a long time to make and is not easy to obtain. Well-rotted garden compost is a good alternative, and this should ideally be incorporated into the soil a few weeks in advance of planting.

Prior to planting, add a small amount of bonemeal into the planting hole, mixing it with the soil thoroughly – bonemeal is rich in phosphorus and will aid good root establishment. If the soil is poor, you may want to incorporate a little general N:P:K fertilizer, such as Growmore.

For most species and cultivars, the planting distance of 30cm (12in) is optimal, though this should be reduced for smaller varieties (such as *P.* 'Little Star', *P.* 'Merlin') and increased for vigorous/large forms (like *P. mollis*, *P.* 'Barfield Regalia').

Check that the soil level is correct while planting – it should normally be the same as the level of compost in the pot. However, the growing point in the centre of the rosette must not be buried and rhizomes or roots should not be exposed, although *P. angustifolia*, *P. angustifolia* hort. and *P. longifolia* can cope with deeper planting. After filling in the hole, firm the soil well – one of the commonest reasons for failure is that planting was not firm enough.

WATERING

Newly planted pulmonarias must be thoroughly watered throughout the first season; afterwards, they should be watered well during dry spells. Try not to let the soil dry out completely, but if it does happen, give the plants a thorough soak. If possible, avoid wetting the leaves, especially in direct sunlight. In countries that have hot summers, timely and generous watering will help maintain growth of pulmonarias that otherwise may go dormant.

DEAD-HEADING

Removing faded flowerheads will prolong flowering and prevent self-seeding, as well as giving the plants a neater look, but it can be a daunting task if you grow a large number of specimens.

MULCHING

An annual mulch of leaf mould will keep the weeds down, conserve moisture and improve soil structure. Well-rotted garden compost is also suitable as long as it is completely decomposed. Composted bark looks natural in a woodland setting, but beware of fresh bark which, when decomposing, will rob the soil of nitrogen. A relatively novel type of mulch is cocoa shell – it is light, attractive and pleasant-smelling, but not cheap. In certain situations, such as rock gardens and low maintenance gardens, a gravel mulch may be preferred, although in my opinion it does not go with pulmonarias as well as other kinds.

DIVISION AND REPLANTING

Vigorous forms (*Pulmonaria* 'Barfield Regalia', *P. longifolia* 'Blue Crown', *P. rubra*) should be divided every four to five years and less vigorous ones (*P.* 'Blauhimmel', *P. rubra* 'David Ward') can be left longer (see Propagation, below, for details of division technique). The sections can be either potted up or planted straight back into the ground. The success rate is likely to be higher if the divisions are potted first, and planted out once the roots have filled the container.

FEEDING

Feeding should be unnecessary for plants grown in the ground if they are well mulched with leaf mould or other organic material annually. On very poor soils, or for plants recovering from damage by pests or diseases, a moderate amount of general-purpose fertilizer, like Growmore, can be beneficial. *Pulmonaria rubra* seems to appreciate feeding more than others. If growing pulmonarias in containers, mix some slow-release fertilizer into the top layers of compost in spring, and it will provide sufficient nutrients throughout the season.

PROPAGATION
SEED
Many pulmonarias self-sow very happily in the garden, although there is a number of sterile varieties that are unable to produce viable seed. Many of the existing cultivars have originally occurred as individual self-sown seedlings, noted for the distinct flower or leaf colour.

It is usually assumed that *Pulmonaria* species and varieties hybridize very freely, and that the numerous variants of flower and leaf colour known in this genus arise as a result of such crossing. However, Joe Sharman is of the opinion that hybridization among pulmonarias might be less common than previously thought, and many of the so-called 'hybrids' are simply very variable seedlings originating from same-species parents.

Self-sown seedlings may grow into good garden plants, but only very rarely do they turn out to be distinct and worthwhile forms. Far too often in the past, inferior plants, or those almost identical to the already existing forms, were given cultivar names and status. This does not do much good to the reputation of this fantastic group of plants among gardeners, and adds further confusion to the already complicated classification of *Pulmonaria* species and cultivars.

Unless you are seriously intent on discovering new, exciting varieties, or are not concerned about the quality of the seedlings that are liable to be popping up around the garden, it is probably better to dead-head the existing plants to prevent self-seeding.

However, it may be appropriate to collect and sow your own *Pulmonaria* seed for two reasons: firstly if you grow only pulmonarias of one species, which are unlikely to have cross-fertilized with other forms, and secondly if you are deliberately setting out to produce a new cultivar.

Remember that there are no *Pulmonaria* cultivars that will come true from seed, and if you do sow seed of a variety, not a species, the resultant plants must not be given the name of their parent, and the seeds of these plants should not enter seed-exchange schemes. A couple of forms – *P. saccharata* Argentea Group and *P. officinalis* Cambridge Blue Group – will produce a number of seedlings similar in their characteristics to the parent, but there is likely to be a considerable variation in the colour and markings. It is permissible to distribute such plants under a 'group' name, which indicates that the plants may be variable, as opposed to a cultivar in which all plants should (at least in theory) be identical.

Pulmonaria seed starts ripening approximately one month after flowers are fertilized. There may be up to four nutlets per flower, but often there are just one or two. Collecting the nutlets is a daunting task, as it is necessary to inspect each calyx flower for the presence of ripe seed. Unless you look for seed on a daily basis throughout the period of ripening, you may find that mice have already harvested most of the nutlets.

To safeguard the seed from such predators or from simply falling onto the ground, tie a fine net bag over the seedheads – you could customize an old stocking for this purpose, although this may lead to rotting if it is left on for too long.

Seed is rarely available from commercial suppliers, though one or two common species may be occasionally found in some of the larger seed catalogues. *Pulmonaria* seed is often listed in seed-exchange schemes of specialist societies, but since cultivars will not come true to type, and the species are likely to hybridize, it is hardly worth obtaining these, unless you have plenty of garden space and wish to experiment with a large number of potentially poor-quality seedlings.

Ideally the seed should be sown as soon as ripe, in containers of good-quality seed compost, covered lightly and placed in a lightly shaded position, in a garden frame or cold greenhouse. It is possible to sow in an outdoor seed bed, but losses will be higher in these conditions. If you collect the seed later than midsummer, there is a risk that, if sown at this time, the seedlings will be too small to overwinter without the protection of a cool greenhouse, so it may be wiser to leave the sowing until spring. In areas where hot summer weather would endanger germinating seedlings, spring sowing is also preferable.

Pulmonaria seed usually germinates well, and it is possible to achieve almost 100 per cent germination. If the seedlings are potted up in 7cm (2³/₄in) or 9cm (3¹/₂in) pots, and overwintered in a frame or cold greenhouse, they may flower the following spring. Left in the open ground, they will probably require at least another season to reach flowering size.

DIVISION
The easiest method of propagation for many pulmonarias is by division of rhizomes. This is usually done after flowering, or in autumn, for plants grown in the open ground. Monksilver

Nursery have found that their stock plants grown in pots are best divided in late July or August. Some pulmonarias, like 'Blue Ensign', 'Blue Pearl' and 'Mawson's Blue', split readily into many pieces, while others, such as *P. officinalis* and 'Barfield Regalia', may still only have one crown at the end of the season, which cannot be divided. Make sure that the soil is moist before lifting the plants carefully. Wash the soil or compost off the rootstock (though this is not strictly necessary), and pull the plant into pieces, with at least one leaf rosette per section of rhizome. Discard the oldest parts, leaving healthy-looking, strong stock. Especially during warm weather, it is a good idea to trim the leaves to about 7–10cm (3–4in) above the crown level, to reduce water loss. Either pot up the sections and keep in a shaded position, watering well, until autumn, when they can be planted or, for larger divisions, plant into the ground straight away.

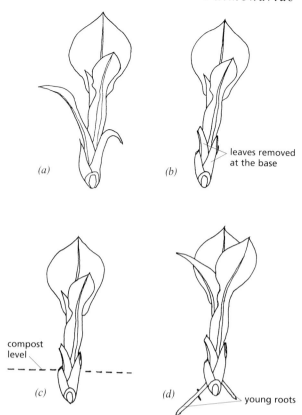

Pulmonaria propagation by shoot cuttings. (a) unprepared cutting; (b) prepared cutting; (c) inserted cutting; (d) rooting.

SHOOT CUTTINGS

Some pulmonarias, such as *P. rubra* and its forms, and *P.* 'Blue Ensign', can also be propagated by means of so-called 'top'-cuttings. This method produces a larger number of young plants than division, but is more difficult and mainly used in commercial propagation. The parts of the plant used are the new vegetative shoots that can be found at the base of the flowering stem just before or during flowering, so the timing for this propagation method will coincide with the blooming period of pulmonaria species or cultivar. These shoots should be carefully cut off the parent plant with a sharp knife. Remove the basal leaves, trying not to damage the base of the cutting, then dip it in hormone rooting powder and insert in a sand bed covered with clear polythene. It is essential to keep the cuttings out of direct sunshine, as too much heat will kill them. Shade the frame if necessary, and provide ventilation on warm, sunny days. Rooting should take place within about 3–4 weeks, but is dependent on temperature. If the sand is too cold, the cuttings may fail to root. Unusually, in *P. rubra* varieties and *P.* 'Blue Ensign', even flowering shoots can be used as cuttings.

ROOT CUTTINGS

Some pulmonarias, especially those that originate from *P. longifolia* and *P. saccharata*, are suitable for propagation by root cuttings. The best time to do this is the autumn/fall. Use the strong-looking young roots of the current season's growth. These should be cut into sections about 5cm (2in) long, and inserted in deep trays of freely draining compost mix. The trays should be overwintered in a cool but frost-free

place, like a greenhouse or garage. A cold frame is also suitable as long as the cuttings are given some extra protection, such as bubble polythene or straw, in frosty weather. It is important to note that the variegated cultivar, *P. rubra* 'David Ward', will not come true from root cuttings, but will revert back to the original plain green-leaved form.

TISSUE CULTURE/MICROPROPAGATION

Large commercial nurseries, such as Terra Nova Nurseries (USA), have been propagating pulmonarias successfully by tissue culture, which allows rapid multiplication of newly raised hybrids. It involves taking a single shoot tip, which is then proliferated in a test tube, eventually producing hundreds or even thousands of young plantlets.

However, it has been noted that during micropropagation, sports may arise that are not true to type (Gill Payne, Pulmonaria Group Newsletter, Nov 2000), so careful monitoring of the micropropagated stock is essential to maintain the purity of each cultivar.

Because this method requires the use of costly equipment and high technological expertise, only

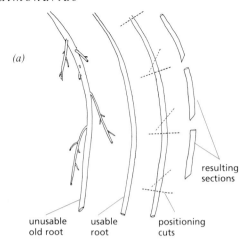

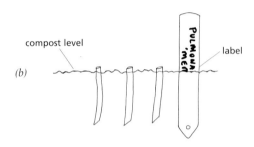

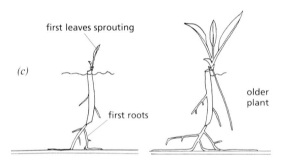

Propagation by root cuttings. (a) selecting and preparing cuttings; (b) cuttings inserted in tray of compost; (c) developing young plants.

some of the newer, 'trendy' cultivars are normally propagated by this method, which reduces the time between the selection of a new plant and introducing it to the mass market because of the vast number of young plantlets that can be raised from a single shoot.

PROBLEMS

Pulmonarias are relatively resistant to most pests and diseases, but a small number of organisms, described below, may cause trouble.

PESTS

Insects Sawfly larvae may feed on pulmonaria foliage and, if they go unnoticed, can destroy whole leaves.

Adult sawflies cause no harm to the plants, but the larvae can be a great nuisance. Superficially, they resemble caterpillars, but can be distinguished by having at least six pairs of fleshy prolegs on their abdomens. Collect them by hand as soon as you see them. Contact insecticides, such as malathion, can be used for control.

At least two kinds of aphids attack pulmonarias in the UK, one of them causing purple-red spots on foliage. However, the most serious damage incurred by aphids is the transmission of viruses. These insect pests can be controlled with a proprietary pesticide, such as malathion, permethrin, derris or insecticidal soap.

Pulmonarias may be attacked by moth caterpillars, which tear holes in the center or edges of leaves, with the most damage occurring in November to December (in England). The caterpillars should be picked off by hand when seen, or, if the infestation is severe, a proprietary insecticide can be used.

Thrips cause brown patches on leaves – on close inspection, each patch is an interveinal space sucked dry. They feed at night on upper leaf surface. *Pulmonaria angustifolia* hort. and *P.* 'Blue Ensign' seem especially susceptible.

Molluscs Slugs and snails do feed on pulmonarias, especially the young growth, but their hairy foliage is probably less palatable than the smooth leaves of hostas and other such delicacies, so if there is a choice between the two, the molluscs are more likely to leave pulmonarias alone. Of course, the young seedlings of any plant are vulnerable to slug damage and should always be protected, especially when they are grown in the open. *Pulmonaria rubra* 'David Ward' is more susceptible to slug and snail damage than any other form.

These pests can be collected by hand, or trapped. If you choose to use a molluscicide, the type based on aluminium sulphate is more environmentally friendly than methaldehyde pellets.

Birds Birds sometimes attack and disfigure pulmonaria flowers in spring, especially those of white-flowered cultivars. Lengths of black cotton sewing thread stretched horizontally around the plants should prevent this problem.

Mammals The most common mammal pests – deer and rabbit – are not usually a problem, as pulmonarias appear to be somewhat unpalatable to these animals, probably because of their rather rough, hairy leaves. However, in harsh winter weather, when there is little food around, almost no plant is safe. In the USA, woodchucks have been known to eat pulmonarias, and fencing may be needed to keep these rodents away. If you like to collect seed from your plants, mice could be a nuisance if they get there before you.

DISEASES

Fungi Powdery mildews are caused by ascomycete fungi of the family Erysiphaceae. A powdery white coating on leaves and occasionally other parts disfigures the plant. In severe cases, most of the plant may be covered with mildew. Mildew does not usually kill pulmonarias, but it restricts gaseous exchange in the affected area, and decreases the efficiency of photosynthesis, thus weakening the plant. Good air circulation and regular watering (avoiding leaves) may work as preventive measures. If the disease does strike, fungicides will have little effect. If the problem is severe, you can try cutting off the worst-affected leaves or even all of them, if necessary. Pulmonarias that lose a number of leaves through this should be fed with a general-purpose fertilizer, which will help them recover.

Pulmonarias vary in the degree of susceptibility to the mildew fungus. Some of the more or less mildew-resistant *Pulmonaria* cultivars include 'Margery Fish', 'Mary Mottram' and 'Wendy Perry', and among the more susceptible ones are 'Mrs Moon', 'Prestbury Pink' and 'Sissinghurst White'.

Viruses Viral infection can be transmitted from plant to plant by aphids or nematodes. There appear to be two main types of virus that attack pulmonarias. One results in deformed leaves which acquire a narrow, pointed shape with jagged edges. Another one causes pale green distorted patches on leaves. In either case, the plants may fail to flower and the whole plant is weakened, becoming more susceptible to other ailments, including mildew. Viral diseases in these plants are incurable, and the only thing we can do is to destroy the affected plants by burning, to prevent the spread of virus.

OTHER DISORDERS

Chlorosis Some pulmonarias, such as *P. angustifolia* hort. and *P. rubra*, and their cultivars, may suffer from chlorosis on soils that are strongly alkaline. If growing these on chalk, mix plenty of leaf mould or moss peat into the soil before planting, and mulch annually with these organic materials. Proprietary soil-acidifying chemicals are available in garden centres, but must be applied repeatedly to maintain lower pH.

Wind damage Though low-growing, pulmonarias are susceptible to damage by wind, which causes unsightly brown bruising on leaves. Growing these plants under the canopy of trees or shrubs – in their natural conditions – should prevent this problem. If they have to be positioned in the open, try to provide some shelter by means of a windbreak.

Frost damage Though most pulmonarias are very hardy and grow happily in Zones 3 and 4, some frost damage is possible, especially if freezing temperatures combine with little or no snow cover. The earliest flowers, on such as *P. rubra*, are often frosted, though the plant withstands the temperatures without any lasting harm. In very cold climates – Zones 1 or 2 – pulmonarias will require some type of protection, like a cover of straw or bracken, or, alternatively, could be planted in containers and transferred under cover for the winter months.

Drought During prolonged periods of dry weather, pulmonarias are likely to suffer. If the basal leaves are beginning to wilt, it is essential to water plentifully – allow at least 10 litres (a large bucketful) per square metre/yard of soil. It is best to water in the evening to minimize moisture loss through evaporation.

PULMONARIAS IN FOLKLORE

A few pulmonaria common names reflect some old beliefs about these plants. As already mentioned, the spotted leaves were taken to indicate the plant's use for diseased lungs, which gave rise to the rather unattractive name of 'lungwort'. A more appealing legend is that the silvery-white spots originated from the Virgin Mary's milk falling on them – hence the vernacular names for *P. officinalis* such as 'Mary-spilt-the-milk' and 'Virgin Mary's milkdrops'. This biblical connection is also reflected in the names like 'Jerusalem cowslip' or 'Bethlehem sage'.

Those interested in Zodiac signs may like to know that lungwort, *P. officinalis*, is one of four plants (the others are bilberry, meadowsweet and rose hips) that are associated with *Pisces* (19 February–20 March). Jupiter is said to 'own' this herb.

'Flower essence' practitioners believe that the lungwort 'opens psychic airways' and prescribe it to 'loosen the blockage of repressed emotion'.

Pulmonaria species in cultivation

Pulmonaria affinis Jordan in F. W. Schultz
Ref: *Arch. Fl. Fr. Allem.* 321 (1854)
A clump-forming deciduous *Pulmonaria*, growing to around 30cm (12in) high, with rough, hairy, usually white-spotted leaves and purple to violet-blue, funnel-shaped flowers in early spring. Rather rare in cultivation.

NAMING
Etymology: *affinis* – related.
Synonym: *Pulmonaria saccharata* subsp. *affinis* (Jord.) Nyman.

DESCRIPTION
General: A hairy clump-forming perennial.
Stems: 30cm (12in) high.
Leaves: Oval, pointed, to 18cm (7in) long and 9cm (3½in) wide, usually white-spotted, abruptly narrowing into a winged stalk to 18cm (7in) long. Leaves feel rough to touch, being covered with many short bristly hairs, with some scattered long hairs and glandular hairs.
Inflorescence: With unequal bristly hairs and long glandular hairs.
Flowers: *Corolla* purple to blue-violet, tube interior hairless, tube below the ring of hairs.
Nutlets: To 4 x 2.5mm.
Chromosomes: 2n=22.

GEOGRAPHY AND ECOLOGY
Distribution: Central and western France and northern Spain.
Habitat: Woods at 400–2100m (1300–6890ft) in Spain.
Flowering: March to May in the wild, occasionally into June.
Associated species: In La Garrotxa in Spain, grows in moist forests of oak (*Quercus robur* and *Q. humilis*) that also contain ash (*Fraxinus excelsior*), field maple (*Acer campestre*), smooth-leaved elm (*Ulmus minor*), small-leaved lime (*Tilia cordata*) and box (*Buxus sempervirens*). Other plants found in this habitat are wayfaring tree (*Viburnum lantana*), male fern (*Dryopteris filix-mas*), early dog-violet (*Viola reichenbachiana*), barren strawberry (*Potentilla sterilis*) and greater stitchwort (*Stellaria holostea*). In El Vidranès, *P. affinis* inhabits forests of beech (*Fagus sylvatica*), growing in company with both green and stinking hellebores (*Helleborus viridis* and *H. foetidus*), wood anemone (*Anemone nemorosa*),

Hepatica nobilis, cowslip (*Primula veris*), wood spurge (*Euphorbia amygdaloides*), spurge laurel (*Daphne laureola*) and Pyrenean squill (*Scilla liliohyacinthus*).

CULTIVATION
Hardiness: Zone 4.
Sun/shade aspect: More tolerant of full sun than most other pulmonarias.
Soil requirements: Any well-drained soil that does not dry out.
Garden uses: As for the genus.
Spacing: 30cm (12in).
Propagation: As for the genus.
Availability: 4. Uncommon in cultivation, but may be found in some specialist herbaceous nurseries.

RELATED FORMS AND CULTIVARS
Pulmonaria affinis may hybridize with *P. longifolia* in the wild, and the resultant hybrids have been named *Pulmonaria ovalis* (Bastard) Boreau. We do not know whether this hybrid has ever appeared in cultivation. Only one cultivar of *P. affinis* is known: 'Margaret' (see page 50).

Pulmonaria angustifolia Linnaeus
Ref: *Sp. Pl.* ed. 1 135 (1753)
A hairy, deciduous *Pulmonaria* about 20–30cm (8–12in) high, with rosettes of long, narrowly lanceolate, plain dark green leaves, and smallish, tubular, intensely blue flowers, which are produced much later in spring than in most other species. The true *P. angustifolia* is probably very rare in cultivation, but if available, it would make a good plant for either a partially shaded border or a woodland garden.

NAMING
Etymology: *angustifolia* – narrow-leaved.
Synonyms: *Pulmonaria azurea* Besser; *Pulmonaria angustifolia* subsp. *azurea* (Besser) Gams; *Pulmonaria maculata* Stokes; *Pulmonaria stiriaca* Kern.
Nomenclature note: Alan Leslie and Joe Sharman, who have observed and identified the true *P. angustifolia* in the wild, believe that the plant known in cultivation as *P. angustifolia* is distinct from that species, and is either a different species or a hybrid. However, as this plant continues to be listed as *P. angustifolia* in the *RHS Plant Fnder* and other major reference books, we will use this name here, to avoid confusion, but will refer to it as *P. angustifolia* hort. (see separate description below) until its true status is clarified.

Common names: *English*: Blue Lungwort, Cowslip Lungwort, Blue Cowslip; *Danish*: Himmelblå Lungeurt; *German*: Schmalblättriges Lungenkraut; *Norwegian*: Blå lungeurt; *Russian*: medunitza uzkolistnaya; *Slovak*: pl'úcnik úzkolisty; *Swedish*: Smalbladig lungört.

(Some of these names may refer to *P. angustifolia* hort. – see separate description below)

Description

General: A rather softly haired perennial, usually forming a few large rosettes.

Stems: To 30cm (12in) high, hairy.

Leaves: Basal leaves to 40cm (16in) long and 5cm (2in) wide, narrowly lanceolate, with the base of leaf blade narrowing gradually into the stalk. Stem leaves narrowly lanceolate to narrowly elliptical, stalkless. All leaves usually plain green, unspotted, covered with hairs of more or less uniform length, occasionally slightly glandular.

Inflorescence: Bristly-hairy, sparingly glandular.

Flowers: *Calyx* short, slender. *Corolla* bright blue, rather narrow, with the limb spreading only slightly; tube hairless inside below the ring of hairs in throat.

Nutlets: To 4.5 × 3.5mm.

Chromosomes: 2n=14, 28.

Geography and ecology

Distribution: Austria, Denmark, France, Germany, Sweden, Hungary, Italy, Poland, Czech Republic, Russia, the Baltic, Ukraine, Belarus.

Habitat: Woods, woodland margins and meadows, usually on acid soils, to 2600m (8500ft) above sea level.

Flowering: April to July in the wild – this is the latest-flowering pulmonaria.

Associated plants: In southern Sweden, in dry hills of Vastergotland, grows with fountain grass (*Stipa pennata*), large self-heal (*Prunella grandiflora*) and northern dragonhead (*Dracocephalum ruyschianum*). In western Poland, it inhabits woods dominated by sessile oak (*Quercus petraea*), where it grows with white cinquefoil (*Potentilla alba*), bloody crane's-bill (*Geranium sanguineum*), multiflowered buttercup (*Ranunculus polyanthemos*), angular spurge (*Euphorbia angulata*), German greenweed (*Genista germanica*), white asphodel (*Asphodelus alba*), variegated iris (*Iris variegata*) and mountain sedge (*Carex montana*). In southern Italy, it can be found with *Helleborus multifidus* subsp. *istriacus*, *Cyclamen hederifolium*, *Hepatica nobilis*, cowslip (*Primula veris*) and *Crocus vernus* subsp. *albiflorus*.

Cultivation

Hardiness: Zone 3.

Sun/shade aspect: As for the genus.

Soil requirements: In the wild, *P. angustifolia* prefers acid to neutral soil, so such conditions are likely to suit it in cultivation.

Garden uses: As for the genus.

Planting: 30cm (12in) apart.

Problems: Limited availability and confusion between the true species and the spurious '*P. angustifolia*' hort. As this species is totally deciduous, it provides no interest in winter, unlike other pulmonarias. It is likely to suffer from chlorosis in alkaline soils.

Propagation: As for the genus.

Availability: We believe that the true *P. angustifolia* is rarely, if ever, available. Monksilver Nursery currently holds small stocks of the true species and this may be on offer in due course.

Related forms and cultivars

We believe that at present there is only one named pulmonaria form in cultivation that originates from the true *P. angustifolia*:

'Alba' (see page 50).

The remaining cultivars commonly assigned to *P. angustifolia* belong to a different species we call here *P. angustifolia* hort.

Pulmonaria angustifolia hort. ♀

A semi-deciduous, sparsely hairy *Pulmonaria* 25–30cm (10–12in) high, forming many small clumps of elliptical-lanceolate to narrowly oval, plain green leaves, and with large funnel-shaped bright blue flowers in spring. Sold as *P. angustifolia*, but differs from the true species by a number of characteristics (see below). A wonderful shade of blue and is worth the space in any garden. This plant gained the Royal Horticultural Society's Award of Garden Merit in 1993.

'Elliptic leaves, bristly and of plain dark green opening with the first flowers. These are carried in little sprays, pink in tight bud, opening a pure rich blue.' (Graham S. Thomas)

Naming

Etymology: *angustifolia* – narrow-leaved.

Common names: At least some of the names given for the true *P. angustifolia* L., are likely to have been applied to this plant.

Nomenclature note: There is little doubt that this is not the true *P. angustifolia* described by Linnaeus, in spite of being listed as such in all catalogues and reputable books, including *The RHS Plant Finder*.

Pulmonaria angustifolia hort. (Howard Rice)

This is clearly an altogether different species, or possibly a hybrid, the identity of which is unclear.

DESCRIPTION

The plant known in cultivation as '*P. angustifolia*' differs from the true species by a number of characteristics:

• It is only sparsely, and not as softly hairy, as the true *P. angustifolia*.

• The leaves are never as long, and are wider – lanceolate to narrowly oval, as opposed to narrowly lanceolate in the true *P. angustifolia*.

• The corolla is larger (about 18mm (³/₄in) long), with a wider expanded limb, and flowers are earlier than in the true species – early to mid-spring, rather than from mid-spring into summer.

• The habit of this plant is also different, forming small clumps/tufts rather than a few large rosettes.

GEOGRAPHY AND ECOLOGY

Distribution: Not known, as this plant is yet to be identified.

Habitat: Not known, but we can probably assume that, in the wild, this species would be found in habitats similar to other pulmonarias, i.e. woodland and scrub.

Flowering: In cultivation in the UK, usually flowers in March to April.

CULTIVATION

Date of introduction: 1731 – this date has been given for the true *P. angustifolia*, but it is likely to be this 'spurious' species.

Hardiness: Zone 3.

Sun/shade aspect: As for the genus.

Soil requirements: As for the genus. On alkaline soils, it may suffer from chlorosis.

Garden uses: As for the genus.

Spacing: 30cm (12in) apart.

Propagation: As for the genus. The clumps readily fall apart into small pieces, which makes this the most convenient method of propagation.

Problems: The confusion with the identity of this species. The plant is liable to lose its leaves in harsh winters and sometimes during dry summers, and is susceptible to chlorosis on lime-rich soils.

Availability: 18. Available as *P. angustifolia* from many herbaceous and general nurseries.

RELATED FORMS AND CULTIVARS

The forms listed below were assumed to originate from the true *P. angustifolia* L., but we are confident that the parent for some of them is a different species, which here we call *P. angustifolia* hort.

'Azurea' (subsp. *azurea*)   (see page 49)
'Blaues Meer'   (see page 51)
'Munstead Blue'   (see page 51)

Pulmonarias 'Beth's Blue' (see page 52), 'Beth's Pink' (see page 52) and 'Blue Pearl' (see page 53) were sometimes listed under this species but are not related to it, while 'Blue Ensign' (see page 52), though listed as a hybrid, is clearly affiliated to this species.

Pulmonaria longifolia (Bastard) Boreau

Ref: *Fl. Centre Fr.* ed. III. 2: 460 (1857)

A semi-evergreen clump-forming *Pulmonaria*, to 30cm (12in) high, with narrowly lanceolate, usually conspicuously white-spotted leaves, and rather dense clusters of funnel-shaped flowers turning from pink to violet or bright blue. One of the most popular pulmonaria species, and deservedly so because, at its

Pulmonaria longifolia. (Howard Rice)

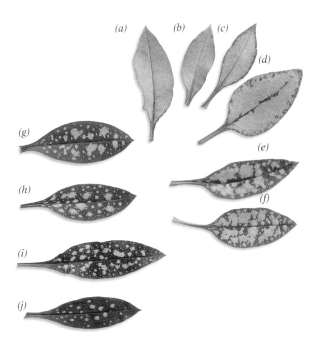

Leaves of *P. longifolia* and *P. saccharata* forms. (Howard Rice) (a) *P.* 'Majesté'; (b) *P.* 'Cotton Cool'; (c) *P.* 'Excalibur'; (d) *P. saccharata* Argentea Group; (e) *P. longifolia*; (f) *P.* 'Merlin'; (g) *P.* 'Blue Crown'; (h) *P.* 'Little Star'; (i) *P.* 'Roy Davidson'; (j) *P.* 'Blauer Hugel'.

best, it is an extremely attractive plant both in foliage and flower. Brightly spotted leaves will light up a semi-shaded border or a woodland garden.

NAMING
Etymology: *longifolia* – long-leaved.
Synonyms: *Pulmonaria angustifolia* invalid, in part, not L.; *Pulmonaria angustifolia* subsp. *longifolia* (Bastard) P. Fourn.; *Pulmonaria vulgaris* invalid, in part; *Pulmonaria vulgaris* Merat in part.
Common names: *English:* Narrow-leaved Lungwort; *Portuguese:* Pulmonaria. *Welsh:* Llys yr Ysgyfaint Culddail.

DESCRIPTION
General: A bristly-hairy glandular perennial, forming congested clumps.
Stems: Upright, 20–40cm (8–16in), not scaly at base.
Leaves: Basal leaves to 40–60cm (16–24in) long and up to 6cm (2¹/₂in) wide, narrowly lanceolate, gradually narrowed into a stalk, upper surface usually spotted with white or pale green, or sometimes unspotted (found especially in W and C France). Stem leaves lanceolate, or oval-lanceolate, stalkless, half-clasping the

stem. Hairs covering the leaves are bristly, all more or less the same length, mixed with some glandular hairs.
Inflorescence: Cymes short, scarcely elongating after flowering, with long bristly hairs, slightly glandular.
Flowers: *Calyx* lobes lanceolate, pointed. *Corolla* 8–12mm (³/₈–¹/₂ in) long, pink turning blue or violet, tube interior hairless below the ring of hairs in throat.
Nutlets: To 4 × 3mm, strongly compressed, ovate, crested, shining.
Chromosomes: 2n=14.

GEOGRAPHY AND ECOLOGY
Distribution: Western Europe including Britain, France, Spain, Portugal. From Sweden to France. Native in the UK in S Hants (New Forest), Dorset and on the Isle of Wight, where it is found very locally.
Habitat: Semi-shaded habitats, including woodland and scrub, sometimes among rocks, usually on rather heavy clay soils, to 2000m (6550ft) above sea level.
Flowering: April to May in the wild.
Associated plants: In the New Forest (UK), grows in same habitat as wild daffodil (*Narcissus pseudonarcissus*), bluebell (*Hyacinthoides non-scripta*),

Pulmonaria longifolia subsp. *cevennensis*. (Dan Heims)

wood-sorrel (*Oxalis acetosella*), bastard balm (*Melittis melissophyllum*), wood anemone (*Anemone nemorosa*) and columbine (*Aquilegia vulgaris*). In Spain, inhabits forests of *Quercus pubescens* and *Q. humilis* with Scots pine (*Pinus sylvestris*), or beech (*Fagus sylvatica*) woods.

CULTIVATION

Hardiness: Zone 5.

Sun/shade aspect: Tolerates sun better than most other pulmonarias.

Soil requirements: As for the genus, but somewhat more tolerant of dryness. Can grow on very heavy clay soil.

Garden uses: As for the genus.

Spacing: 45cm (18in).

Propagation: As for the genus.

Problems: Unlike most other pulmonarias, it will not grow well in heavy shade.

Availability: 29. Widely available in nurseries and garden centres.

RELATED FORMS AND CULTIVARS

A number of subspecies are known in *P. longifolia*. They include **P. longifolia subsp. *cevennensis*** M. Bolliger, from France and Spain (see page 58), **subsp. *delphinensis*** M. Bolliger, from France, and **subsp. *glandulosa*** M. Bolliger, from Portugal and Spain. Of these three subspecies, only the first one is known to be in cultivation. The following cultivars are recognized as *P. longifolia* forms:

'Ankum' ('Coen Jansen')   (see page 59)
'Bertram Anderson'   (see page 59)
'Dordogne'   (see page 60)

Many other cultivars, usually listed as hybrids, are closely related to this species, for example 'Merlin', 'Coral Spring', 'Roy Davidson' and 'Weetwood Blue'.

Pulmonaria mollis Wulfen in Hornemann

Ref: *Hort. Hafn.* 1: 179 (1813)

A deciduous *Pulmonaria*, growing to a clump of 45–60cm (18–24in) high and wide, with very long, plain dark green, softly hairy leaves that are held upright at first, and large funnel-shaped blue flowers that often fade to purplish-pink. The largest of all pulmonaria species, it makes good ground cover in any lightly shaded spot, but loses its leaves in winter.

'The largest-growing species with velvety, long deep green leaves. The flowers are of rich deep blue, fading to purple and coral-red, though by far the most remain blue. Prolific and admirable.' (Graham S. Thomas)

NAMING
Etymology: *mollis* – soft, referring to leaves.
Synonyms: *Pulmonaria montana* subsp. *mollis* Gams;
Pulmonaria montana invalid, not Lej.; *Pulmonaria dacica* invalid, not Simonkai; *Pulmonaria mollissima* A. Kerner.
Common names: *Chinese*: Xian mao fei cao; *Finnish*: pehmoimikkä; *French*: Pulmonaire molle; *German*: Weiches lungenkraut; *Italian*: Polmonaria morbida; *Polish*: miodunca miekkowlosa; *Russian*: medunitza myagkaya; *Slovak*: pl'úcnik mäkky.

DESCRIPTION
General: A large, softly hairy, glandular clump-forming perennial, arising from a rather thick short black rhizome.
Stems: To 45cm (18in) high.
Leaves: Basal leaves oval to elliptical-lanceolate, up to 60cm (24in) long and 12cm (4¹/₂in) wide, normally unspotted, narrowed gradually into a stalk that can be shorter or as long as the blade. The upper surface distinctly soft to touch, with a mixture of dense, slender, short hairs, scattered long hairs and glandular hairs.
Inflorescence: Densely glandular-hairy, sticky to touch.
Flowers: *Corolla* violet to blue violet, tube densely hairy inside below the ring of hairs.
Nutlets: To about 5 × 4mm, ovoid, nearly hairless.
Chromosomes: 2n=18.

GEOGRAPHY AND ECOLOGY
Distribution: Central and south-eastern Europe, central Asia (Siberia), Mongolia, China (Shanxi).
Habitat: Woodland, scrub, other shaded habitats, rocky places.
Flowering: Spring.
Associated plants: In Siberia and central Urals grows in mixed silver birch (*Betula pendula*) and Scots pine (*Pinus sylvestris*) woodland, also containing rowan (*Sorbus aucuparia*), wild cherry (*Prunus avium*), *Crataegus sanguinea* and *Chamaecytisus ruthenicus*. Here *P. mollis* shares its habitat with globeflower (*Trollius europaeus*), tormentil (*Potentilla erecta*), wild strawberry (*Fragaria vesca*), germander speedwell (*Veronica chamaedrys*), cat's-foot (*Antennaria dioica*), twayblade (*Listera ovata*), bird's-nest orchid (*Neottia nidus-avis*) and creeping lady's-tresses (*Goodyera repens*).

  In Hungary, it is found in oak (*Quercus* species) woodland with *Acer tataricus*, service tree (*Sorbus torminalis*), dogwood (*Cornus sanguinea*), spindle (*Euonymus europaeus*), *E.verrucosus*, wild pear (*Pyrus pyraster*) and blackthorn (*Prunus spinosus*).

Pulmonaria mollis. (Dan Heims)

Herbaceous species accompanying it can include black pea (*Lathyrus niger*), blue gromwell (*Buglossoides purpurocaeruleum*), wood avens (*Geum urbanum*), *Tanacetum corymbosum* and *Polygonatum latifolium*.

CULTIVATION
Date of introduction: 1816.
Hardiness: Zone 4.
Sun/shade aspect: Full or light shade.
Soil requirements: As for the genus. Does not like drying out, but withstands it better than some other pulmonarias.
Garden uses: As for the genus. Especially good for ground cover due to its size.
Planting: Space 45–60cm (18–24in) apart.
Problems: Needs plenty of space to achieve its full potential, and dies down in winter, unlike many other pulmonarias.
Propagation: As for the genus.
Availability: 19. Mainly from specialist herbaceous nurseries.

RELATED FORMS AND CULTIVARS
Pulmonaria mollis **subsp.** *alpigena* W. Sauer is native to Germany, Austria and Switzerland but is probably not in cultivation. Two cultivars are grown:
'Royal Blue' (see page 60)
'Samobor' (see page 60)

Pulmonaria obscura Dumort

Ref: *Bull. Soc. Bot. Belg.* 4: 341 (1865)

A clump-forming semi-evergreen *Pulmonaria*, to 30cm (12in) high, rather similar to the better-known *P. officinalis* but usually with plain green rather than spotted, heart-shaped leaves. Flowers purple-pink turning purplish-blue, in spring. An uncommon species of interest to a *Pulmonaria* collector, but not as spectacular as some other species and cultivars.

NAMING

Etymology: *obscura* – dark, probably refers to darker green leaf colour than in the similar *P. officinalis*; could also refer to the indistinct spotting of leaves.
Synonyms: *Pulmonaria officinalis* subsp. *obscura* (Dumort.) Murb.; *P. officinalis* invalid, *Pulmonaria officinalis immaculata*.
Common names: *English*: Suffolk lungwort, Unspotted lungwort; *Danish*: Almindelig lungeurt; *Finnish*: Imikkä, Lehtoimikkä; *French*: Pulmonaire sombre; *German*: Echtes lungenkraut, Dunkelgriunes lungenkraut; *Italian*: Polmonaria scura; *Norwegian*: Lungeurt; *Polish*: miodunka c´ma; *Russian*: medunitza neyasnaya, medunitza tyomnaya; *Slovak*: pl'úcnik tmavy´; *Swedish*: (vanlig) lungört.

DESCRIPTION

General: A bristly, semi-evergreen, clump-forming perennial, rather similar to *P. officinalis*.
Stems: 15–30cm (6–12in), densely covered in fine prickles, and sparsely with coarse bristle-like hairs, with few or no glands.
Leaves: Blade of basal leaves 20–30cm (8–12) long and 2–5cm (1–2in) wide, narrowly oval, more or less unspotted (rarely weakly discoloured), with heart-shaped or truncate base. Leaves usually shorter than stalks. Stem leaves oval-lanceolate, 4–7cm ($1\frac{1}{2}$–$2\frac{3}{4}$in) long and 1. 5–4cm ($\frac{1}{2}$–$2\frac{3}{4}$in) wide, stalkless, wedge-shaped at base.
Inflorescence: With sparse glandular hairs
Flowers: *Pedicels* 3–5mm (to 10mm in fruit), densely hairy and bristly. *Calyx* 8–11 mm bristly-hairy with few or no glands. *Corolla* is violet at first, becoming more or less blue, tube 10–15mm, hairless within except for ring of hairs, limb 7–10mm in diameter.
Chromosomes: 2n=14.

GEOGRAPHY AND ECOLOGY

Distribution: Much of Europe, including Britain, France, Germany, Switzerland, Austria, Denmark, Sweden, Finland, Czech Republic, Slovakia, Poland, Bulgaria, Hungary, Romania, the Balkans, Russia, the Baltic, Belarus, Ukraine, Moldova; also Turkey and

Leaves of *Pulmonaria officinalis* forms and cultivars. (Howard Rice) (a) *P.* 'Dora Bielefeld'; (b) and (d) *P. officinalis* 'Blue Mist'; (c) *P.* 'Oliver Wyatt's White'; (e) *P.* 'Marlene Rawlins'; (f) *P.* 'Glacier'; (g) *P.* 'Sissinghurst White'; (h) *P.* 'Nürnberg'; (i) *P. officinalis* Cambridge Blue Group; (j) *P. officinalis*; (k) *P.* 'Blauhimmel'.

Siberia in Asia. In England it has been confirmed native in Suffolk woodland.
Habitat: Coniferous and deciduous woodland, streamsides, up to abut 500m (1650ft) above sea level.
Flowering: Early to late spring.
Associated plants: In central Russia, grows in taiga of Norway spruce (*Picea abies*) and *P.* x *fennica*, also containing some lime (*Tilia*), with plants such as sweet woodruff (*Galium odoratum*), chickweed wintergreen (*Trientalis europaea*), greater stitchwort (*Stellaria holostea*), yellow archangel (*Lamium galeobdolon*), May lily (*Maianthemum bifolium*), lady fern (*Athyrium felix-femina*) and broad buckler fern (*Dryopteris dilatata*). Near Lake Ladoga, it is found in Scots pine (*Pinus sylvestris*) woods, with wood-sorrel (*Oxalis acetosella*), spring vetchling (*Lathyrus vernus*) and hepatica (*Hepatica nobilis*). In Turkey, grows on streamsides in hornbeam (*Carpinus*) and oak (*Quercus*) woodland.

CULTIVATION
Hardiness: Zone 4.

Sun/shade aspect: As for the genus.
Soil requirements: As for the genus.
Garden uses: As for the genus. Like *P. officinalis,* this species has been an important medicinal herb and used as a food.
Spacing: 30cm (12in).
Propagation: As for the genus.
Problems: A plant of humble appearance, *P. obscura* is outshone by most other pulmonarias both in leaf and flower.
Availability: 1. Can be found in some specialist herbaceous nurseries.

Pulmonaria officinalis Linnaeus

Ref: *Sp. Pl.* ed. 1. 135 (1753)
A clump-forming semi-evergreen *Pulmonaria* to 25–30cm (10–12in) high, with roughly hairy, variously white-spotted, oval to heart-shaped leaves, and clusters of bright rose-pink flowers turning to purple-blue, in spring. The oldest pulmonaria in cultivation, grown for many centuries as a medicinal herb (see Medicinal Uses, page 32). At its best it deserves a place in any garden where a suitable partially shaded spot can be found.

> *'... how welcome are those jaunty sprays of innocent flowers in blue and pink, often rising above the snow, and fluttering unconcernedly in the bitter winds of March.' (Margery Fish)*

NAMING

Etymology: *officinalis* – sold as a herb.
Synonyms: *Pulmonaria maculosa* Liebl.; *Pulmonaria officinalis* subsp. *maculosa* (Hayne) Gams; *Pulmonaria maculata* F.G. Dietr.; *Pulmonaria mollis* Tenore, not Wulfen.
Common names: *English*: Lungwort, Jerusalem cowslip, Jerusalem sage, Spotted dog, Soldiers and sailors, Boys and girls, Bloody butcher, Hundreds and thousands, Joseph and Mary, Mary-spilt-the-milk; *Danish*: Hvidplettet Lungeurt; *Dutch*: Gevlekt Longkruid; *French*: Pulmonaire officinale; *German*: Echtes Lungenkraut, Gemeines lungenkraut, Gewhnliches lungenkraut; *Italian*: Polmonaria maggiore; *Norwegian*: Lækjelungeurt; *Polish*: miodunka plamista; *Russian*: medunitza lekarstvennaya; *Slovak*: pl'úcnik lekársky; *Swedish*: Fläcklungört, fläckbladig lungört, fläckig lungört, skånsk lungört, vitfläckig lungört. *Welsh*: Llys yr Ysgyfaint.

DESCRIPTION

General: A very variable, semi-evergreen, bristly perennial, arising from rather long, thin rhizomes.

Stems: 10–30cm (4–12in) high, bristly with sparse glandular hairs, with brown scales at the base.
Leaves: Basal leaves to 16cm (6^1/$_4$in) long and 10cm (4in) wide, oval, pointed, often with a heart-shaped base, abruptly narrowed into a winged stalk 5–15cm (2–6in) long; usually conspicuously spotted white, rough with stiff hairs of more or less equal length. Stem leaves smaller, stalkless, partly clasping the stem.
Inflorescence: Short paired, few-flowered cymes, densely hairy, with stiff bristles and some glandular hairs.
Flowers: *Calyx* 6–7mm (to 10mm or longer in fruit), nearly cylindrical or narrowly bell-shaped, coarsely bristly; lobes, a third to a quarter the length of the calyx, triangular, bluntish or pointed. *Corolla* pink-red turning violet or purple-blue; tube interior glabrous below the ring of hairs in throat, limb bell-shaped, 7–10mm across.
Nutlets: To 4 × 3mm with two sharp narrow heels, black and downy.
Chromosomes: 2n=16.

GEOGRAPHY AND ECOLOGY

Distribution: Much of Europe: Holland, France, Switzerland, Austria, Germany, Belgium, Denmark, southern Sweden, northern Italy, Czech Republic, Slovakia, Bulgaria, Romania, Hungary, Albania, the former Yugoslavia, Russia, Caucasus. Naturalized in Britain (scattered throughout) and the USA (Wisconsin).
Habitat: Shaded and semi-shaded habitats on deep, humus-rich soils, generally over limestone, to 1900m (6230ft) above sea level. In the UK naturalized in woods, on hedgebanks in scrub, rough ground.
Flowering: March to May in the wild.
Associated plants: On Pavlov Hills in Moravia (Czech Republic), *P. officinalis* grows in forests of large-leaved lime (*Tilia platyphyllos*), hornbeam (*Carpinus betulus*) and sycamore (*Acer pseudoplatanus*), in company with garlic mustard (*Alliaria petiolata*), wild ginger (*Asarum europaeum*), spring vetchling (*Lathyrus vernus*), nettle-leaved bellflower (*Campanula trachelium*) and northern wolfsbane (*Aconitum lycoctonum*).
In the northern parts of former Yugoslavia, it is found in the same habitats as *Helleborus atrorubens, Cyclamen purpurascens,* willow gentian (*Gentiana asclepidacea*), perennial honesty (*Lunaria rediviva*), hepatica (*Hepatica nobilis*), alpine barrenwort (*Epimedium alpinum*) and *Lamium orvala,* while in the south of the same region it may grow with *Helleborus torquatus, Hepatica nobilis, Corydalis solida,* primrose (*Primula vulgaris*), sweet violet (*Viola odorata*), alpine squill (*Scilla bifolia*) and dog's-tooth violet (*Erythronium dens-canis*).

CULTIVATION
Date of introduction: Before 1597.
Hardiness: Zone 4.
Sun/shade aspect: As for the genus.
Soil requirements: As for the genus.
Garden uses: As for the genus.
Spacing: 30–45cm (12–18in) apart.
Propagation: As for the genus.
Availability: 25. Easy to find in herbaceous perennial and herb nurseries, and some garden centres.

RELATED FORMS AND CULTIVARS
The following cultivars of *P. officinalis* are grown:

'Alba'	(see page 61)
'Blue Mist' ('Bowles Blue')	(see page 61)
Cambridge Blue Group	(see page 61)
'Stillingfleet Gran'	(see page 62)
'White Wings'	(see page 62)

Pulmonaria 'Sissinghurst White' (see page 67) is probably a hybrid of this species and *P. saccharata*.

Leaves of *Pulmonaria rubra* forms and cultivars. (Howard Rice) (a) unnamed hybrid; (b) var. *albocorollata*; (c) 'David Ward'; (d) 'Barfield Pink'; (e) 'Ann'; (f) 'Warburg's Red'; (g) 'Barfield Ruby'; (h) 'Prestbury Pink'.

Pulmonaria rubra 'David Ward'. (Dan Heims)

Pulmonaria rubra Schott ♓

Ref: *Bot. Zeit.* 9: 395 (1851)
A semi-evergreen, tufted *Pulmonaria* 30–50cm (12–20in) high, with rosettes of light green, softly hairy, more or less oval leaves and coral-red flowers in late winter and spring. Received the Award of Garden Merit from the Royal Horticultural Society in 1993. Plant in a moist, shady spot.

> *'It is a fine reward for facing the draughty air of a February day to come face to face with little clusters of tight coral flowers tucked into cups of soft green foliage.'* (Margery Fish)

NAMING
Etymology: *rubra* – red, referring to the colour of flowers.
Synonyms: *Pulmonaria transsilvanica* Schur; *Pulmonaria angustifolia* 'Rubra'. Sometimes listed as synonymous with *P. montana* Lej., but this is not likely.
Common names: *English*: Red lungwort, Bethlehem sage, Christmas cowslip; *Russian*: medunitza krasnaya.

DESCRIPTION
General: A hairy perennial arising from a thin rhizome.
Stems: 20–50cm (8–20in) high, usually with long hairs and stiff bristles, glandular-hairy above.
Leaves: Basal leaves to 15cm (6in) long and 7cm (2³/₄in) wide, oval or narrowly so, pointed, with a more or less rounded base; stem leaves to 8cm (3¹/₄in) long and 2.5cm (1in) wide, oblong, upper ones stalkless.
Inflorescence: Cymes slightly elongating in fruit, paired or in small corymbs, with many long glandular hairs.

Flowers: *Pedicels* finely glandular. *Calyx* more or less tubular, 12mm long, shortly bristly and more or less glandular, with pointed lobes. *Corolla* red, never turning blue, tube is hairy inside below the ring of hairs, limb about 10mm across.
Nutlets: To 4.5 × 3mm.

GEOGRAPHY AND ECOLOGY
Distribution: The Balkans (the former Yugoslavia, Albania), Bulgaria, Romania, Carpathians, Middle East. Naturalized in Britain – scattered in northern England, central and southern Scotland.
Habitat: Moist woodland, mostly deciduous, such as beech, but sometimes ascending to coniferous forests; usually grows on slopes. In Britain, naturalized in grassy places, hedges and scrub.
Flowering: February to April in the wild.
Associated plants: In the wild, may be found growing with such plants as *Helleborus purpurascens*, *Pulsatilla halleri* and *Trollius europaeus*.

CULTIVATION
Date of introduction: Prior to 1914.
Hardiness: Zone 5.
Sun/shade aspect: As for the genus.
Soil requirements: As for the genus. Does not tolerate drying out. On poor soils, it benefits from feeding.
Garden uses: As for the genus.
Spacing: 60cm (24in) apart.
Propagation: As for the genus. Gives good results with top cuttings.
Availability: 22. Widely available from many nurseries and garden centres.

RELATED FORMS AND CULTIVARS
Pulmonaria rubra* var. *albocorollata is a naturally occurring white-flowered form (see page 63). The following cultivars of *P. rubra* are grown:

'Ann'	(see page 63)
'Barfield Pink'	(see page 63)
'Barfield Ruby'	(see page 64)
'Bowles' Red'	(see page 64)
'David Ward'	(see page 64)
'Prestbury Pink'	(see page 64)
'Rachel Vernie'	(see page 64)
'Redstart'	(see page 64)
'Warburg's Red'	(see page 65)

Pulmonaria saccharata Mill. , 1768
Ref: *Gard. Dict.* ed. 8 no. 3 (1768)
A semi-evergreen *Pulmonaria* to about 30cm (12in) high, forming dense clumps of elliptical leaves strikingly

Pulmonaria rubra 'Redstart'. (Howard Rice)

spotted or mottled silver-white. The flowers appear in early and mid-spring, opening from pink or reddish-purple buds and gradually changing to sky blue.

> *'Handsome, long, elliptic leaves, more or less heavily spotted or almost wholly grey. The flowers are conspicuous, pink buds emerging from purplish velvety calyces, bright pink turning to blue. An extremely handsome plant in its best forms.' (Graham S. Thomas)*

NAMING
Etymology: *saccharata* – literally, 'sprinkled with sugar', referring to silver-white markings on leaves.
Synonyms: *Pulmonaria officinalis* subsp. *saccharata* (Mill.) Arcang.; *Pulmonaria picta* Rouy; *Pulmonaria tuberosa* invalid, not Schrank; *Pulmonaria saccharata* 'Picta'.
Common names: *English*: Bethlehem sage, Jerusalem sage, Bethlehem lungwort.

DESCRIPTION

General: A bristly-hairy perennial.

Stems: To 45cm (18in) high.

Leaves: Basal leaves to 27cm (10³/₄in) long and 10cm (4in) wide, elliptical, pointed, with the base narrowing gradually into a stalk to 15cm (6in) long; usually conspicuously spotted silver-white, covered with more or less dense, unequal short hairs and long, slender bristles and glandular hairs.
Stem leaves ovate-oblong, stalked or nearly stalkless.

Inflorescence: Bristly-hairy, slightly glandular.

Flowers: *Calyx* purplish. *Corolla* to 18mm long, from pink or reddish-purple changing to blue or violet, occasionally white; tube hairy inside below the ring of hairs in the throat.

Nutlets: 4 x 3mm.

Chromosomes: 2n=22.

GEOGRAPHY AND ECOLOGY

Distribution: South-eastern France, northern Italy (Apennines). Naturalized in Belgium and New York state, USA.

Habitat: Woodland and scrub.

Flowering: March to May.

Associated plants: In Italy, grows in woods of beech (*Fagus sylvatica*) and hazel (*Corylus avellana*), with hornbeam (*Carpinus betulus*) and black hornbeam (*Ostrya carpinifolia*), in company of such plants as lords-and-ladies (*Arum maculatum*), butcher's-broom (*Ruscus aculeatus*), *Allium pendulinum*, twayblade (*Listera ovata*), early purple orchid (*Orchis mascula*), yellow birthwort (*Aristolochia lutea*), wood anemone (*Anemone nemorosa*), hepatica (*Hepatica nobilis*), spurge laurel (*Daphne laureola*) and wood spurge (*Euphorbia amygdaloides*).

CULTIVATION

Date of introduction: Before 1863.

Hardiness: Zone 3.

Sun/shade aspect: As for the genus.

Soil requirements: As for the genus.

Garden uses: As for the genus.

Spacing: 30–45cm (12–18in) apart.

Propagation: As for the genus. Root cuttings work well with this species

Availability: 17. This species is commonly available from many nurseries and garden centres (though sometimes it may be misnamed), but losing its popularity to some of the highly ornamental cultivars.

Pulmonaria saccharata 'Frühlingshimmel'. (Howard Rice)

RELATED FORMS AND CULTIVARS

In our opinion, many cultivars attributed to
P. saccharata are more likely to be of *P. officinalis*
parentage, or hybrids. On the other hand, *P.* 'Margery
Fish', often listed as a cultivar of the rare *P. vallarsae*,
has probably originated from this species.

We believe the cultivars listed below have been
selected from this species:

'Alba'	(see page 65)
Argentea Group	
('White Windows')	(see page 65)
'Brentor'	(see page 65)
'Diana Chappell'	(see page 65)
'Glebe Cottage Blue'	(see page 66)
'Jill Richardson'	(see page 66)
'Leopard'	(see page 66)
'Mrs Moon'	(see page 66)
'Pink Dawn'	(see page 66)
'Reginald Kaye'	(see page 63)
'White Leaf'	(see page 67)

A large number of cultivars sometimes erroneously
attributed to *P. saccharata* are either hybrids or
belong to different species. They are:

'Blauhimmel'	(see page 52)
Cambridge Blue Group	(see page 61)
'Cotton Cool'	(see page 54)
'Dora Bielefeld'	(see page 55)
'Frühlingshimmel'	(see page 65)
'Highdown' = 'Lewis Palmer'	(see page 57)
'Majesté'	(see page 58)
'Tim's Silver'	(see page 68)
'White Barn' = 'Beth's Blue'	(see page 52)

Pulmonaria vallarsae A. Kerner
Ref: Monog. Pulm.33 (1878)
A deciduous, softly hairy *Pulmonaria* with dark green,
wavy-margined leaves, spotted whitish or paler green,
and violet-blue flowers in early spring. There are
many sticky glands on leaves and especially in
inflorescence. Rare in cultivation.

NAMING
Etymology: *vallarsae* – named after Vallarsa, east of
Lake Garda, Italy, where this species grows.
Synonyms: None

DESCRIPTION
General: A deciduous, softly hairy perennial.
Stems: 15–33cm (6–13in) high, hairy.
Leaves: Basal leaves to 20cm (8in) long and 10cm
(4in) wide, usually with wavy margins, rather abruptly
narrowed into a winged stalk to 18cm (7in) long,
with distinct bright green or whitish, often merging,
spots or mottling or, rarely, almost unspotted. Leaf
surface is very soft to touch, covered with very dense,
short, fine hairs, mixed with long hairs and glandular
hairs.
Inflorescence: Densely glandular hairy, sticky to
touch, with long and also short hairs.
Flowers: *Calyx* often stick with glandular hairs,
sometimes tinged reddish-brown. *Corolla* violet or
purple-blue, tube hairy inside below the ring of hairs
in throat.
Nutlets: 4.5 × 3.5mm.
Chromosomes: 2n=22.

GEOGRAPHY AND ECOLOGY
Distribution: Italy.
Habitats: Woodland and woodland edges.
Flowering: Early spring
Associated plants: In S Antonio forest, mainly of
beech (*Fagus sylvatica*) with some chestnut (*Castanea
sativa*) and conifers, grows nearby plants such as
Doronicum columnae, bugle (*Ajuga reptans*), wood
anemone (*Anemone nemorosa*), *Helleborus multifidus*
subsp. *bocconei*, *Viola eugeniae*, *Phyteuma
scorzonerifolium* and mouse plant (*Arisarum
proboscideum*). In Partenio region, is found in oak
(*Quercus*) woodland with sycamore (*Acer
pseudoplatanus*) and field maple (*A.campestre*), where
it may grow in association with such plants as elder
(*Sambucus nigra*), roses (*Rosa canina* and
R.sempervirens), herb robert (*Geranium robertianum*),
stinking hellebore (*Helleborus foetidus*), *Campanula
foliosa*, sticky sage (*Salvia glutinosa*), wild strawberry
(*Fragaria vesca*), sweet violet (*Viola odorata*), alpine
squill (*Scilla bifolia*), *Lilium croceum* and polypodies
(*Polypodium* species).

CULTIVATION
Hardiness: Zone 6?.
Sun/shade aspect: Light to full shade.
Soil requirements: As for the genus.
Garden uses: As for the genus.
Spacing: 30cm (12in) apart.
Propagation: As for the genus.
Availability: 1. Not generally available, but grown by
some plant collectors.

RELATED FORMS AND CULTIVARS
None. *Pulmonaria* 'Margery Fish' (see page 59) is
frequently listed as a cultivar of this species, but it is
more likely to be a hybrid, closer to *P. saccharata*
than *P. vallarsae*.

Pulmonaria species rare or not in cultivation

Pulmonaria dacica Simonkai
Ref: Simonkai 1887

Synonyms: *Pulmonaria molissima* M.Popov not
A.Kerner; *P.mollis* Wulfen in Hornem.

Common names: *German*: Sibirisches Lungenkraut;
Slovenian: Dacijski pljucnik

Nomenclature note: There have been suggestions
that, possibly, the plant we call '*P. angustifolia*' hort.
may be in fact this species, and that some hybrids,
such as 'Blue Ensign', have this in their parentage.
The recent *Flora of Siberia* (MALYSHEV, 1997)
considers this plant to be the same as *P. mollis*
Wulfen in Hornem.

Description: A softly hairy perennial, 30–50cm
(12–20in) high, with stems covered in glandular hairs
and with some bristly hairs above. Leaves to 30–50cm
(12–20in) long and 8–12cm (3$^{1}/_{4}$–4$^{1}/_{2}$in) wide,
unspotted, softly hairy with sparse bristly hairs and
stalked glandular hairs above, more sparsely and
shortly hairy underneath, narrowing into a stalk.
Upper stem leaves 5–8cm (2–3$^{1}/_{4}$in) long and
1.2–2.5cm ($^{1}/_{2}$–1in) wide, rounded or nearly heart-
shaped at base. Flowers borne on glandular-hairy
pedicels 8–13mm long, with *calyx* 10–14mm long,
divided to about one third of their length and blue
corolla with tube 12–20mm long, interior shortly
hairy, limb 8–12mm across.

Distribution: Romania, northern Balkans, southern
Russia, Caucasus, central Asia east to eastern Siberia,
Turkey (north-eastern Anatolia).

Habitats: A variety of habitats, most frequently in
woodland and on woodland edges, although in central
Asia it also grows in alpine meadows and close to
snowline in the mountains.

Flowering: April to June in the wild.

Associated plants: In Siberia, grows in the forests of
pine (*Pinus sibirica*), fir (*Abies sibirica*) and spruce
(*Picea obovata*), with such plants as mezereon
(*Daphne mezereum*), May lily (*Maianthemum
bifolium*), mountain melick (*Melica nutans*), *Clematis
speciosa*, herb paris (*Paris quadrifolium*), *Viola
selkrikii*, alpine enchanter's-nightshade (*Circaea
alpina*), wood vetch (*Vicia sylvatica*), ground-ivy
(*Glechoma hederacea*), red baneberry (*Actaea
erythrocarpa*), wintergreen (*Pyrola rotundifolia*),
twinflower (*Linnaea borealis*). In aspen (*Populus
tremula*) woods, it shares its habitat with northern
wolfsbane (*Aconitum lycoctonum*), moschatel (*Adoxa
moschatellina*), *Dryopteris australis*, *Lilium
pilosiusculum*, *Lathyrus gmelinii*, *Paeonia anomala*,

Delphinium elatum, chickweed wintergreen
(*Trientalis europaeus*). In silver birch (*Betula
pendula*) woods, its companions may include meadow
crane's-bill (*Geranium pratense*), *Bupleurum aureum*,
Viola mirabilis, solomon's seal (*Polygonatum
odoratum*), *Trollius asiaticus*, *Adenophora lilifolia*,
spring vetchling (*Lathyrus vernus*) and *Vicia unijuga*.

Availability: Not known. Not offered by any
suppliers, but some plants grown under different
names may turn out to be this species or perhaps its
hybrids.

Pulmonaria filarszkyana Javorka
Ref: Jav., *Bot. Cozl.* 15: 51 (1916)

Synonym: *Pulmonaria rubra* subsp. *filarszkyana*
(Jav.) Domin. *Flora of the USSR* regards this as the
same plant as *P. dacica* Simk., but this does not
seem likely.

Description: A perennial 20–40cm (8–16in) high.
Similar to *P. rubra*, but has thinner and longer
rhizomes and longer basal leaves that are narrowed
gradually into a wide, winged stalk, which is shorter
than in *P. rubra*. The leaves are always unspotted,
covered in shorter and softer hairs, up to 30cm
(12in) long including the stalk and 6cm (2$^{1}/_{2}$in)
wide. Stem leaves oblong-lanceolate, the lowermost
ones are almost as large as the basal leaves, tapering
almost to the base; upper leaves distinctly decurrent,
the uppermost glandular hairy. Calyx to 15mm, with
longer lobes than in *P. rubra*, corolla red. Nutlets
4.5–5mm, ovoid, slightly downy.

Chromosomes: 2n=14

Distribution: Romania, Carpathians. A rare plant,
included in *The IUCN Red List of Threatened Plants*.

Habitat: Subalpine region, in krummholz (the zone
where the trees are stunted due to high altitude
conditions, and above which they are unable to
grow).

Flowering: June to July in the wild.

Associated plants: In the Carpathians, grows nearby
such plants as yellow gentian (*Gentiana lutea*),
spotted gentian (*G.punctata*), *Aconitum firmum*,
Anemone narcissiflora, alpine avens (*Geum
montaum*), verticillate lousewort (*Pedicularis
verticillata*) and *Narcissus angustifolius*.

Availability: Probably unavailable. The plant sold by
Monksilver nursery a few years ago as *P.* aff.
filarszkyana is unlikely to have been this species.

Pulmonaria helvetica M.Bolliger

Ref: Bolliger, *Pulmonaria in Westeuropa* (*Phanerogam.Monogr.*, 8):99 (1982).

Common names: *French*: Pulmonaire de Suisse; *German*: Schweizerisches Lungenkraut; *Italian*: Polmonaria elvetica.

Description: A hairy perennial 20–60cm (8–24in) high. Leaves indistinctly spotted pale green, or plain green, to 30cm (12in) long in summer. Corolla large, 15–24mm ($^5/_8$–1in) long, at first pink-red, then turning blue and finally violet; interior of tube is hairy below the ring of hairs. This species is said to have arisen quite recently through natural hybridization, with the possible parentage of *P. officinalis* and *P. montana* or *P. mollis.*

Chromosomes: 2n=24

Distribution: Endemic to Switzerland.

Habitat: Woodland, at 400–800m (1300–2600ft).

Flowering: March–May in the wild.

Associated plants: Grows in beech (*Fagus sylvatica*) woods, sometimes with lime (*Tilia*) and maple (*Acer*), in company of such plants as lords-and-ladies (*Arum maculatum*), ramsons (*Allium ursinum*), dog's-mercury (*Mercurialis perennis*) and winter horsetail (*Equisetum hyemale*).

Availability: Probably unavailable.

Pulmonaria kerneri Wettstein von Westersheim

Ref: Wettst., *Verh. Zool.-Bot. Ges. Wien*, 38: 559 (1888)

Description: A perennial with thick, somewhat leathery lanceolate leaves, to 60cm (24in) long and 10cm (4in) wide, very gradually narrowed to the base, usually with prominent white spots or, rarer, more or less unspotted, with stiff bristles and sparse short glandular hairs. Stem leaves lanceolate, stalkless, with broad base. Corolla bright blue, tube hairless inside below the ring of hairs. Nutlets 5 × 3mm.

Chromosomes: 2n=26.

Distribution: Austria (north-eastern Alps)

Habitat: Woodland.

Flowering: Spring.

Availability: Probably unavailable.

Pulmonaria montana Lejeune

Ref: *Fl.Spa*, 1: 98 (1811)

Synonyms: *Pulmonaria tuberosa* invalid, not Schrank; *P. vulgaris* Merat, in part; *P. vulgaris* invalid, in part; *P. angustifolia* subsp. *tuberosa* Gams, in part. Sometimes listed as synonymous with *P.rubra* Schott (in *The New RHS Dictionary*) or *P. mollis* Wulf. (*Flora of the USSR*), but this does not seem likely.

Common names: *English*: Mountain lungwort; *Dutch*: Smal Longkruid; *French:* Pulmonaire des montagnes; *German*: Knollen-Lungenkraut, Berg-lungenkraut.

Description: A hairy perennial, to 45cm (18in) high, somewhat resembling *P. mollis,* but leaves are rougher to touch and inflorescence is less sticky. Basal leaves to 50cm (20in) long and 12.5cm (5in) wide, oval to elliptical-lanceolate, pointed, base gradually narrowing into a stalk, usually plain green, unspotted, or occasionally with pale green blotches; covered with unequally distributed long hairs and sparse glandular hairs above. Inflorescence is densely covered in long hairs and has some glandular hairs. Corolla violet or blue, tube interior hairless below. Nutlets to 4.5 × 3.5mm.

Distribution: Western and west central Europe (Belgium, France, Germany, Switzerland).

Habitat: Woodland.

Flowering: Spring.

Associated plants: In France, this plant can be found in oak (*Quercus robur* and *Q. petraea*) woodland, with beech (*Fagus sylvatica*) and ash (*Fraxinus excelsior*). Other plants found in these habitats, include wood avens (*Geum urbanum*), cowslip (*Primula elatior*), sanicle (*Sanicula europaea*), alpine enchanter's-nightshade (*Circaea alpina*), *Ranunculus platanifolius*, sweet woodruff (*Galium odoratum*), whorled Solomon's-seal

Pulmonaria dacica in Turkey. (Erich Pasche)

(*Polygonatum verticillatum*), early purple orchid (*Orchis mascula*), greater butterfly orchid (*Platanthera chlorantha*), lesser spotted orchid (*Dactylorhiza fuchsii*) and *Poa chaixii*. In Belgium, its associates may include wild gooseberry (*Ribes uva-crispa*), stinking hellebore (*Helleborus foetidus*), lesser celandine (*Ranunculus ficaria*), hairy violet (*Viola hirta*), dog's-mercury (*Mercurialis perennis*), lords-and-ladies (*Arum maculatum*), meadow saffron (*Colchicum autumnale*), hairy wood-rush (*Luzula pilosa*), Solomon's-Seal (*Polygonatum multiflorum*), wood melick (*Melica uniflora*) and herb paris (*Paris quadrifolia*)

Availability: Not known – it is possible that some plants may be in cultivation under a different name.

Pulmonaria stiriaca A.Kerner

Ref: Monogr. Pulm. 36 (1878)
Synonym: *Pulmonaria saccharata* subsp. *stiriaca* (A.Kerner) Nyman
Description: A hairy perennial with leaves up to 22cm (8^1/$_2$in) long and 8cm (3^1/$_4$in) wide, covered with unequal bristles and glandular hairs above, with stalk up to 12cm (4^1/$_2$in). Inflorescence with short bristles and long glandular hairs. Corolla bright blue, with the tube indistinctly hairy inside, below the ring of hairs. Nutlets 3 x 2mm.
Chromosomes: 2n=18.
Flowering: Spring
Distribution: Austria, the former Yugoslavia.
Habitat: Woodland.
Associated plants: In Bela Krajina in Slovenia, grows in oak (*Quercus robur*) and hornbeam (*Carpinus betulus*) woods, with such plants as *Aconitum variegatum*, alpine barrenwort (*Epimedium alpinum*), false stitchwort (*Pseudostellaria Europaea*) and ivy (*Hedera helix*).
Availability: Probably unavailable.

Pulmonaria visianii Degen & Lengyel

Ref: Degen, Fl. Velebit., ii. 569 (1937).
Synonyms: *Pulmonaria angustifolia* invalid, in part, not L.; *P. media* Host, in part.
Description: This plant is said to be similar to *P. angustifolia*, with the basal leaves shorter (to 30cm/12in) and wider (6cm/2^1/$_2$in). Leaves are unspotted or sometimes with indistinct paler green spots, covered in unequal bristles and sparse glandular hairs. Inflorescence bristly and glandular, with deep blue-violet or blue flowers. Nutlets 3–3.5mm.
Chromosomes: 2n=20.

Flowering: Spring.
Distribution: Central and eastern Alps, northern and western Balkans (Austria, France, Switzerland, Italy, former Yugoslavia).
Habitat: Woodland.
Availability: Probably unavailable.

Pulmonaria forms, cultivars and hybrids

Pulmonaria 'Abbey Dore Pink'

Synonym: *Pulmonaria* 'Abbey Dore Pale Pink'.
Parentage: It has been suggested that this is a hybrid between *P. affinis* and *P. mollis*, but it is more likely that the parents are *P. officinalis* and *P. rubra*.
Origin: From Abbey Dore Court Garden, Herefordshire (UK).
Description: 30cm (12 in). Vigorous, more or less deciduous. Flowers large, soft pale pink or shell pink, without any blue. Leaves somewhat yellowish-green, with pale green spots. Self-pollinates and does not cross. *Pulmonaria* 'Vera May' is similar, but a superior plant.
Availability: 2.

Pulmonaria affinis 'Margaret'

Parentage: Selection from wild-collected *P. affinis*.
Origin: Found by Margaret Nimmo-Smith at Monksilver Nursery (UK) in 1998.
Description: 30cm (12in). Flowers purple-blue with paler grey-purple stripe down each corolla lobe. Leaves finely pointed, with large spots.
Availability: 0 (last listed in 1999).

Pulmonaria 'Alan Leslie'

This is listed by some US nurseries, but does not appear to be a valid name – Dr Alan Leslie himself is puzzled as to why it may be named after him. Possibly this could have originated from *P. longifolia* subsp. *cevennensis*, which Alan introduced to the UK, and which may have appeared in the USA later. It is described as 'Dark blue flowers over elliptical foliage that is spotted silvery white, 12in tall'.

Pulmonaria angustifolia 'Alba'

Synonym: *Pulmonaria angustifolia alba*. As it is possible that other white-flowered clones may be introduced, it should perhaps not be given a cultivar status, to avoid confusion, and instead be referred to as *P. angustifolia* var *alba*.
Parentage: *P. angustifolia* (true species).
Origin: This is an old form, originating from the true

P. angustifolia, unlike other varieties normally assigned to this species. Joe Sharman and Alan Leslie have found a virtually identical plant except for blue flowers in a type locality for *P. angustifolia*.
Description: 20cm (8in). Deciduous. Flowers small, narrow, pure white, in mid- or late spring. Leaves plain green, as in the normal species. Drought-tolerant.
Availability: 1.

Pulmonaria angustifolia subsp. azurea
Synonym: *Pulmonaria angustifolia* 'Azurea'
Parentage: *P. angustifolia* hort., not the true wild *P. angustifolia*.
Origin: Of wild origin, although it is uncertain whether the plants currently in cultivation are identical to the original clone.
Description: 25–30cm (10–12in). Deciduous. Flowers bright blue, opening from pink buds. Leaves dark green, unspotted.
Availability: 31.

Pulmonaria angustifolia 'Beth's Blue' see *Pulmonaria* 'Beth's Blue'

Pulmonaria angustifolia 'Beth's Pink' see *Pulmonaria* 'Beth's Pink'

Pulmonaria angustifolia 'Blaues Meer' ♥
Parentage: *P. angustifolia* hort., not the true wild *P. angustifolia*.
Origin: From Ernst Pagels (Germany), possibly originating from a plant collected in Yugoslavia.
Description: 30cm (12 in). Deciduous. Flowers large, bright blue, freely produced. Leaves about 30cm (12in) long, unspotted. May have been confused with *P.* 'Blue Pearl'. This cultivar gained the RHS Award of Garden Merit in 1998.
Availability: 13.

Pulmonaria angustifolia 'Blue Pearl' see *Pulmonaria* 'Blue Pearl'

Pulmonaria angustifolia 'Munstead Blue'
Parentage: *P. angustifolia* hort., not the true *P. angustifolia*.
Origin: From Gertrude Jekyll.
Description: 15cm (6in). Deciduous. Flowers clear blue, early. Leaves smallish, rather dark green, but not as dark as some other cultivars in this group, unspotted.
Availability: 23.

Pulmonaria angustifolia 'Rubra' see *Pulmonaria rubra*

Pulmonaria angustifolia 'Variegata' is listed in *The New RHS Dictionary* and described as 'leaves narrow, variegated white'. To our knowledge, such a plant does not exist.

Pulmonaria 'Apple Frost'
Parentage: Unknown
Origin: From Dan Heims of Terra Nova Nurseries (USA), introduced in 1997.
Description: 25–30cm (10–12in). Flowers rose-pink. Leaves apple-green with heavy silver mottling. Said to be resistant to mildew. Dan Heims describes it as 'charming silver appliqué over Granny Smith apple-green leaves'.
Availability: 5.

Pulmonaria 'Barfield Regalia'
Parentage: Unknown.
Origin: Named by Richard Nutt, who originally obtained it from the garden of Margery Fish (Barfield is Richard Nutt's garden near High Wycombe); introduced by Monksilver Nursery in 1992.
Description: 30–45cm (12–18in). Vigorous. Flowers 'pin', rather large, violet-blue to dark blue, borne in branched clusters on upright stems, in early spring. Leaves to 45cm (18in) on petioles to 20cm (8in), narrow, with a small number of more or less regularly spaced pale green spots, or sometimes almost unspotted. Was voted the favourite purple/blue flowered cultivar by the Pulmonaria Group of the Hardy Plant Society. A very distinct plant.
Availability: 13.

Pulmonaria 'Benediction'
Parentage: Unknown
Origin: Named by Jerry Fintoff after Loie Benedict (USA).
Description: 20–23cm (8–9in). Flowers deep blue, in loose clusters. Leaves narrow, dark green with pale green to silvery, widely spaced spots.
Availability: 6.

Pulmonaria 'Berries and Cream'
Parentage: Said to be a hybrid between *Pulmonaria rubra* 'Redstart' and *P.* 'Excalibur', but is possibly a seedling of one of these.
Origin: Raised by Dan Heims of Terra Nova Nurseries (USA) and introduced by them in 1997.
Description: 25–30cm (10–12in). Flowers are said to be raspberry-pink, fading slightly to purplish. Leaves silvery, with wavy margins, and Dan Heims describes the foliage as 'undulating and shimmering silver'.
Availability: 1.

Pulmonaria **'Beth Chatto'** see *Pulmonaria* 'Beth's Blue'

Pulmonaria 'Beth's Blue'
Synonyms: *Pulmonaria angustifolia* 'Beth's Blue'; *P.* 'Beth Chatto'; *P.* 'White Barn'; *P.* 'Cedric Morris'; *P.* 'Special Blue'; '*Pulmonaria linifolia*'.
Parentage: Possibly a hybrid between *P. longifolia* and *P. affinis*, though previously assigned to *P. angustifolia*, which it is not.
Origin: From Beth Chatto's garden, possibly given to her by Sir Cedric Morris.
Description: 30cm (12in). Forms a compact clump. Flowers 'pin', rich blue with no hint of pink, in rather tight clusters on upright stems. Leaves to 32cm (12¹⁄₂in) long and 12cm (4¹⁄₂in) wide, dark green, sparsely spotted with pale green. A very distinct plant.
Availability: 6.

Pulmonaria 'Beth's Pink'
Synonym: *P. angustifolia* 'Beth's Pink', *P.* 'Beth Chatto's Red'
Parentage: Possibly a *P. affinis* hybrid, but not *P. angustifolia* though it has been listed as such in the past.
Origin: From Beth Chatto's garden, possibly given to her by Sir Cedric Morris as a seedling.
Description: 25cm (10in). Flowers 'pin', coral red, in mid- to late spring, a slight tinge of violet with age. Leaves oval, to 25cm (10in) long and 10cm (4in) wide, narrowing abruptly to petioles, heavily spotted with whitish- or very pale green.
Availability: 6.

Pulmonaria 'Blauer Hugel'
Parentage: *P. longifolia*.
Origin: From Ernst Pagels (Germany).
Description: 30cm (12in). Forms a compact clump. Flowers clear blue, freely produced. Leaves long, narrow, with a few evenly spaced spots. Similar to *Pulmonaria* 'Little Star'.
Availability: 4.

Pulmonaria 'Blauhimmel'
Synonyms: *Pulmonaria saccharata* 'Blauhimmel'; *P. saccharata* 'Frühlingshimmel'; *P. saccharata* 'Spring Beauty' (RHS Trials have it separate from Frühlingshimmel').
Parentage: Has been frequently listed as a *P. saccharata* cultivar, but this is incorrect. More likely to be related to *P. officinalis*, possibly with another parent being *P. angustifolia* hort.
Origin: Germany.
Description: 15cm (6in). Flowers 'pin', clear blue, with purple calyx. Leaves distinctively white-spotted. Monksilver Nursery catalogue describes it as 'one of the most beautiful but not the strongest'. Needs shade. This is probably the same plant as *Pulmonaria* 'Fruihlingshimmel', though the latter is said to have a darker 'eye'.
Availability: 6.

Pulmonaria 'Blue Buttons'
Parentage: Seedling from *Pulmonaria* 'Munstead Blue'.
Origin: Named and introduced by P. W. Plants (UK) in 2000.
Description: 30cm (12 in). Flowers deep blue, produced in abundance. Leaves dark green. Said to be an improvement on *P.* 'Blue Ensign', but time will tell.
Availability: 7.

Pulmonaria 'Blue Crown'
Synonym: *Pulmonaria longifolia* 'Blue Crown'.
Parentage: *P. longifolia* hybrid.
Origin: From Piet Oudolf (Holland); in the UK introduced by Elizabeth Strangman of Washfield Nurseries, and listed by Monksilver Nursery in 1993.
Description: 25–30cm (10–12in). Semi-evergreen, vigorous. Flowers dark blue or violet-blue, freely produced. Leaves narrow, dark green, with a few bold, large spots. Very tough and long-lived. It doesn't flower well in shade, and some growers find it prone to mildew.
Availability: 18.

Pulmonaria 'Blue Ensign' ♀
Parentage: Not known, but bears similarities with *P. angustifolia* hort.
Origin: Found in Bowles' Corner, RHS Garden Wisley; introduced by Monksilver Nursery (UK) in the early 1990s.
Description: 25–30cm (10–12in). Vigorous, deciduous. Flowers 'pin', large, rich blue or blue-violet flowers in rather compact clusters, borne on more or less upright stems. Leaves rather broad, to 25cm (10in) long, very dark green, unspotted. One of the showiest cultivars, and one of the best dark blues. Was voted the favourite dark/royal blue pulmonaria by the Pulmonaria Group of the Hardy Plant Society, and won the Award of Garden Merit from the Royal Horticultural Society in 1998. Tolerant of sun but can be susceptible to mildew.
Availability: 49.

Pulmonaria **'Blue Mist'** see *Pulmonaria officinalis* 'Blue Mist'

Pulmonaria 'Blue Ensign'. (Howard Rice)

***Pulmonaria* 'Blue Moon'** see *Pulmonaria officinalis* 'Blue Mist'

Pulmonaria 'Blue Pearl' ♈

Synonym: *Pulmonaria angustifolia* 'Blue Pearl'.
Parentage: *P. angustifolia* hort., not *P. angustifolia* Linnaeus.
Origin: Unknown. One of the older cultivars.
Description: 20–30cm (8–12in). Flowers 'thrum', pure blue, a little lighter than in the similar *P.* 'Mawson's Blue'. Leaves rounded, unspotted. Awarded the AGM by the Royal Horticultural Society in 1998.

RHS trials note that most plants sold in the trade as 'Mawson's Blue' are in reality 'Blue Pearl'.
Availability: 5.

***Pulmonaria* 'Boughton Blue'** see *Pulmonaria* 'Lewis Palmer'

***Pulmonaria* 'Bowles' Red'** see *Pulmonaria rubra* 'Bowles' Red'

***Pulmonaria* 'Brentor'** see *Pulmonaria saccharata* 'Brentor'

Pulmonaria 'British Sterling'

Parentage: *P. saccharata* Argentea Group.
Origin: Selected by Henry Ross (USA) from seedlings of *P. saccharata* Argentea Group given to him by Adrian Bloom (UK).
Description: 25cm (10in). Flowers pink turning blue. Leaves almost completely greenish-silver, with green margins. Sometimes said to be an unstable cultivar.
Availability: 2.

Pulmonaria Cally hybrid

Parentage: *P. mollis* hybrid.
Origin: From Michael Wickenden of Cally Gardens (UK).
Description: 20cm (8in). Flowers rich, bright blue, borne in many clusters. Leaves dark green, unspotted, much smaller than in *P. mollis*.
Availability: 6.

***Pulmonaria* 'Cambridge Blue'** see *Pulmonaria officinalis* Cambridge Blue Group

***Pulmonaria* 'Cedric Morris'** see *Pulmonaria* 'Beth's Blue'

Pulmonaria 'Chintz'

Parentage: *P. officinalis*.
Origin: Appeared in Jennifer Hewitt's garden as a seedling; introduced by Elizabeth Strangman of Washfield Nurseries (UK) in the early 1990s.
Description: 15–25cm (6–10 in). Flowers 'thrum', white with each corolla lobe having a broad pink stripe that turns purple and finally blue, in tight clusters on short stems. Usually there is a white vein in the centre of the stripe, thus dividing it into two. Leaves mid-green with sparse or moderate pale green spotting; petioles of basal leaves are approximately equal in length to blades. Monksilver Nursery found this to be quite a weak plant, but micro-propagation is said to have improved its vigour.
Availability: 10.

Pulmonaria 'Cleeton Red'

Parentage: Probably a hybrid between *P. rubra* 'Bowles' Red' and *P. longifolia*.
Origin: Originated as a seedling in Jennifer Hewitt's garden (UK).
Description: 30cm (12in). Flowers 'pin', rich pinkish- or coral-red, with brown calyces, borne in branched, rather lax nodding clusters, on somewhat lanky stems, from upright to sprawling. Leaves long, narrow, with a few large pale green spots; petioles about a third of the length of the blades; overall leaf length to 50cm (20in), together with petioles. Similar to *Pulmonaria* 'Esther', *P.* 'Elworthy Rubies' and *P.* 'Middleton Red'.
Availability: 4.

Pulmonaria 'Coral Springs'

Parentage: Possibly *P. longifolia*.
Origin: From Dan Heims of Terra Nova Nurseries (USA).
Description: 25–30cm (10–12in). Flowers coral-pink opening from purple-pink buds. Leaves described by Dan Heims as 'fountains of *P. longifolia* foliage spotted in mint-green'.
Availability: 6.

Pulmonaria 'Coralie'

Synonym: Has been sold as *P. longifolia* red form.
Parentage: Possibly *P. longifolia* form or hybrid.
Origin: Raised by David Ward at Beth Chatto's Gardens (UK).
Description: Similar to 'Elworthy Rubies', 'Cleeton Red' and 'Esther', but flowers are coral-pink and leaves a little narrower and slightly lighter green.
Availability: 0.

Pulmonaria 'Corsage'

Parentage: Unknown.
Origin: From Louise Vockins of Foxgrove Plants (UK).
Description: 30cm (12in). Flowers light blue. Leaves rounded, ovate with more or less heart-shaped bases, rather light green, moderately spotted in paler green.
Availability: 2.

Pulmonaria 'Cotton Cool'

Synonyms: None, but has been known to be mistakenly distributed as *Pulmonaria* 'Cotton Wool'.
Parentage: *P. longifolia*.
Origin: From Diana Grenfell of Apple Court (UK), introduced in the USA in 1999 by Terra Nova Nurseries.
Description: 20–30cm (8–12in). Flowers blue and pink in dense clusters, tending to be darker blue in damper soil. Leaves held more or less upright, and

Pulmonaria 'Cotton Cool'. (Dan Heims)

are long, rather narrow, tapered, completely silver, with or without a dark green margin. Voted the favourite pulmonaria for its foliage by the Hardy Plant Society Pulmonaria Group, and noted for its good foliage by the RHS trials. The Heronswood Nursery catalogue describes it as 'possessing elegant long and narrow foliage of pure platinum that reflects the dim light of the woodland where it grows.' A good, long-lived, drought-tolerant plant.
Availability: 17.

Pulmonaria 'Crawshay Chance'

Parentage: Unknown.
Origin: From Jerry Webb of Hardy Plant Society (UK).
Description: 45cm (18in). Vigorous and can spread to 90cm (3ft) wide. Flowers 'pin', large, light blue, violet-tinged, in rather large clusters. Leaves with heart-shaped bases, blades and stalks approximately equal, about 15cm (6in) long, bright green with many prominent, bright white spots, evenly spaced. Noted by the RHS trials for its good foliage.
Availability: 2.

Pulmonaria 'Dark Vader'

Parentage: Unknown.
Origin: From Dan Heims of Terra Nova Nurseries (USA), introduced in 1999.
Description: 30cm (12 in). Flowers blue-purple to rose pink. Leaves very dark green with silver spotting.
Availability: 1.

Pulmonaria 'Dawn Star'

Synonyms: *Pulmonaria* 'Jennifer's Pale Pink'; *Pulmonaria* 'Jennifer's Pale Blue'.
Parentage: Possibly *P. longifolia* hybrid.
Origin: From Jennifer Hewitt's garden (UK).
Description: 30cm (12 in). Flowers pale pink opening from pink buds, later changing to pale violet and violet-blue. Leaves long, rather narrow, pointed, dark green, with many whitish-green spots; stem leaves similar. Noted by the RHS trials for its good foliage.
Availability: 0.

Pulmonaria 'De Vroomen's Pride'

Parentage: Unknown.
Origin: Holland.
Description: 25–40cm (10–16in). Flowers blue fading to pink. Leaves near-white with a narrow green edge. *Pulmonaria* cultivars 'Jill Richardson', 'Paul Aden', 'Reginald Kaye' and 'Silver Mist' are all quite similar to 'De Vroomen's Pride' and, in the opinion of both Vanessa and Joe Sharman, should probably all be grouped under the name 'Reginald Kaye'.
Availability: 17.

Pulmonaria 'Crawshay Chance'. (Dan Heims)

Pulmonaria **'Diana Chappell'** see *Pulmonaria saccharata* 'Diana Chappell'

Pulmonaria **'Diana Clare'**

Parentage: *P. longifolia* hybrid.
Origin: From Bob Brown of Cotswold Garden Flowers (UK), who named it after his wife.
Description: 20–30cm (8–12in). Flowers large, violet blue, with a longitudinal purple stripe between the corolla lobes. Leaves long, pointed, silver-green, very decorative. Jennifer Hewitt considers this Pulmonaria to be among the best for foliage and flower.
Availability: 35.

Pulmonaria **'Dora Bielefeld'**

Synonym: *Pulmonaria saccharata* 'Dora Bielefeld'. Sometimes misspelt as *Pulmonaria* 'Dora Bielefild' or 'Dora Bielefeldt'.
Parentage: Unknown. Often listed as a cultivar of *P. saccharata*, which it is not.
Origin: Germany.
Description: 25–30cm (10–12in). Flowers 'pin', clear pink opening from coral buds. Leaves bright green, lightly spotted pale silvery-green. The best pink-flowered pulmonaria among the older cultivars.
Availability: 44.

Pulmonaria **'Duke's Silver'**

Parentage: Unknown.
Origin: First found in the Dukes Garden at Kew, later thought to have been lost but has been re-introduced by Jenny Spiller of Elworthy Cottage Plants, who obtained a surviving plant from Angela Whinfield, Dorset (UK).
Description: 30cm (12 in). Flowers rather dark violet-blue opening from pink buds. Leaves very silver, rather dark with no green margin. Not a strong grower.
Availability: 1.

Pulmonaria 'Dora Bielefeld'. (Howard Rice)

Pulmonaria **'Elworthy Rubies'**

Parentage: Possibly *P. longifolia*.
Origin: From Jenny Spiller of Elworthy Cottage Plants (UK), named in 1998.
Description: 30cm (12in). Robust and vigorous. Flowers salmon-red. Leaves long, spotted. One of the earliest-flowering pulmonarias, midwinter to mid-spring. Could be grouped together with 'Esther' and 'Cleeton Red' as very similar cultivars.
Availability: 2.

Pulmonaria **'Emerald Isles'**

Parentage: Unknown.
Origin: From Dan Heims of Terra Nova Nurseries (USA), introduced in 1999.
Description: 30cm (12in). Flowers blue. Leaves silver with green spots, which, according to Dan Heims, are 'like emerald isles floating in a sea of silver'.
Availability: 1.

Pulmonaria **'Esther'**

Parentage: Possibly *P. longifolia*.
Origin: Raised and introduced by Coen Jansen (Holland), named for his wife.
Description: 25–30cm (10–12in). Flowers 'thrum', reddish, with brown calyces. Leaves long, rather narrow, to 30cm (12in) long, dark green with a few light green spots and blotches. Similar to 'Cleeton Red', 'Elworthy Rubies' and 'Middleton Red', and is the best among them.
Availability: 17.

Pulmonaria 'Excalibur'. (Dan Heims)

Pulmonaria **'Excalibur'**

Parentage: *Pulmonaria* 'Margery Fish'.
Origin: From Dan Heims of Terra Nova Nurseries (USA), raised and introduced in 1992.
Description: 20–30cm (8–12in). Rather vigorous. Flowers opening from coral-pink buds gradually turning to light blue, freely produced. Leaves silver, with a narrow dark green margin. Good mildew resistance, but can scorch in sun and does not tolerate drying out – needs shade and moist soil.
Availability: 22.

Pulmonaria **'Fiona'**

Parentage: Unknown.
Origin: Unknown.
Description: 22–30cm (9–12in). Flowers 'pin', large, at first coral-red gradually changing to violet and then blue, borne in large, branched, lax clusters on brownish stems. Leaves widening rather abruptly at the base, mid- to dark green, with many bright whitish spots; petiole and blade are approximately equal in length, about 18cm (7in), the blade about twice as long as it is wide.
Availability: 3.

Pulmonaria **'Glacier'**

Parentage: *P. officinalis*.
Origin: Found by Dr Alan Leslie's mother, Julia, in a friend's Surrey garden; named and introduced by Monksilver Nursery (UK) in 1990.
Description: 20–30cm (8–12in). Flowers 'pin', large, opening from pink buds, a very pale blue, sometimes also palest pink or white on the same plant, flowers borne on spreading stems. Leaves widen abruptly at base, blades about 15cm (6in) long and 8cm (3¹/₄in)

wide; petioles about twice as long, lightly spotted pale yellowish green. Can tolerate sun if the soil is moist.
Availability: 24.

Pulmonaria **'Glebe Blue'** or **'Glebe Cottage Blue'** see *Pulmonaria saccharata* 'Glebe Cottage Blue'

Pulmonaria **'Golden Haze'**

Parentage: *Pulmonaria longifolia* 'Bertram Anderson'.
Origin: From Walters Gardens, introduced by Terra Nova Nurseries (USA) in 1999.
Description: 23cm (9in). Golden-edged form of *Pulmonaria* 'Bertram Anderson'. Flowers blue. Leaves narrow, edged in yellow. This is a protected cultivar, and a licence is required for its propagation and distribution. 'The flowers are lighter than in "Bertram Anderson", and the foliage is overlaid with a golden haze and an irregular border of gold,' according to Dan Heims. The Heronswood Nursery catalogue concludes: 'This is a superb departure in Pulmonarias and will certainly give rise to future editions of exciting foliage plants.'
Availability: 1.

Pulmonaria **'Hascombensis'** see *Pulmonaria* 'Lewis Palmer'

Pulmonaria **'Hazel Kaye's Red'**

Parentage: Possibly *P. saccharata*.
Origin: From Hazel Kaye's nursery (UK).
Description: 18–22cm (7–8¹/₂in). Vigorous. Flowers 'pin', pinkish-red, with dark calyces, profuse. Leaves, both basal and stem, medium green with many pale green spots and blotches. Stalks of basal leaves are longer than the blades. Superficially similar to 'Esther', but flowers are pinker, stems shorter, leaves lighter green and broader. Extremely free-flowering.
Availability: 4.

Pulmonaria **'Highdown'** see *Pulmonaria* 'Lewis Palmer'

Pulmonaria **'Ice Ballet'**

Synonyms: Terra Nova Nurseries (USA) have originally sold this as *Pulmonaria* 'White Wings', which was incorrect. This plant may be identical to *P.* 'Sissinghurst White'.
Parentage: Probably *P.* 'Sissinghurst White'.
Origin: Introduced by Dan Heims of Terra Nova Nurseries (USA) in 1997.
Description: 30cm (12 in). Similar to *P.* 'Sissinghurst White' and could be the same, but is said by Dan Heims to have 'larger pure white blooms, better vigour and nicer spotting'. Differs from *P.* 'White Wings', with which it

has been confused, by the lack of a pink eye and larger flowers.
Availability: 0.

Pulmonaria 'Joan's Red'
Parentage: Probably a hybrid between *P. rubra* and *P. saccharata* Argentea Group.
Origin: From Joan Grout (UK).
Description: Jennifer Hewitt: 'A very striking plant with dark green basal leaves and paler stem leaves, all boldly spotted and blotched bright silvery white. The basal leaves are quite large with blades about one and half times the length of the petioles and a third as wide as they are long; they widen and narrow gradually. It makes a compact clump. The flower heads are fairly closely clustered, on brown stems up to 22cm (9in) tall, and dark reddish-brown calyces hold bright pink to red flowers with short styles.'
Availability: 4.

Pulmonaria 'Lewis Palmer' ♀
Synonyms: *Pulmonaria* 'Highdown'; *P. saccharata* 'Highdown'; *P. longifolia* 'Lewis Palmer'; *P.* 'Boughton Blue'; *P.* 'Hascombensis'.
Parentage: Possibly *P. longifolia*.
Origin: UK. Named after the Hon. Lewis (Luly) Palmer, a famous plantsman.

Pulmonaria 'Glacier'. (Howard Rice)

Description: 30–35cm (12–14in). Vigorous. Flowers large, rich violet-blue, facing mostly upward and outward. Leaves long, rather narrow, dark green, strongly spotted and blotched (up to half the leaf surface), spots conspicuous, greenish-white. Blades about three times as long as wide, and about twice as long as the petioles. One of the best pulmonarias. Gained the Royal Horticultural Society Award of Garden Merit in 1993.
Availability: 41.

Pulmonaria 'Little Blue'
Parentage: Unknown.
Origin: Terra Nova Nurseries, USA.
Description: 30cm (12in). Flowers small, dark blue and prolific. Leaves lanceolate, silver-spotted. Does well in regions that are too warm for most pulmonarias (Zone 9). Susceptible to chlorosis.
Availability: 1.

Pulmonaria 'Little Star'
Parentage: *Pulmonaria longifolia* 'Bertram Anderson'. Said to be a cross of this with *P. angustifolia*, but it is not likely.
Origin: From Dan Heims of Terra Nova Nurseries (USA), introduced in 1993.
Description: 15–20cm (6–8in). Flowers dusky pink in bud changing to cobalt blue, borne on short stems, produced very freely and over a long period of time. Pedicels are shorter than in its parent, *P. longifolia* 'Bertram Anderson'. Leaves narrowly lanceolate, rather short, with a few small silvery spots. Very similar to *P.* 'Blauer Hugel'.
Availability: 12.

Pulmonaria longifolia 'Ankum'
Synonym: *Pulmonaria longifolia* 'Coen Jansen'.
Parentage: *P. longifolia*.
Origin: From Coen Jansen (Holland), in the UK first introduced by Washfield Nurseries, who sold it as 'Coen Jansen'.
Description: 30–35cm (12–14 in). A compact plant. Flowers small, bright violet-blue. Leaves narrow, to 30cm (12in) long, with wavy margins, very silvery. The foliage merits more attention than the flowers, forming a good mound.
Availability: 33.

Pulmonaria longifolia 'Bertram Anderson'
Synonym: *P. longifolia* 'E. Bertram Anderson'.
Parentage: *P. longifolia*.
Origin: Named after the Gloucestershire plantsman B. Anderson.
Description: 25–30cm (10–12in). Flowers smallish,

Pulmonaria 'Lewis Palmer'. (Dan Heims)

vivid blue. Leaves long, narrow, spotted silver. Dan Heims: 'Small cobalt bells held tightly together over delicious dark green strap leaves'.
Availability: 25.

Pulmonaria longifolia subsp. *cevennensis* M. Bolliger

Parentage: *P. longifolia*.
Origin: Wild, from near Cevennes, France.
Description: 30cm (12in). Flowers blue. Leaves very long, almost reaching 60cm (2ft), and heavily silvered.
Availability: 15

Pulmonaria longifolia 'Coen Jansen' see *Pulmonaria longifolia* 'Ankum'

Pulmonaria longifolia 'Dordogne'

Parentage: *P. longifolia*.
Origin: From Elizabeth Strangman of Washfield Nurseries (UK), collected in the wild in France.
Description: 30–45cm (12–18in). Flowers blue, borne on upright stems. Leaves lanceolate, larger than the species, well-spotted in silver-white.
Availability: 20.

Pulmonaria 'Lovell Blue'

Parentage: Possibly *P. longifolia* hybrid.
Origin: Raised by Trevor Bath, a seedling from a vigorous blue-flowered hybrid.
Description: 22–30cm ($8^{1}/_{2}$–12in). Flowers 'pin', large, bright purplish-blue, with dark calyces, borne in rather open clusters on upright stems, rather late. Leaves rather narrow, bright green, moderately spotted in whitish, blades somewhat shorter than the petioles. New shoots in spring emerge brown-tinged.
Availability: 3.

Pulmonaria 'Majesté'

Synonym: *Pulmonaria saccharata* 'Majesté'. Sometimes misspelt as 'Majesty' or 'Majestee'.
Parentage: Unknown.
Origin: From La Ferme Fleurie Nursery, France (1986).
Description: 20–30cm (8–12in) Flowers pink, turning blue. Leaves almost completely silver, with 'shiny finish' and a fine green margin. Monksilver Nursery finds this difficult to grow on drier soil. It does best in shade, but with ample moisture will stand more sun.
Availability: 45.

Pulmonaria 'Margery Fish' ♥
Synonyms: *Pulmonaria vallarsae* 'Margery Fish'; *P. saccharata* 'Margery Fish'.
Parentage: Probably *P. saccharata*, or may be *P. affinis*, although consistently listed as *P. vallarsae*, which is incorrect.
Origin: From Margery Fish. Known from 1974.
Description: 30cm (12in). Flowers open from pink buds, turning blue, early. Leaves long and narrow, spotted to almost completely silver. Mildew resistant. A parent of many recent US introductions. Received the RHS Award of Garden Merit in 1998.
Availability: 37.

Pulmonaria 'Marlene Rawlins'
Parentage: *P. saccharata*.
Origin: Named for Ronnie Rawlins' wife.
Description: 30cm (12 in). Vigorous, healthy. Flowers large, blue and pink striped, freely produced. Leaves spotted with many small bright silver spots. It is difficult to grow and is not yet in general cultivation.
Availability: 0.

Pulmonaria 'Mary Mottram'
Synonyms: *Pulmonaria* 'Wendy Perry' is considered to be identical to this form by Vanessa Cook, but the current *RHS Plant Finder* lists them as separate cultivars. This plant was originally distributed as *P. mollissima*, which is incorrect.
Parentage: Possibly *P. saccharata*, although *P. mollis* has also been suggested.
Origin: Grown by Mary Mottram in her Devon nursery (UK).
Description: 30cm (12in). Flowers 'pin', large, blue-violet, with brown calyces, early. Jennifer Hewitt describes the colour of the blooms as follows: 'Flowers are violet in the centre of the corolla lobes; the violet extends and becomes more blue as the flower ages though the rim and tube remain red.' Leaves large, to 30cm (12in) long, about three times as long as they are wide, petioles to 22cm (8½in); becoming almost completely covered in silver, apart from a narrow green margin (stem leaves densely spotted). Tolerates full sun.
Availability: 28.

Pulmonaria 'Mawson's Blue'
Synonym: *Pulmonaria* 'Mawson's Variety'.
Parentage: Probably a hybrid of *P. angustifolia* hort., and has been listed under *P. angustifolia* L.

Pulmonaria 'Marlene Rawlins'. (Howard Rice)

Origin: Appeared at Mawson Brothers' Nursery in Cumbria (UK), in the 1930s when this seedling was selected.
Description: 20–30cm (8–12in). Flowers 'thrum', rich dark blue, rather late flowering. Leaves plain dark green, unspotted. The original 'Mawson's Blue' was a different plant from the one that is grown under this name now; the former was more like *P. angustifolia* 'Blauer Meer' or 'Munstead Blue', and the latter resembles 'Blue Ensign'.
Availability: 21.

Pulmonaria 'May Bouquet'
Parentage: Unknown.
Origin: From Dan Heims of Terra Nova Nurseries (USA), 1999.
Description: 27–30cm (10¾–12in). Dan Heims: 'Our absolute winner in the 1999 trials for profusion of bloom and individual flower size. This plant was

stunning here for over a month in masses and masses of blue flowers that change to pink.' This is a protected variety and a licence is required for its propagation and distribution.
Availability: 1.

Pulmonaria 'Merlin'

Parentage: *P. longifolia.*
Origin: Obtained by Joe Sharman of Monksilver Nursery (UK) from the garden of Valerie Finnis, the widow of Sir David Scott and the founder of the Merlin Trust; named by Joe after Sir David's late son Merlin in 1992.
Description: 20–30cm (8–12in). Neat, compact habit. Flowers 'pin', pale pink to pale blue. Leaves more or less typical of *P. longifolia*, variably spotted with bright silvery-green, sometimes almost completely covered. Jennifer Hewitt finds this a weak grower.
Availability: 6.

Pulmonaria 'Milchstrasse'

Synonym: *Pulmonaria* 'Milky Way'.
Parentage: *P.* 'Margery Fish' cross.
Origin: From Dan Heims of Terra Nova Nurseries (USA), introduced in 1996.

Pulmonaria 'Mawson's Blue'. (Howard Rice)

Description: 30cm (12in). Flowers blue opening from wine-red buds, then fading to pink. Leaves large, lanceolate, heavily spotted with silver.
Dan Heims: 'Tony A. of Juniper Level says it's the best form for the south.'
Availability: 2.

Pulmonaria **'Milky Way'** see *Pulmonaria* 'Milchstrasse'

Pulmonaria mollis 'Royal Blue'

Parentage: *P. mollis* hybrid.
Origin: From Hidcote, 1977.
Description: Flowers of the original cultivar were described as blue, but plants currently in cultivation under this name have dull purple flowers. Leaves softly hairy, relatively small compared to it parent.
Availability: 2.

Pulmonaria mollis 'Samobor'

Synonym: None, but sometimes misspelt as *Pulmonaria mollis* 'Samabor'.
Parentage: Wild *P. mollis.*
Origin: From Elizabeth Strangman of Washfield Nurseries (UK), who collected it in the former Yugoslavia.
Description: 45–50cm (18–20in). Flowers first pink, turning violet to violet-blue. Leaves plain mid- to dark green, softly hairy.
Availability: 4.

Pulmonaria 'Monksilver'

Synonym: Monksilver Nursery had been selling this plant as *P.* 'Mournful Purple', until it was pointed out by Vanessa Cook that this was a distinct clone; this was named in 1999.
Parentage: *P. affinis* and *P. longifolia.*
Origin: Originally obtained by Langthorns Planthery from Beth Chatto as 'Mournful Purple', later sold by Monksilver Nursery under this name until 1999 when the correct name was published.
Description: 20–30cm (8–12in). Flowers 'pin', dull bluish to rich, dark purple. Leaves long, broad, spotted. Flowers are larger than in 'Mournful Purple', and the leaves are wider and more brightly spotted.
Availability: 3.

Pulmonaria 'Moonstone'

Parentage: Has been described as a 'sport' of *Pulmonaria saccharata* 'Glebe Cottage Blue' (which is a hybrid of *P. saccharata* and *P. longifolia*), but is more likely to be a seedling.
Origin: From Carol Klein of Glebe Cottage Plants (UK).
Description: 30cm (12in). Flowers first white,

becoming very pale blue-grey. Leaves heavily spotted, very decorative.
Availability: 6.

Pulmonaria 'Mournful Purple'
Synonyms: *Pulmonaria longifolia* 'Mournful Purple'; *P. longifolia* 'Mourning Widow'; *P. longifolia* 'Mournful Widow'.
Parentage: *P. longifolia* hybrid.
Origin: From Sir Cedric Morris who possibly collected it in Portugal or the Pyrenees; first named and sold by Beth Chatto (UK).
Description: 30cm (12in). Flowers 'amethyst purple'. Leaves lanceolate, dark green with pale green blotches. Until 1999, Monksilver Nursery sold a pulmonaria under this name. It has since been shown to be a distinct clone and is now correctly known as *P.* 'Monksilver'. There may still be more than one plant sold under this name.
Availability: 8.

Pulmonaria 'Mrs Kittle'
Parentage: Possibly *P. longifolia*.
Origin: Obtained as an unnamed form from a small amateur nursery at Kittle on the Gower (Wales), by Sheila and Bob Mousley, who wanted to name it for the nursery owner but could not remember her real name. Became well-known when introduced into Holland by Coen Jansen in 1980s.
Description: 22–30cm (8½–12in). Neat, more or less upright. Flowers 'pin', very pale pink, opening from pink buds, and gradually turning pale lavender. Leaves rather narrow, about 15cm (6in) long with the petioles of about the same length, well-spotted and blotched in silver white. One of the older pale pink forms. Dislikes dry soil.
Availability: 25.

Pulmonaria 'Netta Statham'
Synonym: Occasionally misspelt as *Pulmonaria* 'Netta Stratham'.
Parentage: Unknown.
Origin: Given by Margery Fish to the late Netta Statham, who had a garden at Erway Farm House, Shropshire (UK) and was generous with plants.
Description: 30–40cm (12–16in). Flowers gentian blue, on upright stems, over a long period. Leaves narrow, with spaced pale green spots.
Availability: 2.

Pulmonaria 'Nürnberg'
Parentage: *P. officinalis*.
Origin: From Dr Hans Simon (Germany); introduced into the UK by Monksilver Nursery.

Description: 20–30cm (8–12in). A slightly running habit. Flowers 'pin', pink and blue. Leaves very well spotted in summer, including stem leaves. Mid- to late spring. Needs shade and moist soil. The leaves are as good as in the similar *P. saccharata* 'Leopard', the flowers are even better, but habit is somewhat inferior.
Availability: 8.

Pulmonaria 'Ocupol' (PBR)
Synonym: *Pulmonaria* Opal.
Parentage: Possibly a *P. saccharata* cross.
Origin: First appeared in the garden of Sue Cupitt (UK), the Hardy Plant Society stalwart.
Description: 25cm (10in). Flowers opening from palest pink buds turning to luminous pale blue. Leaves rather narrow, well-spotted. Protected against commercial propagation by Plant Breeder's Rights (PBR). A very good plant.
Availability: 39.

Pulmonaria officinalis 'Alba'
Parentage: *P. officinalis*.
Origin: The true white form of *P. officinalis* may be of wild origin, but this is probably unavailable.
Description: 25–30cm (10–12in). Flowers white. Plants sold under this name are nearly always likely to be *P.* 'Sissinghurst White' or *P.* 'White Wings'.
Availability: 4.

Pulmonaria officinalis 'Blue Mist' ♡
Synonyms: *Pulmonaria officinalis* 'Bowles' Blue'; *P.* 'Blue Mist'; *P.* 'Blue Moon'.
Parentage: *P. officinalis*.
Origin: From Amy Doncaster, but introduced by Elizabeth Strangman of Washfield Nurseries (UK).
Description: 15–20cm (6–8in). Flowers 'pin', large, very pale clear blue opening from buds tinged lavender. Leaves spotted whitish green. There may be more than one plant sold under this name. Similar to *P. officinalis* Cambridge Blue Group, but flowers are larger and earlier. It gained the RHS Award of Garden Merit in 1998.
Availability: 29.

Pulmonaria officinalis **'Bowles Blue'** see *Pulmonaria officinalis* 'Blue Mist'

Pulmonaria officinalis Cambridge Blue Group
Synonyms: *Pulmonaria saccharata* Cambridge Blue Group; *P. saccharata* 'Cambridge Blue'.
Parentage: *P. officinalis*, although previously often listed as *P. saccharata*.

Origin: From Amy Doncaster, introduced by Elizabeth Strangman of Washfield Nurseries (UK).
Description: 25–30cm (10–12in). Flowers 'thrum', pale blue, tinted pink in bud, borne in profusion. Leaves heart-shaped, spotted. Tolerant of sun as long as the soil does not dry out. Was for a long time the best pale blue. Some nurseries claim to hold the true original 'Cambridge Blue', but others offer a number of similar strains, which is why they are classified as Cambridge Blue Group.
Availability: 16.

Pulmonaria officinalis **'Sissinghurst White'** see *Pulmonaria* 'Sissinghurst White'

Pulmonaria officinalis 'Stillingfleet Gran'

Parentage: *P. officinalis.*
Origin: A clone growing in Vanessa Cook's mother's garden for many years, and introduced by Stillingfleet Lodge Nurseries.
Description: 30cm (12in). Vigorous and sterile. Flowers small pink and blue. Leaves well-spotted with pale green spots. This is a sterile clone, and not a very exciting plant.
Availability: 3.

Pulmonaria officinalis 'White Wings'

Synonym: *Pulmonaria* 'White Wings'.
Parentage: *P. officinalis.*
Origin: Unknown.
Description: 25–30cm (10–12in). Flowers 'pin', white with a distinct pink eye, late-flowering. Leaves silver-spotted. Sometimes confused with *P.* 'Sissinghurst White', but differs by the pink eye, pinkier calyces, smaller more compact habit. It is less susceptible to mildew.
Availability: 20.

Pulmonaria 'Oliver Wyatt's White'

Synonym: *Pulmonaria* 'Nutt's White'.
Parentage: Unknown.
Origin: From Richard Nutt, who originally obtained this from his schoolteacher, Oliver Wyatt; named and introduced by Monksilver Nursery (UK) in 1995.
Description: 30cm (12in). Similar to *P.* 'Sissinghurst White', but with more strongly silver-spotted leaves. The RHS trials have noted this cultivar for good foliage. Possibly the best white-flowered pulmonaria.
Availability: 5.

Pulmonaria 'Opal' see *Pulmonaria* 'Ocupol'

Pulmonaria 'Paul Aden'

Parentage: Unknown.

Origin: From Baldwin, *Hosta* breeder (USA), introduced in the UK by Glebe Garden Nursery in 1996.
Description: Flowers pink turning blue. Leaves heavily spotted silver. This is similar to *Pulmonaria* 'De Vroomen's Pride', *P.* 'Jill Richardson' and *P.* 'Silver Mist', and all of these, according to Vanessa Cook, should all be grouped under *P.* 'Reginald Kaye'.
Availability: 1.

Pulmonaria **'Peter Chappell's Pink'** see *Pulmonaria saccharata* 'Diana Chappell'

Pulmonaria 'Pewter'

Parentage: Possibly *P. saccharata.*
Origin: Found by A. Cary, a stalwart of the Hardy Plant Society, in a small local nursery in Herefordshire (UK), believed to have been introduced from Holland.
Description: 22cm (9in). Flowers deep pink in bud, opening reddish-pink and becoming quite dark blue-violet. Leaves entirely silver, broadly lanceolate. Not a strong plant.
Availability: 1.

Pulmonaria 'Polar Splash'

Parentage: Unknown.
Origin: From Dan Heims of Terra Nova Nurseries (USA), introduced in 1998.
Description: 20–23cm (8–9in). Flowers blue fading to pink. Leaves rather rounded, dark green, with contrasting bright silver spots. In winter leaves may be tinged purple.
Availability: 5.

Pulmonaria 'Purple Haze'

Parentage: Unknown.
Origin: From Dan Heims of Terra Nova Nurseries (USA), introduced in 1997.
Description: 20cm (8in). Vigorous but compact. Flowers purple, freely produced. Leaves with short petioles, mid-green, well-spotted.
Availability: 2.

Pulmonaria **'Rachel Vernie'** see *Pulmonaria rubra* 'Rachel Vernie'

Pulmonaria 'Raspberry Splash'

Parentage: *P. longifolia.*
Origin: From Dan Heims of Terra Nova Nurseries (USA), introduced in 1998.
Description: 30cm (12in). Flowers dark pink, freely produced. Leaves held upright, pointed, dark green, with distinct silver spots. Superficially resembles 'Esther' and other similar cultivars. This is a

protected cultivar and a licence is required for its propagation and distribution.
Availability: 4.

Pulmonaria 'Red Freckles'
Parentage: Unknown.
Origin: From Mrs Pat Jones of Sunnybank Nursery, Herefordshire (UK).
Description: 30cm (12 in). Vigorous. Flowers clear coral-red. Leaves oval, dark green with a moderate number of very bright silver spots. Noted for its good foliage by the RHS trials.
Availability: 0 (last available 1995).

Pulmonaria 'Regal Ruffles'
Parentage: Unknown.
Origin: Terra Nova Nurseries (USA), 1995.
Description: 30cm (12in). Dan Heims: 'A breakthrough on several fronts; first, it has ruffled flowers, secondly, the clasping vestigial leaves of the stalk form a halo around the flower cluster. The plant is compact and well silvered.' Some English nurseries have found this difficult to keep, especially in damp conditions.
Availability: 0 (last available 1995).

Pulmonaria 'Reginald Kaye'
Synonyms: *Pulmonaria saccharata* 'Reginald Kaye'; *P.* 'Silver Mist'; *P.* 'Jill Richardson'.
Parentage: Though often listed as *P. saccharata*, this is closer to *P. officinalis*.
Origin: Given by Gill Richardson, who obtained it as an unnamed seedling from Reginald Kaye's nursery, to Beth Chatto, who introduced it in the UK in 1990. Introduced to the USA in 1997 by Terra Nova Nurseries.
Description: 20–30cm (8–12in). Flowers opening pink and turning light blue, in dense clusters. Leaves large, rounded, with larger silver spots in the centre, and smaller spots at border; very good in summer. It is likely that more than one clone is sold under this name.
Availability: 35.

Pulmonaria 'Roy Davidson'
Parentage: *P. longifolia* 'Bertram Anderson' seedling.
Origin: From Roy Davidson, famous US plantsman.
Description: 30cm (12in). Flowers 'pin', pale to mid-blue (fading to pink), borne in dense clusters. Leaves rather narrow, but somewhat broader than in its parent *P. longifolia* 'Bertram Anderson', lightly spotted silver. An extremely good plant. The 'blue' is unlike any other, so very distinct.
Availability: 36.

Pulmonaria 'Roy Davidson'. (Howard Rice)

Pulmonaria rubra var. albocorollata
Synonyms: *Pulmonaria rubra* 'Albocorollata'; *P. rubra* var. *alba*.
Parentage: *P. rubra*.
Origin: Wild origin.
Description: 20–30cm (8–12in). Flowers small, pure white, 'thrum', borne in nodding clusters, very early. Leaves pale apple green, unspotted. Has been included on National Council for Conservation of Plants and Gardens Pink Sheet (list of plants rare or endangered in the garden), but now is common. Seedlings of this variety turn out red-flowered.
Availability: 22.

Pulmonaria rubra 'Ann'
Parentage: *P. rubra*.
Origin: From John Metcalf of Four Seasons Nursery; named 1993, introduced by Monksilver Nursery (UK).
Description: 30cm (12in). Vigorous. Flowers 'pin', bright pink-red with white, similar in colour to *P. rubra* 'Barfield Pink'. Leaves with light green spotting. Much more free-flowering and vigorous than 'Barfield Pink' with larger flowers.
Availability: 19.

Pulmonaria rubra 'Barfield Pink'
Synonyms: None. Graham S. Thomas lists this under *P. officinalis* in his *Perennial Garden Plants*, but this is incorrect.
Parentage: *P. rubra*.
Origin: Richard Nutt named this *c*.1980, having

obtained it as an unnamed seedling from the garden of Margery Fish, in Somerset, UK.
Description: 25–30cm (10–12in). Flowers 'thrum', brick-red, with white margins and white veins. Leaves soft green, unspotted.
Availability: 23.

Pulmonaria rubra 'Barfield Ruby'

Parentage: Seedling of *P. rubra* 'Bowles' Red'.
Origin: Seedling in Richard Nutt's garden; introduced by Monksilver Nursery (UK). Name published 1992.
Description: 30cm (12in). Flowers 'pin', large – the biggest amongst the forms of *P. rubra*; typical coral-red. Leaves green, unspotted. Vanessa Cook of Stillingfleet Nurseries does not consider this to be significantly different from the species.
Availability: 4.

Pulmonaria rubra 'Bowles' Red'

Parentage: *P. rubra*.
Origin: From Edward A. Bowles (UK).
Description: 30cm (12in). Flowers coral-red. Leaves faintly spotted pale green. Similar to *P. rubra* 'Redstart'.
Availability: 20.

Pulmonaria rubra 'David Ward' ♀

Parentage: This is a sport of *Pulmonaria rubra* 'Redstart'.
Origin: Beth Chatto's nursery (UK), *c*.1986; named after the nursery's propagator, David Ward, who first found it. Introduced to the USA in 1996 by Terra Nova Nurseries.
Description: 30cm (12in). The first variegated pulmonaria cultivar. Less vigorous than the species. Flowers 'thrum', coral-red. Leaves long, pale green with white margins, in large rosettes. Not easy, the leaves will scorch in sun or wind; must be shaded and not allowed to dry out. Very susceptible to slug and snail damage. At Monksilver Nursery it persists well. Dan Heims: 'Occasional all-green or all-white leaves pop-up, but this plant is a certified knock-out.' This cultivar received the RHS Award of Garden Merit in 1998.
Availability: 54.

Pulmonaria rubra 'Prestbury Pink'

Parentage: *P. rubra*.
Origin: Found in the wild in Czechoslovakia and named by John Anton-Smith (UK). Name published in 1993.
Description: 20–30cm (8–12in). Weak plant, compared to other forms of *P. rubra*. Flowers very soft pale pink, slightly smaller than in the type. Leaves plain green. Susceptible to mildew.
Availability: 1.

Pulmonaria rubra 'Rachel Vernie'

Synonym: *Pulmonaria* 'Rachel Vernie'.
Parentage: *P. rubra*.
Origin: Seedling from Jennifer Hewitt's garden, named for her younger daughter.
Description: 30cm (12in). Flowers 'pin', coral-red. Stem leaves with a white margin as in *P. rubra* 'David Ward'. Basal (summer) leaves mainly soft grey-green with darker streaks and margins, and some white edges. Importantly, leaves do not scorch in full sun, unlike those of 'David Ward'. This plant is also more vigorous and has darker red flowers. Monksilver Nursery have encountered problems with reversion, with variegated foliage giving way to spotted green leaves, but this is not the experience of Jennifer Hewitt and other growers. Prefers moist soil.
Availability: 1.

Pulmonaria rubra 'Redstart'

Parentage: *P. rubra*.
Origin: During the RHS trials in 1970s, this was selected as a good form among J. Archibald's plants of *P. rubra* (UK); he was asked to name this clone, though he doesn't think that this is different from the species.

Pulmonaria rubra 'Barfield Ruby'. (Howard Rice)

Description: 30–45cm (12–18in). Vigorous but with a compact habit. Flowers large, coral-red. Leaves plain fresh green. Vanessa Cook does not consider this form significantly different from the species to warrant a cultivar status. This plant is the parent of the first variegated pulmonaria, *P. rubra* 'David Ward'.
Availability: 42.

Pulmonaria rubra 'Warburg's Red'

Parentage: *P. rubra.*
Origin: From Primrose Warburg (UK), noticed outside her front door by many 'galanthophiles' (snowdrop-lovers). Named by Monksilver Nursery in 1995.
Description: 30cm (12in). Flowers somewhat smaller and darker red (especially in bud) than in other forms of *P. rubra*; with brownish-green calyces. Leaves lightly spotted pale green.
Availability: 2.

Pulmonaria saccharata 'Alba'

Parentage: Probably not the true *P. saccharata.*
Origin: Unknown.
Description: 30cm (12in). Flowers large, larger than those of *P. officinalis*, snow-white. Leaves well-marked.
Availability: 6.

Pulmonaria saccharata
Argentea Group ♀

Synonym: *P. saccharata* 'Argentea'.
Parentage: *P. saccharata.*
Origin: Margery Fish was one of the first to sell this plant.
Description: 25–30cm (10–12in). Flowers pink at first, turning violet-blue, borne in open clusters, early spring. Leaves become wholly silvery-grey, at their best in late summer. Comes more or less true from seed, but the seedlings are variable. This plant gained the RHS Award of Garden Merit in 1993.
Availability: 25.

Pulmonaria saccharata 'Blauhimmel' see *Pulmonaria* 'Blauhimmel'

Pulmonaria saccharata 'Brentor'

Synonyms: *Pulmonaria officinalis* 'Brentor'; *P.* 'Brentor'.
Parentage: Possibly a hybrid of *P. rubra* and *P. saccharata.*
Origin: Introduced by Rowden Gardens, Devon (UK).
Description: Flowers magenta-red. Leaves large, dark green, lightly silver-spotted.
Availability: 3.

Pulmonaria rubra 'David Ward'. (Howard Rice)

Pulmonaria saccharata **Cambridge Blue Group** see *Pulmonaria officinalis* Cambridge Blue Group

Pulmonaria saccharata **'Cotton Cool'** see *Pulmonaria* 'Cotton Cool'

Pulmonaria saccharata 'Diana Chappell'

Synonyms: *Pulmonaria* 'Diana Chappell'; *P.* 'Peter Chappell's Pink'.
Parentage: Probably *P. saccharata.*
Origin: A seedling found by Peter Chappell, owner of Spinners Nursery, (UK), who named it after his wife.
Description: 30cm (12in). Flowers large, deep pink ageing to violet, on upright stems. Leaves large, brightly spotted.
Availability: 1.

Pulmonaria saccharata **'Dora Bielefeld'** see *Pulmonaria* 'Dora Bielefeld'

Pulmonaria saccharata 'Frühlingshimmel'

Synonym: *Pulmonaria saccharata* 'Spring Sky'.
Parentage: According to Joe Sharman, this is not a *P. saccharata* but possibly a *P. officinalis* hybrid; and is likely to be the same as *P.* 'Blauhimmel'.

Pulmonaria 'Blauhimmel'. (Howard Rice)

Origin: Germany.
Description: 20–25cm (8–10in). Flowers open wide from pale pink buds into clear blue; calyces reddish-brown. Leaves silver-spotted. Slightly running habit. Very similar to *P.* 'Blauhimmel', if not the same plant, and said to differ by slightly deeper-coloured flowers, with a darker throat.
Availability: 21.

Pulmonaria saccharata 'Glebe Cottage Blue'
Synonyms: *Pulmonaria* 'Glebe Cottage Blue';
P. 'Glebe Blue'.
Parentage: *P. saccharata* x *P. longifolia*.
Origin: From Carol Klein of Glebe Cottage Plants (UK).
Description: Flowers blue, borne in dense clusters. Leaves well-spotted.
Availability: 5.

Pulmonaria saccharata 'Jill Richardson'
Parentage: Listed as *P. saccharata* but could be *P. officinalis*.
Origin: Named after plantswoman Gill Richardson (UK), misspelt as 'Jill'.

Description: Similar to *P.* 'Reginald Kaye', 'De Vroomen's Pride', 'Paul Aden' and 'Silver Mist', which should all be grouped under the first name, according to Vanessa Cook and other *Pulmonaria* experts.
Availability: 2.

Pulmonaria saccharata 'Leopard' ♀
Parentage: *P. saccharata*.
Origin: Occurred in G.S. Thomas's garden *c*.1970 and named in 1977.
Description: 20–30cm (8–12in). Flowers reddish-pink fading to lavender. Leaves regularly spotted in silver-white. Similar cultivars are *Pulmonaria* 'Beth's Pink', *P.* 'Diana Chappell' and *P.* 'Nürnberg'. Gained the RHS Award of Garden Merit in 1998.
Availability: 29.

Pulmonaria saccharata 'Majesté' see *Pulmonaria* 'Majesté'

Pulmonaria saccharata 'Mrs Moon'
Parentage: *P. saccharata*.
Origin: Though the origin of this plant (or plants) is not clear, there have been suggestions that it may be named after the orchid grower H.F.C. Sander's daughter, who was married to Henry George Moon (1857–1905).
Description: There are many contradicting descriptions of this plant in the literature, but it is not clear whether the original plant, whose name has been known since at least the 1930s, is still in cultivation. The flowers are described in various sources as 'lilac tinted red', 'magenta buds with flowers turning blue', and 'pink to violet'. Leaves with silver spotting, and prone to mildew. Many plants in cultivation under this name are probably seedlings, which further contributes to the confusion.
Availability: 22.

Pulmonaria saccharata 'Picta' see *Pulmonaria saccharata*

Pulmonaria saccharata 'Pink Dawn'
Parentage: *P. saccharata*.
Origin: Unknown
Description: 25–30cm (10–12in). Flowers large, deep pink, with no blue but ageing to violet. Leaves long, elliptical, way-margined, light green with paler green spots.
Availability: 7.

Pulmonaria saccharata 'Reginald Kaye' see *Pulmonaria* 'Reginald Kaye'

Pulmonaria saccharata 'Spring Sky' see
Pulmonaria 'Frülingshimmel'

Pulmonaria saccharata 'White Barn' see
Pulmonaria 'Beth Chatto'

Pulmonaria saccharata 'White Leaf'
Parentage: *P. saccharata.*
Origin: Abbey Dore Court (UK).
Description: 22–25cm (8½–10in). Flowers rather
large, pink to violet, in lax clusters on upright stems.
Leaves light green, almost completely covered with
pale green to silvery-white blotches, except for
narrow, spotted margins.
Availability: 1.

Pulmonaria 'Saint Ann's'
Synonym: *Pulmonaria mollis* hort.
Parentage: Possibly *P. angustifolia* hort.
Origin: Traced back as far as possible, the plant
originally came from St Ann's near Nottingham.
Description: 30cm (12in). Flowers dull purplish-pink
to dull blue, early. Leaves plain green, much smaller
than in true *P. mollis,* with which this has been
confused in the past.

   The plant that Monksilver Nursery has under this
name is virtually identical to *P. angustifolia* ssp. *azurea*
hort.
Availability: 6.

Pulmonaria 'Silver Mist'
Parentage: Possibly *P. officinalis.*
Origin: Jennifer Hewitt has traced it to a pulmonaria
enthusiast in Derbyshire (UK).
Description: Flowers violet-blue changing to pink.
Very similar to *P.* 'De Vrooman's Pride' and, together
with this and other cultivars ('Paul Aden' and 'Jill
Richardson'), should all be assigned to 'Reginald
Kaye', according to Vanessa Cook and other experts.
Availability: 1.

Pulmonaria 'Silver Streamers'
Origin: From Dan Heims of Terra Nova Nurseries
(USA), introduced in 1997.
Description: 20–25cm (8–10in). Flowers medium
violet-blue, opening from pink buds, with dark
calyces. Leaves lanceolate, with wavy margins, silver,
almost white.
Availability: 1.

Pulmonaria 'Sissinghurst White' ♀
Synonyms: *Pulmonaria officinalis* 'Sissinghurst
White'; *P. saccharata* 'Sissinghurst White'.
Parentage: This plant is closer to *P. officinalis* than

P. saccharata, to which it has frequently been
assigned in the past, but is now considered a hybrid.
Origin: Supplied to the RHS Trial 1973–76 by
Sissinghurst Gardens (UK), but did not originate
there.
Description: 30cm (12in). Evergreen. Flowers pure
white, early. Leaves silver-spotted.

   This old cultivar sometimes gets confused with
P. officinalis 'White Wings', but is more vigorous
and lacks pink eye. Susceptible to mildew. Gained
the RHS Award of Garden Merit in 1993.
Availability: 55.

Pulmonaria 'Smoky Blue'
Synonyms: *P. angustifolia* 'Smokey Blue', *P.* 'Smokey'.
Parentage: Probably *P. officinalis.*
Origin: Holland.
Description: 25–30cm (10–12in). Flowers soft dusky
(or, to put it less kindly, muddy) pink to purple blue.
Leaves dark green, purple-tinged, with light silver
spotting. RHS trials in 1996–98 comment on it as
being 'vigorous, healthy, but unremarkable for flower
or leaf', and Monksilver's catalogue bluntly describes
it as 'boring'.
Availability: 23.

Pulmonaria 'Special Blue' see *Pulmonaria* 'Beth's
Blue'

Pulmonaria 'Spilled Milk'
Synonyms: None, but may be misspelt as *Pulmonaria*
'Spilt Milk'.
Parentage: Unknown.
Origin: Terra Nova Nurseries (USA), 1992.

Pulmonaria 'Silver Streamers'. (Dan Heims)

Description: 20–30cm (8–12in) Compact habit. Flowers open blue fading to pink, in tight clusters. Leaves wide, mostly silver, with green margins and a few green blotches.
Availability: 8.

Pulmonaria 'Tim's Silver'
Synonym: *Pulmonaria saccharata* 'Tim's Silver'.
Parentage: *P. saccharata*.
Origin: Found by a horticulture student at Beth Chatto's Gardens (UK), where he worked part-time.
Description: 30cm (12in). Flowers mid-blue. Leaves almost completely silver-coated, with a satiny sheen and a fine green rim. Difficult to grow well.
Availability: 4.

Pulmonaria 'Trevi Fountain'
Parentage: Unknown.
Origin: From Terra Nova Nurseries (USA), 1999.
Description: 30cm (12in). Dan Heims: 'Superb, brightly silver-spotted long leaves give rise to profuse clusters of large cobalt-blue flowers in spring. A stunner!' This is a protected cultivar and a licence is required for its propagation and distribution.
Availability: 4.

Pulmonaria 'Ultramarine'
Parentage: Possibly *P. longifolia*, perhaps crossed with *P. affinis*.
Origin: Alan Leslie bought it in 1970s as *P. officinalis* 'Ultramarine'; re-introduced by Monksilver Nursery in 1997.
Description: 20cm (8in). Flowers soft violet to dusky blue. Leaves pointed, regularly silver-spotted.
Availability: 2.

Pulmonaria 'Sissinghurst White'. (Howard Rice)

Pulmonaria vallarsae 'Margery Fish' see
Pulmonaria 'Margery Fish'

Pulmonaria 'Vera May' * ♀
Parentage: Possibly *Pulmonaria* 'Sissinghurst White' hybrid.
Origin: Occurred in the garden of Vera May, mother of Michael Bowyer (UK), who named it after her.
Description: Compact plant. Flowers pale pink, freely produced. Similar to *Pulmonaria* 'Abbey Dore Pink', but leaves a little darker green and have fewer spots. *The RHS Award of Garden Merit was given to this plant in 1998 subject to it becoming available in the trade.
Availability: 2.

Pulmonaria 'Victorian Brooch'
Parentage: Unknown.
Origin: From Dan Heims of Terra Nova Nurseries (USA), introduced in 1998.
Description: 30cm (12in). Flowers magenta/rose-pink, facing outwards, with dark red calyces, on upright stems, produced over a long period. Leaves nearly round, well-spotted with large silver spots. In 1998, Dan Heims rated this as his best *Pulmonaria*.
Availability: 8.

Pulmonaria 'Weetwood Blue'
Parentage: More likely to originate from *P. longifolia* than, as it has been thought previously, *P. angustifolia*.
Origin: From the garden of Harold and Joan Bawden in Devon (UK).
Description: 20cm (8in). Flowers 'pin', smallish, open from dark purplish-pink buds into intense clear blue, borne on upright or spreading stems. Leaves rather long, narrow, with petioles shorter than blades, deep green, unspotted or with a few pale green or white spots. A very good plant. Gained the RHS Award of Garden Merit in 1998.
Availability: 6.

Pulmonaria 'Wendy Perry'
Synonyms: Some growers consider this synonymous with *Pulmonaria* 'Mary Mottram'.
Parentage: Unknown.
Origin: Raised by Wendy Perry of Bosvigo Plants (UK), and named by RD Plants, Devon, after its originator.
Description: Vigorous, clump-forming. Leaves almost completely silver. Flowers pink and blue. Vanessa Cook considers this plant to be the same as *P.* 'Mary Mottram'. However, other gardeners believe it different. Sue Ward (Pulmonaria Group newsletter,

October 1999) says that her 'Wendy Perry' flowers two weeks earlier than 'Mary Mottram', has about one-third more flowering stems, which are tall and strong in comparison with smaller and weaker 'Mary'. She says that during summer the plants look very similar but at the end of October, in *P.* 'Mary Mottram' 'leaves start to deteriorate very quickly whereas *P.* 'Wendy Perry' is still looking good up until Christmas'. She suggests that the latter cultivar could be a seedling of the former.
Availability: 6.

Pulmonaria **'White Leaf'** see *Pulmonaria saccharata* 'White Leaf'

Pulmonaria **'White Wings'** see *Pulmonaria officinalis* 'White Wings'

Pulmonaria **'Wisley White'**
Parentage: Unknown
Origin: Possibly from the RHS Gardens at Wisley.
Description: Flowers 'thrum', white, in somewhat drooping clusters, borne on upright stems. Leaves light to mid-green, with a moderate number of well-spaced spots that are pale green and vary in size. Leaf blades and petioles are approximately the same length.
Availability: 2.

REFERENCES
BIRD, 1993; BLAMEY and GREY-WILSON, 1989; BOWN, 1995; BRICKELL, 1996; CLAPHAM *et al.*, 1952; HEWITT, 1994; HUXLEY, 1992; KOSHEYEV and KOSHEYEV, 1994; LORD, 2001; LORD, 2002; MABEY, 1997; MATHEW, 1982; OBERRATH *et al.*, 1999; PHILLIPS and RIX, 1994; POPOV, 1992; RICE & STRANGMAN, 1993; RICHARDS, 1993; SHARMAN, 1998; SHISKIN & BOBROV, 1967–77; SHPILENYA and IVANOV, 1989; STACE, 1991; THOMAS, 1993; TUTIN *et al.*, 1972; *Pulmonaria* Group newsletters; RHS Trial of *Pulmonaria* 1996–98; Plant Finder

Pulmonaria 'Trevi Fountain'. (Dan Heims)

CHAPTER 2
OTHER GENERA IN CULTIVATION

This chapter deals with Boraginaceae genera other than *Pulmonaria* that are fairly commonly found in cultivation. In most cases, each genus will contain quite a few species familiar to gardeners, e.g. *Myosotis*, *Cynoglossum* and *Symphytum*, while others, such as *Nonea*, may only be represented by one well-known species, with the rest yet to be introduced and widely accepted as good garden plants.

Anchusa species with *Adonis aestivalis* in central Turkey. (Erich Pasche)

The genus *Anchusa* Linnaeus, 1753
Alkanet and Bugloss

NAMING

Etymology: *Anchusa* – from Greek *anchousa*, paint for the skin, as the roots of some species are used for this purpose.

Synonyms: This generic name has also been applied to *Brunnera* Stev., *Buglossoides* Tausch. and *Pentaglottis* Tausch. *Anchusa barrelieri* has been recently allocated a separate genus, *Cynoglottis* (Gusuleac) Vural & Kit Tan, but here we include it under *Anchusa*.

GENUS CHARACTERISTICS AND BIOLOGY

Number of species: About 35.

Description: Annual, biennial or perennial, usually bristly herbs, occasionally woody at base. Leaves alternate, simple, with entire or toothed margins. *Calyx* 5–lobed. Flowers usually blue or purple, occasionally white or yellow, borne in lateral or terminal cymes, with bracts. *Corolla* usually blue or violet, sometimes white or yellow, funnel-shaped, with a straight or curved tube, and an abruptly expanding limb, with 5 lobes that may sometimes be unequal, and with 5 hairy scales in the throat. Stamens 5, not protruding from the tube or only slightly so. Style included in the tube, with a headlike stigma. Nutlets 4, erect, egg-shaped to kidney-shaped with a thickened collar at the base, wrinkled or netted, sometimes finely warted/tuberculate.

Distribution and habitats: Europe, west Asia, North and South Africa, mostly in dry, sunny, sometimes grassy habitats.

Life forms: Hemicryptophytes, some therophytes, occasionally chamaephytes (*A. caespitosa*).

Pollination: Anchusas are visited by many insects and are especially attractive to bees.

Fruit dispersal: No specialized mechanism.

Animal feeders: Foodplants for caterpillars of *Coleophora pennella, Ethmia aurifluella, E. bipunctella, Euchrysops crawshayi, Utetheisa pulchella* and painted lady (*Vanessa cardui*). Alkanet (*Anchusa officinalis*) supports a species of weevil *Ceuthorhynchus geographicus* and a leaf beetle *Longitarsus echii*.

USES

GARDEN

This is a versatile genus and offers plants for a wide variety of garden situations: herbaceous and mixed borders (*A. azurea, A. barrelieri*), rock garden (*A. leptophylla* subsp. *incana*), alpine house (*A. caespitosa*), bedding (*A. capensis*), wildlife garden (many species, especially *A. azurea*, are rich in nectar and are very attractive to bees) and 'cornfield patch' (*A. arvensis*).

MEDICINAL

A number of species have been used as medicinal plants. *Anchusa officinalis* has demulcent and expectorant properties, and was used in treatment of cuts and bruises, and taken internally, for coughs and catarrh; also to treat ulcers.

Anchusa azurea is said to have antitussive, depurative, diaphoretic and diuretic properties. When dried, it can be used as a poultice to treat inflammations. It contains a toxic alkaloid, cynoglossin, and should be treated with caution.

CULINARY

The young leaves and young shoots of *A. azurea, A. officinalis* and *A. capensis* can be cooked and eaten like spinach. The flowers, when fresh, make a decorative addition to salad, or can be used as a garnish on other types of food.

WARNING: Like other Boraginaceae, anchusas contain toxic alkaloids and should be used carefully.

ECONOMIC

A red dye can be obtained from the roots of *A. officinalis*, which can be used to colour oils and fats, and that from *A. azurea* has been used in cosmetics (rouge).

CULTIVATION

Most common garden anchusas are very hardy (Zone 3), though *A. capensis* is rather tender (Zone 9), which is not a problem because it is normally grown as an annual.

Anchusas need plenty of sunlight, and do not grow or flower well if shaded. All members of the genus grow on well-drained soil and will not tolerate waterlogging. Most will put up with drought and considerable alkalinity. It is preferable to buy pot-grown plants, to minimize root disturbance at planting time. Tall species and cultivars, like *A. azurea*, need staking, and benefit from being divided every three years or so. In tall and medium-height species, removing spent flower stems may induce a second flowering.

PROPAGATION

These are general notes – see the details for individual species.

Seed Seeds for both perennials and annuals can be

sown from early to late spring, or in autumn, outdoors or in pots, covering the seed lightly. Germination can be erratic. Non-hardy species (see *A. capensis*) should be sown under glass, and rock garden species (*A. cespitosa, A. leptophylla* subsp. *incana*) must be sown in pots in a gritty mix.
Vegetative Root cuttings for perennials, from late autumn to midwinter, in a cold frame. Young plants should be potted up in spring and can be planted the following autumn. Careful division is possible for plants that form more than one rosette.

PROBLEMS
Cucumber mosaic virus produces symptoms of yellow spots on the leaves, and is untreatable. Most species have a taproot and resent disturbance. Some species can be weedy, so should be sited carefully.

REFERENCES
BECKETT, 1993–34; BLAMEY & GREY-WILSON, 1993; BRICKELL, 1996; CLAPHAM *et al.*, 1952; DAVIS, 1978; GRIFFITH, 1964; HUXLEY, 1992; PFAF, 1992–02; POLUNIN, 1987; SLABY, 2001; STACE, 1991; THOMAS, 1993; TUTIN *et al.*, 1972; WALTERS, 2001

Anchusa arvensis (Linnaeus) M. Bieberstein
A weedy, bristly annual, growing 10–60cm (4–24in) high, with lanceolate wavy-margined and toothed leaves, and small bright blue flowers borne in forked cymes. It is suitable for sunny borders in wild gardens or a 'cornfield patch'.

NAMING
Etymology: *arvensis* – of cultivated fields.
Synonym: *Lycopsis arvensis* L.
Common names: *English*: Bugloss, Small Bugloss (USA), Annual Bugloss (USA); *Danish*: Krummhals; *Dutch*: Kromhals; *Finnish*: Peltorasti; *French*: lycopside, Buglosse des champs; *German*: Acker-Krummhals; *Italian*: Buglossa Minore, Lingua di Bove; *Norwegian*: Krokhals; *Polish*: Krzywoszyj polny; *Russian*: krivotzvet polevoi; *Slovak*: phlica roná; *Swedish*: (vanlig) fårtunga, rast; *Welsh*: Bleidd Drem.

DESCRIPTION
General: A bristly annual, with bristles arising from tubercles.
Stems: 10–60cm (4–24in), ascending, bristly.
Leaves: 3–10cm (1¼–4in) long and 5–20mm (¼–¾in) wide, narrowly to broadly lanceolate, with wavy or irregularly toothed margins; the lower stalked, upper unstalked and half-clasping the stem.
Inflorescence: Of several short cymes, with leaflike bracts.

Flowers: *Calyx* 4–5mm, somewhat enlarging in fruit; lobes narrowly-lanceolate, pointed, divided almost to base. *Corolla* tube 4–7mm, blue (or, very occasionally, whitish), curved in the middle; limb to 6mm wide, with 5 unequal lobes. *Stamens* inserted at the middle of the tube, or just below.
Nutlets: to 2 x 5mm, obliquely ovoid, netted.
Chromosomes: 2n=48, 54.

GEOGRAPHY AND ECOLOGY
Distribution: Europe, Asia, North Africa. Classified as a noxious weed in Washington State, USA.
Habitat: Cultivated, fallow and waste ground, bare places, roadsides, vineyards and olive groves, sandy coastal habitats.
Flowering: April to September in the wild in Mediterranean areas.

CULTIVATION
Hardiness: Zone 7.
Sun/shade aspect: Needs full sun.
Soil requirements: Any well-drained soil.
Garden uses: Suitable only for a wild garden, where it can be admitted to a wildflower border, or used in a mixture of cornfield weeds.
Spacing: Thin the seedlings to 20–30cm (8–12in) apart.
Propagation: From seed only, best sown *in situ* in spring or autumn.
Availability: Seed and, occasionally, plants are available from some wildflower merchants.
Problems: To many gardeners this plant probably has little merit as an ornamental. Excessive self-seeding can be prevented by removing the faded flowerheads.

Anchusa azurea Miller
A robust, roughly-bristly perennial, with upright, branched stems to 90–150cm (3–5ft) high, coarse, long-lanceolate leaves and, in summer, large loose panicles of deep blue or violet flowers that are extremely attractive to bees. For a sunny position on well-drained soil and best in a mixed or herbaceous border.

NAMING
Etymology: *azurea* – azure-blue, referring to flower colour.
Synonym: *Anchusa italica* Retzius.
Common names: *English*: Large Blue Alkanet; Italian Bugloss (USA); *Finnish*: Italianrasti; *French*: Buglosse d'Italie; *German*: Italienische ochsenzunge; *Italian*: Buglossa Azzurra; *Russian*: anhusa italyanskaya; *Slovak*: smohla talianska; *Spanish*: Buglosa, Chupamieles, Lenguaza; Lengua de buey; *Turkish*: Sigir dili, Ouzu Dili.

Anchusa azurea 'Loddon Royalist'. (Howard Rice)

DESCRIPTION

General: Coarsely hairy or bristly perennial; hairs may be soft or rigid, spreading, often tubercle-based.

Stems: 20–150cm (8–60in), upright.

Leaves: 5–30cm (2–12in) long and 1–5cm (⁷/₁₆–2in) wide, narrowly-elliptical to lanceolate or oblanceolate, pointed; margin more or less entire or very slightly wavy; covered with dense soft or stiff hairs.

Inflorescence: Consists of many cymes, elongating considerably in fruit; with minute, narrowly-lanceolate bracts.

Flowers: *Pedicels* to 3mm, to 15mm in fruit. *Calyx* about 8–10mm in flower, 12–18mm in fruit, divided almost to base into linear, pointed lobes. *Corolla* violet or deep blue, occasionally paler or even white, tube 6–10mm long; limb 8–15mm across. *Stamens* inserted at top of tube and overlapping the scales.

Nutlets: 6–10 × 2–3mm, oblong to oblong-ovoid, erect.

GEOGRAPHY AND ECOLOGY

Distribution: South Europe, especially the Mediterranean; west Asia, Caucasus, north Africa. Naturalized in parts of northern Europe.

Habitat: Cultivated fields, fallow and waste ground, dry steppe, open garrigue, roadsides, olive groves, from sea level to 2500m (8200ft).

Flowering: March to July in the wild; early to late summer in cultivation.

CULTIVATION

Date of introduction: 1597.

Hardiness: Zone 4.

Sun/shade aspect: Full sun is necessary for good flowering in temperate climate, although in the Mediterranean garrigue and olive groves it sometimes grows in light shade.

Soil requirements: Any reasonably fertile, well-drained soil. Although *A. azurea* will tolerate heavy clay, such soil should, ideally, be improved by incorporating sharp sand and organic matter. Drought-tolerant when established.

Garden uses: A classic herbaceous border plant, *A. azurea* is also suitable for mixed plantings as long as it does not get shaded, and is good for a wildlife garden, the nectar-rich flowers being very popular with insects.

Spacing: 60cm (2ft) for the species and large cultivars; choose dwarfer forms, such as 'Little John', for small spaces.

Propagation: *Seed* – Most cultivars, with the exception of 'Feltham Pride', should not be expected to come true from seed. Sow from mid-spring to early summer in pots or a seed bed about 10mm (⁷/₁₆in) deep. Germination should take place within 2–4 weeks. *Vegetative* – Can be divided, but an easier and more prolific method of multiplying this plant, is from root cuttings, which is especially useful for cultivars that do not come true from seed. Take the root cuttings from late autumn until late winter, inserting in gritty soil-based compost, and pot up or plant out in an outdoor nursery bed in late spring. The young plants should be ready for planting in their permanent position the following autumn.

Availability: The species is listed by a few nurseries but cultivars are much more widely available. Seed is sold by some suppliers, and may also be found through some exchange schemes.

Problems: Is sometimes prone to mildew. Requires quite a lot of maintenance to keep it in good form, as it is short-lived and needs regular division, about every three years. Tall plants usually need staking with twiggy sticks or 'linkstakes'.

RELATED FORMS AND CULTIVARS

'Dropmore' – 150–180cm (5–6ft), with rich deep blue flowers; introduced in 1905.

'Feltham Pride' – more compact than the species, to 90cm (3ft), with clear, bright blue flowers; this is unusual in being grown from seed as a biennial.

'Little John' – dwarf, to 45cm (18in) high and 30cm (12in) wide; long-lived, with deep-blue flowers. This plant is suitable for the front of the border or rock garden.

'Loddon Royalist' – to 90cm (3ft), with bright gentian-blue flowers; valuable for its sturdiness, rarely requiring staking. RHS Award of Garden Merit plant, and by far the most popular cultivar; introduced in 1957.

'Morning Glory' – 100–150cm (40–60in), deep blue flowers.

'Opal' – 90–120cm (3–4ft), light blue; paler than any other form.

'Royal Blue' – 90–120cm (3–4ft), intense gentian-blue, introduced in 1954.

RELATED SPECIES

Anchusa strigosa Labill., from Turkey, Middle East, Iran and Iraq, is a biennial or perennial similar to *A. azurea*, with which it may hybridize, but usually with paler blue or sometimes nearly-white flowers. It is not in general cultivation.

Anchusa capensis Thunberg

A pretty, bushy biennial normally grown as an annual, with narrow, hairy leaves and masses of small, brilliant blue, or sometimes white or pink, saucer-shaped flowers through much of the summer and into autumn. This plant is ideal in sunny beds, borders and containers, and is attractive to bees.

> *'A delightful blue-flowered plant with somewhat rough, hairy leaves. Like the flowers of the perennial varieties, those of annual varieties are also edible and possibly improve complexion to a greater extent than the ancient tinctures.'*
> (Thomas Mansfield)

NAMING

Etymology: *capensis* – of Cape, South Africa, though the plant is found throughout the country.

Common name: *English*: Cape forget-me-not, Summer forget-me-not, Cape bugloss; *Afrikaans*: ystergras, koringblom.

DESCRIPTION

General: A hairy biennial.

Stems: 30–45cm (1–1½ft) high, occasionally to 60cm (2ft), upright.

Leaves: Linear to narrowly lanceolate, to 12cm (4½in) long, with tubercle-based bristly hairs.

Inflorescence: Terminal, open panicle.

Flowers: *Calyx* lobes triangular, blunt. *Corolla* 4–8mm across, bright blue with a white throat, tube equalling calyx.

Nutlets: 1 x 2.5mm, ovoid, horizontal, wrinkled.

GEOGRAPHY AND ECOLOGY

Distribution: South Africa.

Habitat: Dry, sandy places, disturbed ground, roadside verges.

Flowering: In the wild in West Cape, flowers in late spring and early summer. When grown as an annual in the UK, flowers from midsummer until mid-autumn.

CULTIVATION

Hardiness: Zone 9 – not hardy, but this is not a problem because it is usually grown as an annual. If you are growing this species as a biennial, overwinter under glass.

Sun/shade aspect: Full sun is best, although will tolerate very light shade.

Soil requirements: Well-drained soil, neutral is preferred but will tolerate a fair degree of both acidity and alkalinity. Drought-tolerant once established.

Spacing: 20–30cm (8–12in) apart.

Garden uses: Good for use in mixed borders, as bedding and in containers. Attractive to bees and butterflies.

Propagation: By seed only. To grow as an annual, sow the seeds in the open ground in mid- to late spring, covering very lightly, as the seed germinates better in the light. Germination takes 14–28 days in the open, or less at higher temperatures indoors – 7–14 days at 15–18°C (59–64°F). Sow in early summer if they are to be grown as biennials; these will make more robust plants, but will require winter protection under glass or plastic in areas colder than Zone 9. For earlier-flowering pot plants for a glasshouse or conservatory, sow indoors in midwinter, and pot up into 12.5cm (5in) pots of good-quality compost – these should flower in mid- to late spring.

Availability: The species is rather uncommon, but popular seed strains are widely available from seed merchants.

Problems: Take care not to damage taproot when transplanting. Handling the leaves may cause an irritating rash.

RELATED FORMS AND CULTIVARS

The following seed strains are likely to be available:
'Blue Angel' – 20–23cm (8–9in), ultramarine blue.
'Blue Bird' – 45cm (18in), bright indigo-blue flowers.
'Dawn' – 23cm (9in), mixture in pink, lavender, light blue, ultramarine and white.
'Pink Bird' – with pink flowers.

Anchusa cespitosa Lamarck

A choice alpine perennial, forming a prostrate rosette to about 22cm (8½in) across, of long, narrow, bristly leaves, with dense clusters of salver-shaped, deep blue, white-eyed flowers, on very short stems in late spring and early summer. Needs sun and perfect drainage – not easy and probably best in the alpine house.

> *'Quite the most beautiful of [anchusas] is
> A. caespitosa, making starfish-like rosettes of
> deep green, bristly, strap-shaped leaves, and
> bearing many large almost sessile flowers of a
> deep, clear blue, intensified by a white eye.'*
> (Anna N. Griffith)

NAMING

Etymology: *cespitosa* – tufted, referring to the habit of this plant.
Synonyms: *A. leptophylla incana* has sometimes been grown under this name, but is quite a distinct plant. Sometimes misspelt as '*A. caespitosa*'.

DESCRIPTION

General: Small tufted perennial, woody at base, shortly bristly throughout, with many of the hairs having minute tubercles at the base.
Stems: 1–10cm (⁷⁄₁₆–4in).
Leaves: 1–5cm (⁷⁄₁₆–2in) long and 2.5–4mm wide, narrowly linear, blunt, forming a rosette.
Inflorescence: Cymes often almost stalkless, dense, with 1–5 flowers.
Flowers: *Pedicels* very short. *Calyx* 4–6mm, divided almost to the base into linear, blunt oblong-lanceolate lobes. *Corolla* tube 6–7mm long, white; limb deep blue, 10–13mm across; with oblong, scales that are hairy on the margins.
Nutlets: to 3.2 × 3.5mm, obliquely ovoid.

GEOGRAPHY AND ECOLOGY

Distribution: Endemic to Crete. Included on *The IUCN Red List of Threatened Plants* as Rare.
Habitat: Stony slopes of the White Mountains.
Flowering: May to July in cultivation.
Associated plants: In its native habitat, grows with plants such as *Ranunculus brevifolius*, *Ricotia cretica* and *Crocus seiberi*.

CULTIVATION

Date of introduction: 1930s.
Hardiness: Probably Zone 7 or 8.
Sun/shade aspect: Full sun is normally considered essential, though I have observed an apparently happy specimen growing in a semi-shaded woodland border in a Scottish garden.
Soil requirements: Can be tried outside in a very gritty soil mix. In pots, use a mixture of 1 part loam to 1 part leaf mould and 1 part coarse sand, in 15–20cm (6–8in) pots (for mature plants). Alkaline conditions are sometimes recommended for this plant, but it is tolerant of moderate acidity. It should be kept moist throughout the growing season, but almost dry in winter. If grown in the alpine house, re-pot every year or every other year in early spring or after flowering.
Garden uses: Best in the alpine house; difficult outdoors, though could try on a raised bed or in a scree in a sunny spot.
Spacing: 20cm (8in) if planted outside, though usually would be grown in a pot.
Propagation: *Seed* – Sow in early spring in sterilized compost of 1 part loam to 1 part leaf mould and 2 parts sharp sand, or a similarly gritty mixture, covering lightly. *Vegetative* – Detach daughter rosettes in summer, and root in very gritty compost in a lightly shaded, well-ventilated cold frame. Take 5cm (2in) root cuttings in early to mid-spring, inserting them into equal parts of peat/peat substitute and sharp sand. If the young plants are for the raised bed or scree outdoors, plant out in early autumn or mid-spring.
Availability: Only from specialist alpine nurseries; seed sometimes available on exchange lists of alpine societies.
Problems: Difficult to grow outdoors, and hard to transplant without damaging the taproot. Sometimes a different plant, *Anchusa leptophylla* subsp. *incana* (Ledeb.) Chamb., may be incorrectly labelled and sold as this species.

Anchusa leptophylla subsp. *incana* (Ledebour) Chamberlain

An upright, softly hairy perennial with tough stems and long, narrow, dark green leaves, usually around 40cm (16in) tall, but can be higher. Gentian-blue flowers, salver-shaped and 8–12mm (⅜–½in) across, are borne in dense cymes from late spring to midsummer, and sometimes again in autumn. Good for a sunny border, or in a large rock garden.

NAMING

Etymology: *leptophylla* – slender-leaved, *incana* – grey.
Synonyms: *Anchusa angustissima* Koch; *A. aspera*
Boiss.; *A. boissieri* Bornm. & Gusul.; *A. incana* Ledeb.;
A. caespitosa hort.

DESCRIPTION

General: A softly hairy perennial.
Stems: To 70cm (28in), although often not more than
30–40cm (12–16in), upright.
Leaves: linear, entire, 5–11cm (2–4¼in) long and
6–13mm (¼–½in) wide.
Inflorescence: Cymes short, several, crowded, hairy,
elongating in fruit, with minute oval-lanceolate bracts.
Flowers: *Pedicels* to 2mm in fruit. *Calyx* 4–9mm in
flower, elongating in fruit, divided to one-third into
blunt lobes; covered in dense spreading hairy
(sericeous). *Corolla* usually bright azure blue with
white throat, occasionally white to yellow; tube
6–10mm, limb 4–8mm across, with oblong lobes.
Nutlets: 2–3 × 2.5–4mm, obliquely ovoid.

GEOGRAPHY AND ECOLOGY

Distribution: Endemic to Turkey.
Habitat: Rocky slopes, sandy steppe, at 800–3000m
(2600–9850ft).
Flowering: Early to midsummer in the wild; usually
late spring to midsummer in cultivation in the UK.

CULTIVATION

Hardiness: Zone 7
Sun/shade aspect: Needs full sun.
Soil requirements: Any fertile, well-drained soil; heavy
soil can be improved by mixing in sharp sand.
Garden uses: Rock garden or front of the border.
Spacing: 15–22cm (6–8½in) apart.
Propagation: *Seed* – As for *A. cespitosa*. *Vegetative* –
Basal stem cuttings 5cm (2in) long taken in March or
April, inserted in gritty compost in a cold frame.
Rooted cuttings can be potted up or planted out in a
nursery bed about a month later.
Availability: Can be found in some alpine and
herbaceous nurseries. May be mistakenly sold under
the name of *A. cespitosa*, or '*A. caespitosa*'.
Problems: Resents root disturbance, so avoid
transplanting if at all possible.

RELATED SPECIES

The typical species **Anchusa leptophylla** Roemer &
Schultes (syn. *Anchusa ochroleuca* var. *canescens* Boiss.)
is probably not in cultivation. It has a wider distribution –
in Turkey, Bulgaria, Romania, S Russia and Crimea –
and differs by a calyx divided to half to two-thirds of its
length with more sparse hairs, and more branched stems.

Anchusa granatensis. (Masha Bennett)

Anchusa officinalis Linnaeus

An upright, rather coarse perennial or biennial, on
average about 60cm (2ft) tall, with hairy stems,
long leaves and one-sided racemes of purple-violet
flowers in summer. This species is suitable for
planting in a dry, sunny spot in a native wildflower
garden or a herb collection, and is very attractive
to bees.

*'The leaves are covered with short hairs and
when dry emit a rich musky fragrance, rather
like wild strawberry leaves drying.' (Roy
Genders)*

NAMING

Etymology: *officinalis* – sold as a herb.
Synonyms: *Anchusa angustifolia* L.; *A. osmanica*
Velenovsky; *A. procera* Bess in Link; *A. microcalyx*
Vis.; *A. macrocalyx* Hausskn.
Common names: *English*: Alkanet; Common
Bugloss (USA); *Danish*: Læge-Oksetunge;
Dutch: Gewone Ossetong; *Finnish*: Rohtorasti;
French: Buglosse Officinale; *German*: Gewöhnliche
Ochsenzunge, Gemeine ochsenzunge; *Italian*:
Buglossa commune, Lingua Bovina; *Norwegian*:
Oksetunge; *Polish*: Fabrownik lekarski; *Russian*:
anhusa lekarstvennaya; *Slovak*: smohla lekárska;
Swedish: (vanlig) oxtunga; *Turkish*: Sigiidili;
Welsh: Alcanet.

DESCRIPTION

General: An upright, bristly perennial or biennial.
Stems: Usually 20–80cm (8–32in), though can grow to 1m (3ft) and taller, angular, upright, branched.
Leaves: 5–12cm (2–4$^1/_2$in) long and 1–2cm ($^7/_{16}$–$^3/_4$in) wide, lanceolate, the lower stalked.
Inflorescence: Several dense cymes, elongating in fruit.
Flowers: *Pedicels* very short, up to 5mm in fruit. *Calyx* 5–7mm, up to 10mm in fruit, divided almost to the base into lanceolate, pointed lobes. *Corolla* deep blue, violet or reddish, occasionally white or yellowish, tube to 7mm long, limb to 15mm across.
Nutlets: To 2 × 4mm, obliquely ovoid.
Chromosomes: 2n=16.

GEOGRAPHY AND ECOLOGY

Distribution: Balearic Islands and southern France eastwards to Turkey, except Sicily and Crete; Europe, Asia Minor; naturalized in the UK, the USA and Australia; this species is classified as a noxious weed in Washington State and certain areas of Canada.
Habitat: Meadows, banks, roadsides, cultivated ground, gardens.
Flowering: Mid-spring to early autumn.
Associated plants: In Romania, may be found with dwarf elder *(Sambucus ebulus)*, *Delphinium fissum*, *Achillea setacea*, *Centaurea micrantha* and *Berteroa incana*.

CULTIVATION

Hardiness: Zone 5.
Sun/shade aspect: Needs full sun.
Soil requirements: Any well drained soil.
Garden uses: A rather weedy plant, but may be of interest for a herb or a native collection, or a wildlife garden as it is very attractive to bees.
Spacing: 45–60cm (18–24in).
Problems: Weedy, can become a nuisance through self-seeding in favourable conditions.
Propagation: From seed only. Sown in midsummer in outdoor bed or in containers, it will be ready for planting into its final position in early or mid-autumn, for flowering next summer. Germination may be erratic and can take from 1 to 4 weeks. It can also be sown in pots of compost under cover in early spring, and then may flower in the same year.
Availability: Available from some herb and wildflower nurseries.

RELATED FORMS AND CULTIVARS

'Incarnata' is an unusual variation that is generally unavailable, with flesh-pink flowers.

RELATED SPECIES

Barrelier's bugloss, **Anchusa barrelieri** (Allioni) Vitman (syn. *Buglossum barrelieri* Allioni, *Cynoglottis barrelieri* (Allioni) Vural & Kit Tan) – from woods, fields and stony slopes in Italy, northern Balkans, Ukraine and Turkey, is a perennial similar to *A. officinalis*, but can be distinguished by its blue or bluish-violet flowers with the tube only 1.5mm long. Often separated into the genus *Cynoglottis* (Gusuleac) Vural & Kit Tan, which differs by the enlarged calyx at fruiting stage, and non-warty nutlets. It is sometimes offered for sale by herbaceous specialists, is hardy (Zone 4) and has similar requirements to the other taller perennial anchusas.

Anchusa undulata Linnaeus

A hairy, upright annual or biennial to 50cm (20in) tall, with bristly, wavy-margined leaves, and cymes of deep violet funnel-shaped flowers, with prominent, paler-coloured scales in throat. Suitable for a sunny border.

NAMING

Etymology: *undulata* – wavy, referring to leaf margins.
Synonyms: *Anchusa hybrida* Ten.; *A. undulata* ssp. *hybrida* (Ten.) Coutinho; *A. amplexicaulis* Sm.; *A. luschani* Wettst.
Common names: *English*: Undulate alkanet; *French*: Buglosse hybride; *German*: Wellblättrige ochsenzunge; *Italian*: Buglossa ibrida; *Spanish*: Chupamieles ondulada; *Turkish*: Arı çiçegi.

DESCRIPTION

General: A hairy annual or biennial.
Stems: 20–50cm (8–20in), upright, often branched from base.
Leaves: 3–7cm (1$^1/_4$–2$^3/_4$in) long and 5–15mm ($^1/_4$–$^5/_8$in)wide, oblong-lanceolate or elliptic, usually wavy-margined or irregularly toothed, sometimes entire; basal leaves tapering at base, stem leaves stalkless.
Inflorescence: Cymes elongating in fruit; with small, oval or lanceolate bracts.
Flowers: *Calyx* 5–10mm, tubular, divided at most to half their depth inflated in fruit; lobes oblong-lanceolate to triangular. *Corolla* deep violet to deep purple, occasionally white, funnel-shaped, tube 6–10mm, limb 5–10mm across, with oval lobes; scales oval, white to pale purple.
Nutlets: Transversely ovoid, 2–2.5 × 3–4mm, grey, with a rough, netted surface.

GEOGRAPHY AND ECOLOGY

Distribution: Throughout most of the Mediterranean region.

Arnebia pulchra. (Erich Pasche)

Habitat: Cultivated ground, dry hills, sand dunes, waste places, *Pinus brutia* forest (in Turkey) from sea level to 900m (2950ft).
Flowering: Spring, sometimes to midsummer, in the wild, usually summer in cultivation in temperate climates.
Associated plants: The author saw this species growing in Israel in fallow fields with plants such as *Ranunculus asiaticus*, *Anemone coronaria* and *Arum palaestinium*.

CULTIVATION
Hardiness: Zone 8
Sun/shade aspect: Full sun.
Soil requirements: Any well-drained soil, tolerates drought.
Garden uses: Mixed and herbaceous border.
Spacing: 20–30cm (8–12in).
Propagation: As for *Anchusa officinalis*. Dead-head to prevent self-seeding.
Availability: Rarely available, but seed and occasionally plants can be found in some specialist catalogues.
Problems: Can be weedy in ideal conditions.

RELATED SPECIES
Anchusa granatensis (Boiss.) Valdés – sometimes considered a subspecies of *A. undulata*, is an attractive perennial that is unfortunately rarely cultivated.

The genus *Arnebia* Forsskål, 1775
Prophet Flower

NAMING
Etymology: *Arnebia* – from the Arabian name of the plant.
Synonyms: *Echioides* Ortega; *Leptanthe* Klotzsch.; *Munbya* Boiss.; *Toxostigma* A. Rich.; *Ulugbekia* Zakyrov; *Huynhia* Greuter. A number of species are now often allocated to genus *Macrotomia* DC, and one species to *Stenosolenium* Turcz.

GENUS CHARACTERISTICS AND BIOLOGY
Number of species: About 25.
Description: A genus of hairy annuals, biennials or perennials, with roots that often contain a purple dye, and upright or prostrate stems. Flowers borne in cymes with bracts. *Calyx* 5–lobed, divided to base. *Corolla* usually yellow, sometimes purple, funnel-shaped, usually hairy on the outside, with a straight or slightly curved tube. Limb usually shorter than the tube, with spreading lobes. Flowers are often heterostylous – with 'pin' and 'thrum' forms. Style 2- or 4-branched. Nutlets obliquely oval, covered with small swellings.
Distribution and habitats: North Africa, Mediterranean, central and south-west Asia, the Himalayas, China; mostly in mountains, some in semi-deserts and other dry places.
Life forms: Therophytes (annuals) and hemicryptophytes.
Pollination: By various bees. The curious flower pattern of some yellow-flowered species, such as *A. pulchra* and *A. guttata*, where each corolla lobe bears a dark-coloured spot, acts as a signal to bees, indicating which flowers contain most nectar and are ready to be pollinated. Once the flower is fertilized, the dark spots on petals fade and the nectar disappears.
Fruit dispersal: No specialized mechanism.

USES
GARDEN
The genus is poorly represented in cultivation, only *Arnebia pulchra* being to any extent common. However, there are some very attractive species worth trying in rock gardens and mixed borders, where *A. pulchra* is certainly worthy of a place. This species is also suitable for a semi-shaded wall crevice, or a light woodland situation.

MEDICINAL
Arnebia euchroma is an important plant in Chinese medicine, used to treat many ailments including

measles, constipation, burns, frostbite, dermatitis, eczema. The roots contain the compound shikonin, which inhibits pathogenic bacteria, such as *E.coli, Bacillus typhi, B. dysenteriae, Pseudomonas* and *Staphylococcus aureus*. It may also have some contraceptive effect, and can inhibit the growth of cancerous cells. In India the roots of *A. perennis* are bruised and applied to eruptions, and *A. benthamii* is used to treat diseases of the tongue and throat.

Culinary

The root of *A. perennis* is said to be edible and eaten in the western Himalayas.

Cultivation

As only one species is commonly cultivated, cultural information on the genus as a whole is scarce. I will try to make generalizations based on the known aspects of plants' ecology in the wild. Most species are hardy, though probably not easy to grow outdoors in wet climates. *Arnebia pulchra* needs a cool, lightly shaded position, but most other *Arnebia* species probably require full sun, though in the wild they are sometimes found in dappled shade. All species grow on very well-drained, sometimes dry soil. High humus content (especially leaf mould) in the soil would probably be beneficial.

Propagation

Seed The seed may fall on the ground as soon as it is ripe, so check for the ripening seed regularly, or put a cloth underneath the plant to catch the falling seeds. Sow in mid-spring, in containers of gritty compost, under glass at about 15°C (59°F), covering lightly. Germination is likely to be erratic. Take great care watering the seedlings, as they may easily rot – try not to get water on the leaves, so water from below. **Vegetative** For some species, stem cuttings can be tried in midwinter in a heated propagator at 15°C (59°F), or heeled cuttings in late summer or autumn. Careful division in spring or after flowering is possible only for plants that have multiple rosettes. Root cuttings can be tried in sand in a propagator with bottom heat during late winter. At all times prevent the propagation material and young plants from getting too wet.

Problems

Most species are unavailable, and some, such as *A. densiflora*, are very difficult to grow.

References

CHOPRA *et al*, 1986; DAVIS, 1978; DUKE & AYENSU, 1985; ELLIOT, 1935; GENDERS, 1994; HUXLEY, 1992; INGRAM, 1958; OLSEN, 1999; PFAF, 1992–02; POLUNIN & STAINTON, 1984; SINGH & KACHROO, 1976; SLABY, 2001: WALTERS, 2001; WU & RAVEN, 1995; YEUNG, 1985.

Arnebia benthamii (Wallich in G. Don) I.M. Johnston

A striking, upright, very hairy perennial around 60cm (2ft) tall, with a stout leafy stem bearing a tall spike of purple-red flowers with long, grey bracts. Rare in cultivation, but worth seeking out as a curiosity.

> '...a large very dense shaggy-haired cylindrical spike of red-purple flowers showing between much longer linear grey hairy drooping bracts.' (Oleg Polunin & Adam Stainton)

> 'It has attracted attention from everyone who has seen it in my garden. Some think it pretty, others a curiosity, and others think it ugly. I myself think it really pretty.' (O.P. Olsen)

Naming

Etymology: *benthamii* – after George Bentham (1800–84), who compiled *Flora HongKongensis*.
Synonyms: *Macrotomia benthamii* (Wallich) A. DC; *Lithospermum benthamii* (Wallich in G. Don) Johnst.; *Echium benthami* Wall.

Description

General: A hairy perennial, with stout rootstock that is covered with bases of old leaves.
Stems: 15–60cm (6–24in), upright, stout.
Leaves: Linear to narrowly-lanceolate, bristly-hairy, lower leaves to 30cm (12in), upper leaves smaller, numerous.
Inflorescence: In fruit elongating to 30cm (12in); with linear bracts to 5cm (2in) long.
Flowers: *Calyx* lobes linear 23–40mm long, tipped with purple. *Corolla* tube about 25mm, and with 5 triangular lobes spreading to about 10mm across.
Nutlets: Oval, with small swellings.

Geography and ecology

Distribution: From Pakistan to western Nepal. It is included on *The IUCN Red List of Threatened Plants* as an Endangered Species.
Habitat: Open slopes, scrub, at 3000–4300m (9850–14,100ft) in the mountains.
Flowering: May to July in the wild.

Cultivation

Hardiness: Zone 7.

Sun/shade aspect: Full sun will be needed in temperate climates, although in warmer areas it will perhaps do well in dappled shade.
Soil requirements: Gritty, well-drained soil is probably required, with addition of leaf mould.
Garden uses: Although a high mountain plant, this can often grow too large for an average rock garden. Could try in a sunny border.
Propagation: Probably only from seed.
Availability: Generally unavailable commercially, but seed may appear on specialist societies' exchange lists. As this is an endangered plant, make sure that seed/plants come from legitimate sources.

Arnebia densiflora (Nordb.) Ledebour

A beautiful perennial with several dense rosettes of narrowly lanceolate leaves, and flowering stems to 40cm (16in), bearing honey-scented yellow flowers. A connoisseur plant, very difficult to cultivate, but worth every effort.

'According to Clarence Elliott, the plant was discovered in the Taurus Mountains by E.K. Balls in 1934, and the finder considered it to be the most beautiful plant he had ever seen… It bears throughout summer large flower-heads of deep golden-yellow which emit a sweet honey-like perfume, like that of Onosma….' (Roy Genders)

NAMING
Etymology: *densiflora* – dense-flowered.
Synonyms: *Arnebia cephalotes* A. DC; *Macrotomia densiflora* (Ledeb. in Nordm.) Macbride; *M. cephalotes* (A. DC.) Boiss.; *Lithospermum densiflorum* Ledeb. in Nordm.; *Arnebia macrothyrsa* Stapf in Wien.; *Munbya cephalotes* (A. DC.) Boiss.

DESCRIPTION
General: A hairy perennial with stout, erect or horizontal rootstock up to 7cm (2³⁄₄in) long and 4cm (1¹⁄₂in) across, covered with wool bases of old leaves.
Stems: 15–20cm (6–8in), upright, simple or branched above, leafy, velvety and bristly-hairy.
Leaves: Basal leaves to 10–15cm (4–6in) long and 1–2cm (⁷⁄₁₆–³⁄₄in) wide, forming an erect rosette, lanceolate, sharply pointed, narrowing into stalks to about 4cm (1¹⁄₂in) long. Stem leaves narrowly lanceolate, stalkless, all greyish-green, hairy.
Inflorescence: Terminal cluster of cymes forming a dense head 5–12cm (2–4¹⁄₂in) in diameter.
Flowers: Honey-scented. *Calyx* lobes linear, sharply pointed, bristly, 15–20mm (⁵⁄₈–³⁄₄in) in flower, to 20mm (³⁄₄in) in fruit. *Corolla* up to 30mm (1¹⁄₄in)

long, tube twice as long as calyx; limb 20mm (³⁄₄in) across, golden-yellow, unspotted.
Nutlets: Minutely pitted and ridged.

GEOGRAPHY AND ECOLOGY
Distribution: Greece, in the Peloponnese (Aroania Oros); Turkey, from Bursa south to Adana and east to Erzican.
Habitat: Limestone and igneous rocks and cliffs in mountains, at altitudes of 750–2600m (2460–8500ft).
Flowering: May to August in the wild.

CULTIVATION
Hardiness: Zone 5
Sun/shade aspect: Full sun is essential.
Soil requirements: Perfectly drained position is a must.
Garden uses: Probably safest in an alpine house, in gritty compost.
Could be tried outdoors in a crevice.
Spacing: Normally would be in a pot, but if planting outdoors, allow 20–30cm (8–12in) for spread.
Propagation: By seed, or possibly heeled cuttings in autumn.
Availability: Generally unavailable, but seed may appear on alpine and rock plant societies' exchange lists.
Problems: Very difficult to grow: although Roger Phillips collected seed and raised seedlings on several occasions, the plants have always rotted before flowering. Must be kept dry and cold in winter – snow cover is ideal, but in wet climates, like the UK, could try protecting from winter rains with a pane of glass.

Arnebia euchroma (Royle in Benth.) I.M. Johnston

An unusual, hairy perennial, about 30cm (12in) tall, with few upright stems bearing slightly nodding clusters of tubular flowers that open dark blackish-purplish and gradually fade to pink with age, so that there are flowers of different shades in the inflorescence, surrounded by narrow hairy bracts. The flowers smell intensely of peaches. Not in general cultivation, but worth seeking out.

NAMING
Etymology: *euchroma* – good colouring, referring to the dye contained in the roots.
Synonyms: *Macrotomia euchroma* (Royle) Paulsen, *Lithospermum euchromon* Royle.
Common name: *Chinese:* ruan zi cao.

DESCRIPTION
General: A hairy perennial with a stout rootstock, to 2cm (³⁄₄in) in diameter, containing purple dye.

Arnebia euchroma. (Masha Bennett)

Stems: Usually 1 or 2, erect, branched above, sheathed with remaining bases of leaves, 15–40cm (6–16in) tall, covered with white or pale yellow spreading hairs.

Leaves: Basal leaves linear to narrowly lanceolate, pointed, 7–20cm (2³/₄–8in) long and to 1.5cm (⁵/₈in) wide, with a sheathlike base. Stem leaves lanceolate, smaller. All sparsely covered with long bristly hairs, stalkless.

Inflorescence: Cymes terminal, 2–6cm (³/₄–2¹/₂in) across, many-flowered; with lanceolate bracts.

Flowers: Fragrant, smelling of peaches. *Calyx* lobes linear, to 16mm (⁵/₈in), to 30mm (1¹/₄in) in fruit, densely covered in pale yellow hairs on both sides. *Corolla* dark purple, but turning paler red and eventually pink with age, tubular bell-shaped, sometimes sparsely shortly hairy outside; tube to 14mm (¹/₂in); limb 6–10mm (¹/₄–⁷/₁₆in)wide.

Nutlets: About 3.5 x 3mm, black-brown, broadly oval, roughly netted, with few small swellings.

GEOGRAPHY AND ECOLOGY

Distribution: Afghanistan, Pakistan, north west India, central Nepal, China, Kazakhstan, Kyrghyzstan, Tajikistan, Turkmenistan, Uzbekistan, southern Russia.

Habitat: Open slopes, rocks and meadows, at 3300–4500m (10,800–14,750ft) above sea level.

Flowering: June to August in the wild.

Associated plants: In the Tien Shan mountains, the author observed it growing with *Lindelofia stylosa*, *Codonopsis clematidea*, *Tianschaniella umbellulifera* and *Adenophora himalayensis*.

CULTIVATION

Hardiness: Zone 5

Sun/shade aspect: Full sun.

Soil requirements: Well-drained, gritty soil.

Garden uses: Large rock garden or front of the border.

Spacing: 20–30cm (8–12in).

Propagation: As for the genus.

Availability: Not yet available, but may appear on seed-exchange lists.

Arnebia pulchra (Willd. in Roem. & Schult.) Edmondson

A hairy, clump-forming perennial around 30cm (12in) tall, with rosettes of light green leaves, and yellow funnel-shaped flowers with an unusal pattern – a dark brown spot on each petal, which fades gradually. A choice plant for a cool situation in a border or rock garden.

'Forms clumps of rough surfaced, hairy foliage from which rise the characteristic crosiers, displaying widely funnel-shaped, five-petalled flowers of clear yellow, each with a conspicuous black spot. This, which is supposed to represent the Prophet's fingerprint, fades to a dull brown and eventually disappears leaving the flowers, in their last stages, a uniform lemon-yellow.' (Anna N. Griffith)

NAMING

Etymology: *pulchra* – beautiful.

Synonyms: *Macrotomia echioides* Boiss., *Arnebia echioides* (L.) A. D.C, *A. longiflora* C. Koch, 1849, *Echioides longiflorum* (K.Koch) I.M. Johnston; *Lycopsis pulchra* Willd., *L. echioides* Linnaeus.

Common name: *English:* Prophet flower; *Finnish:* profeetankukka; *German:* Prophetenblume; *Swedish:* profetblomma.

DESCRIPTION

Stems: 15–45cm (6–18in) high, few, sprawling, with dense spreading hairs.

Leaves: Basal leaves lanceolate to narrowly oval, 6–15cm (2¹/₂–6in) long and 1–2.5cm (⁷/₁₆–1in) wide, gradually narrowing into a short stalk. Stem leaves stalkless with heart-shaped base, clasping the stem. All leaves stiffly hairy and light green.

Inflorescence: Compact terminal raceme.
Flowers: *Calyx* 10–12 mm ($^{7}/_{16}$–$^{1}/_{2}$in) flower, 13–18mm ($^{1}/_{2}$–$^{3}/_{4}$in) in fruit. *Corolla* bright yellow, 20–24mm ($^{3}/_{4}$–$^{15}/_{16}$in) long and 18–25mm ($^{3}/_{4}$–1in) across, with 5 spreading lobes, each with blackish/dark spot which fades as the flower ages.
Nutlets: To 4.5 × 3.5mm, ovoid to nearly spherical, pointed, with sharp keels, covered with small warts and pits.

GEOGRAPHY AND ECOLOGY

Distribution: Northern Caucasus, northern Iran, north-east Turkey.
Habitat: Rocky places and roadsides at 1525–3000m (5000–9850ft) in Turkey.
Flowering: June to July in Turkey in the wild, mid- to late spring in cultivation in cool temperate climates.
Associated plants: In Georgia may be found growing in association with *Primula algida*, *Rhododendron caucasium*, *Androsace villosa* and *Gentiana angulosa*.

CULTIVATION

Hardiness: Zone 6.
Sun/shade aspect: Needs a sheltered, cool position in light shade, especially in warmer areas. Can tolerate full sun in cooler climates.
Soil requirements: Humus-rich, gritty soil, ideally with plenty of leaf mould.
Garden uses: Rock garden, front of the border, or in an open woodland setting. Can be tried in a north- or east-facing crevice in a wall – ensure there is plenty of organic matter for the plant to root in.
Spacing: 25cm (10in).
Propagation: As for the genus. All methods described could be tried with this species. Seed production is low and divisions are usually only few.
Availability: The only *Arnebia* species generally available from nurseries, but not easily found.

RELATED SPECIES

Other interesting yellow-flowered species include the so-called Arabian primrose, **Arnebia cornuta** Fisch. & Mey. (syn. *A. decumbens* (Vent.) Coss. & Kral.), distributed in parts of Europe, north Africa and western China, an annual with plain yellow flowers.
Arnebia guttata Bunge (syn. *A. thompsoni* C. B. Clarke; *Macrotomia guttata* (Bunge) Farrer) from Afghanistan to Kashmir, China, Mongolia and Central Asia, is biennial to perennial, with dark brown spots on corolla lobes, similarly to *A. pulchra*. These are not in general cultivation, but may crop up in seed-exchange lists, and are worth trying.

The genus *Borago*
Linnaeus, 1753
Borage

NAMING

Etymology: Possibly a corruption of Latin *corago*, from *cor*, the heart, and *ago*, meaning 'I bring', because of its cordial effect. Another possibility is that it may be derived from the Latin *burra*, a flock of wool, referring to the hairiness of the plant, or, alternatively, from *barrach*, a Celtic word meaning 'a man of courage'.
Synonym: *Borrachinea* Lavy. In the past, *B. pygmaea* has been assigned to *Anchusa* L. and even *Campanula* L.! Plants of other genera, such as *Caccinia* Sari and so on, have been included into *Borago*.

GENUS CHARACTERISTICS AND BIOLOGY

Number of species: 3, the third, little-known species being *B. longifolia* Poiret from Algeria and Tunisia.
Description: A genus of annual or perennial herbs. Leaves alternate, hairy. Flowers borne in a branched, loose cyme, usually with bracts. *Calyx* lobed almost to the base, enlarging in fruit. *Corolla* blue, pink or white, star-shaped to bell-shaped, with no tube or only a short one, and lanceolate, widely spreading lobes. Stamens protrude to form a cone-shaped beak; each filament has a long, narrow appendage at the tip. Style included inside the corolla, with a capitate/headlike stigma. Nutlets 4, obovoid, erect, rough.
Distribution and habitats: Europe; dry open places for *B. officinalis* and moist places for *B. pygmaea*.
Life form: Therophyte (non-overwintering individuals of *B. officinalis*) or hermicryptophyte.
Pollination: Visited by bumblebees (*Bombus* species) that carry out the process of so-called 'buzz pollination' – the pollen is shaken off the anthers onto the furry body of the insect by means of vibrations produced by the bee buzzing.
Fruit dispersal: No specialized mechanism.

USES

GARDEN

Borago officinalis is an indispensable ingredient of all herb collections and any garden designed for bees. In 'companion planting', *B. officinalis* is considered a good companion for strawberries, tomatoes, courgettes and most other plants. It is said to deter some pests, including Japanese beetles and tomato hornworm. Both this and the perennial species are good for a mixed border, and *B. pygmaea* may be admitted to a larger rock garden.

Borago officinalis with *Cerinthe major* in an olive grove in southern Italy. (Masha Bennett)

MEDICINAL

Borage has been used in herbal medicine for centuries, mostly as a demulcent and diuretic, but numerous other beneficial properties are attributed to this plant. In particular, it has a reputation for lifting the spirits. John Gerard in *The Herball, or Generall Historie of Plantes* (1597) says that '*Those of our time do use the floures in sallads to exhilerate and make the minde glad. There be also many things made of them, used for the comfort of the heart, to drive away sorrow and increase the joy of the minde. ... Syrrup made of the floures of Borrage comforteth the heart, purges melancholy, and quieteth the phrenticke or lunaticke person.*' According to Dioscorides and Pliny, Borage was the famous Nepenthe of Homer, which when drunk steeped in wine, brought absolute forgetfulness.

Borage is taken internally for fevers, bronchial complaints, mouth infections and a variety of skin conditions; and externally as poultices, gargles, eyewashes and mouthwashes.

The leaves are harvested in late spring and the summer as the plant comes into flower. They can be used fresh or dried, but stored no longer than one year. Flowers should be picked as they open, and oil is extracted from ripe seeds.

The plant (but not oil) contains small quantities of pyrrolizidine alkaloids that may cause liver damage and liver cancer (though this is likely only if very large quantities are consumed). Therefore, it should not be taken by those suffering from liver problems.

Borago officinalis flower. (Masha Bennett)

CULINARY

Borago officinalis leaves have a cucumber-like flavour and can be eaten raw (in salads or with soft cheese) or cooked like spinach. They are also added whole as a flavouring to some drinks, such as Pimm's. The young leaves are rich in potassium and calcium. The flowers can be used raw or candied as a decorative garnish on salads and fruit drinks; they can also be made into syrup.

ECONOMIC

The seed of *B. officinalis* yields 30 per cent oil, 20 per cent of which is gamma-linolenic acid, yielding in total 350–650kg per hectare (310–570lbs per acre). An edible blue dye can be obtained from the flowers, used to colour vinegar. In Russia, fields of borage are grown by apiculturists specially for bees – 1 hectare of borage yielding 200kg (1 acre yielding 175lbs) of excellent-quality honey.

CULTIVATION

The two commonly cultivated *Borago* species are very easy to grow but require rather different growing conditions – see individual species descriptions.

PROPAGATION

Seed Seeds for the annual *B. officinalis* can be sown early to late spring outdoors where they are to flower, or in autumn in mild areas and should be covered lightly with soil. They could also be raised in pots for transplanting later, but care must be taken not to damage the root. Autumn-sown plants, however, will flower and die earlier. Perennial *B. pygmaea* is also easy from seed, and both species self-seed copiously.
Vegetative By division for the perennial species, *B. pygmaea,* during early or mid-spring, replanting the sections immediately or potting them up. Also by cuttings of young shoots in late spring or summer, rooted in the cold frame.

REFERENCES

ALLARDICE, 1993; BLAMEY & GREY-WILSON, 1993; BRICKELL, 1995; BOWN, 1996; FACCIOLA, 1990; HUXLEY, 1992; PFAF, 1992–02; STACE, 1991; THOMAS 1993; TUTIN *et al.,* 1972; WALTERS, 2000.

Borago officinalis Linnaeus

A bristly annual, with robust, branched stems up to 60cm (2ft) tall, rough oval or lanceolate leaves, and star-shaped, bright blue flowers with a whitish centre and a cone of blue-black anthers. One of the most popular and indispensable herbs, and its nectar-rich flowers are a magnet to bees.

'*Pliny calls it Euphrosinum, because it maketh a man merry and joyfull: which thing also the old verse concerning Borage doth testifie:*

> *Ego Borago (I, Borage)*
> *Gaudia semper ago. (Bring alwaies courage.)'*
> (John Gerard)

NAMING

Etymology: *officinalis* – sold as a herb.
Common names: *English*: Borage, Beebread, Star-flower; *Danish*: Hjulkrone; *Finnish*: Purasruoho; *French*: Bourrache (officinale); *German*: Borretsch; *Italian*: Borragine (comune); *Japanese*: Ruri Zisa; *Norwegian*: Agurkurt; *Russian*: burachnik, ogurechnaya trava; *Slovak*: borák lekársky; *Spanish*: Borraja; *Swedish*: Gurkört, borag, stofférblomma; *Turkish*: Hodan; *Welsh*: Bara Gwenyn, Iianwenlys.

Borago officinalis 'Variegata'. (Masha Bennett)

DESCRIPTION
General: A coarse bristly annual.
Stems: 15–60cm (6–24in), upright, robust, bristly, often branched.
Leaves: Basal leaves 5–20cm (2–8in), oval to lanceolate, with rather wavy margins, stalked. Upper stem leaves stalkless, more or less clasping the stem. All leaves bristly.
Inflorescence: Broad, branched cymes, with semi-nodding flowers.
Flowers: *Calyx* 8–15mm ($^3/_8$–$^5/_8$in) at anthesis, up to 20mm ($^3/_4$in) in fruit, lobes narrowly lanceolate, pointed, connivent in fruit. *Corolla* star-shaped, bright blue, rarely white; tube very short or almost absent, lobes 8–15mm ($^3/_8$–$^5/_8$in), lanceolate, pointed. *Stamens* with prominent purple-black anthers forming a cone.
Nutlets: 7–10mm, oblong-obovoid.
Chromosomes: 2n=16, 32.

GEOGRAPHY AND ECOLOGY
Distribution: Originally from southern Europe, now widely naturalized in central, eastern and western Europe.
Habitat: Cultivated and fallow ground, waste places, rough ground, roadsides, usually in dry habitats (though I have observed some very healthy-looking plants standing with their feet in water in southern Spain).
Flowering: March to June in southern Europe; early summer to mid-autumn in cultivation in the UK.

CULTIVATION
Hardiness: Zone 7.
Sun/shade aspect: Full sun is best, will tolerate light shade.
Soil requirements: Any well-drained soil, tolerates a wide range of pH, from very acid to alkaline (4.8 to 8.3). Can grow on stony, chalky, poor soil.
Garden uses: Versatile plant that is useful in a sunny mixed border, herb garden, wildlife garden (where it will provide abundant nectar for bees), or in any gaps in the garden.
Spacing: 30–45cm (12–18in) apart.
Propagation: From seed only, very easy. Seed ripens approximately one month after flowering. Germination takes 2–3 weeks. Dislikes transplanting so best to sow *in situ*, thinning the seedlings to 30–45cm (12–18in).
Availability: A common garden herb, available as seed from most seed merchants, and also as a plant from some herb nurseries.
Problems: Susceptible to powdery mildew, especially towards the end of the season. Prolific self-seeding may be a nuisance.

RELATED FORMS AND CULTIVARS
'Alba' – white-flowered. There seem to be at least two strains in cultivation in the UK, one with more elongated leaves than the other. The former seeds about 50 per cent white and the rest blue, while the shorter leaved form should come 100 per cent true from seed.
'Variegata' (syn. 'Bill Archer') – has yellow-mottled leaves, possibly a result of a viral infection, but a curious plant to grow.

Borago pygmaea De Candolle
A short-lived, somewhat sprawling, hairy, branched perennial, 15–60cm (6–24in) high, with large rosettes of coarse, hairy leaves and clear pale blue, bell-shaped flowers in summer. Best in light shade in moist soil.

> *'Once again the borage family gives us exquisite blue flowers, this time of a pale azure, gracefully nodding in branching sprays, but marred by large rosettes of coarse hairy foliage.'*
> (Graham S. Thomas)

NAMING
Etymology: *pygmaea* – dwarf.
Synonyms: *Borago laxiflora* Poiret; *Anchusa laxiflora* DC, *Campanula pygmaea* DC.
Common names: *English:* Slender borage, Dwarf borage (USA).

DESCRIPTION
General: A hairy, short-lived perennial.
Stems: 15–60cm (6–24in), somewhat sprawling, slender, branched.
Leaves: Lower leaves 5–20cm (2–8in), oblong to obovate, petiolate, upper leaves sessile, amplexicaul.
Inflorescence: Loose branched cymes
Flowers: *Calyx* 4–6mm, enlarging to 8mm in fruit; lobes lanceolate, pointed. *Corolla* bell-shaped, clear blue with a short tube and oval, pointed lobes 5–8mm long.
Nutlets: 3–4mm, obovoid.
Chromosomes: 2n=32.

GEOGRAPHY AND ECOLOGY
Distribution: Corsica, Sardinia, Capri.
Habitat: Damp places.
Flowering: April to July in the wild, in cultivation in the UK from early summer to autumn.

CULTIVATION
Date of introduction: 1813.
Hardiness: Zone 5.
Sun/shade aspect: Tolerates sun or shade; best in light shade.

Soil requirements: Soil should ideally be moist but well-drained. Will tolerate poor conditions and can even grow in gravel.

Garden uses: Grow in a border, rock garden, in light shade of trees and shrubs.

Spacing: 45cm (18in) apart.

Maintenance: Dead-head to avoid self-seeding.

Propagation: As for the genus.

Availability: Plants quite widely available from nurseries, and seed from a few seedsmen and most seed-exchange schemes.

Problems: Prolific seeder and can become invasive.

The genus *Brunnera*
Steven, 1851

NAMING
Etymology: *Brunnera* – after the Swiss botanist Samuel Brunner (1790–1844).

Synonyms: Plants belonging to this genus have previously been assigned to *Anchusa* L. and *Myosotis* L.

GENUS CHARACTERISTICS AND BIOLOGY
Number of species: 3 relict species.

Description: A genus of Perennial rhizomatous herbs, with upright, hairy stems that sometimes may bear glandular hairs. Leaves hairy, basal leaves long-stalked. Flowers borne in a terminal panicle of cymes, without bracts. *Calyx* divided nearly to base, lobes linear-lanceolate, somewhat pointed, enlarged in fruit. *Corolla* blue or purple, small, 'forget-me-not'-like, with a short, straight tube and spreading, oval or rounded lobes; throat more or less closed by 5 hollow, fringed scales. Stamens included in the corolla, inserted at the middle of the tube; filaments short. Style included in the corolla, very short, with head-like stigma. Nutlets 4, straight or somewhat curved, rough, warty, each with a thickened collar-like basal ring.

Distribution and habitats: The eastern Mediterranean, the western Irano-Turanian, the Caucasus and Siberia. Forests and grassy places at high altitudes.

Life form: Hemicryptophytes.

Pollination: Bees.

Fruit dispersal: No specialized mechanism.

GARDEN USES, CULTIVATION AND PROPAGATION
See under the species, *Brunnera macrophylla*.

Brunnera macrophylla in north-eastern Turkey. (Erich Pasche)

REFERENCES
BRICKELL, 1996; DAVIS 1978; HUXLEY, 1992; MALYSHEV 1997; PHILLIPS & RIX, 1991; STACE, 1991; THOMAS, 1993; TUTIN et al., 1972; WALTERS, 2001.

Brunnera macrophylla (Adams) I.M. Johnston
A rhizomatous perennial with rough, softly hairy, heart-shaped leaves, 30–45cm (12–18in) tall. Sprays of blue, forget-me-not-like flowers, 6mm across, appear in late spring. Perfect for light shade under deciduous trees and shrubs. The RHS Award of Garden Merit plant.

'Vivid forget-me-not flowers are held on 46cm (18in) stems in April and May, after which the large heart-shaped leaves assume greater proportions and create attractive greenery through the summer.' (Graham S. Thomas)

NAMING
Etymology: *macrophylla* – large-leaved.

Synonyms: *Anchusa myosotidifolia* Lehm.; *Brunnera myosotidifolia* (Lehm.) Lehm.; *Myosotis macrophylla* Adams in Weber & Mohr.

Common names: *English*: Brunnera, Great forget-me-not, Perennial forget-me-not, False forget-me-not; *Norwegian:* Forglemmegeisøster; *Slovak:* brunera vel'kolistá.

DESCRIPTION

General: A hairy perennial arising from an elongated blackish rhizome.

Stems: 20–50cm (8–20in), simple, bristly-hairy.

Leaves: Basal leaves oval with a heart-shaped base, 5–20cm (2–8in) long and 3.5–10cm (1³/₈–4in) wide, with a stalk to 20cm (8in) long; stem leaves lanceolate or oval to elliptical, stalkless. All hairy, pointed.

Inflorescence: Terminal cymes, forming a lax panicle or corymb-like inflorescence.

Flowers: *Pedicels* 2–5mm, elongating to 8mm in fruit. *Calyx* about 1mm, enlarging to about 2mm in fruit; lobes linear-lanceolate, somewhat pointed. *Corolla* blue, with a tube about 1.5mm long and limb 3–7mm across.

Nutlets: 2.5–4mm, oblong-obovoid, weakly ribbed and warty.

GEOGRAPHY AND ECOLOGY

Distribution: Caucasus, Turkey. Naturalized as a garden escape in the UK.

Habitat: Spruce (*Picea*) forest, grassy banks, at 500–2000m (1650–6550ft). In the UK, sometimes found in woods and on rough ground.

Brunnera macrophylla 'Betty Bowring'. (Howard Rice)

Flowering: March to May in Turkey; usually late spring and early summer in cultivation.

CULTIVATION

Date of introduction: 1713.

Hardiness: Zone 3.

Sun/shade aspect: Light dappled shade is best, but will tolerate both deeper shade and full sun.

Soil requirements: Any well-drained soil, ideally moisture-retentive, with added leaf mould, but is very tolerant of soil conditions.

Garden uses: Excellent ground cover in light shade of trees and shrubs, or in north-facing gardens; especially good in wild and informal settings. Lovely by stream or pond. Sheltered situation is preferable.

Spacing: 45cm (18in) apart.

Propagation: *Seed* – Not very practical as vegetative propagation is faster and easier, but, if desired, the seed can be sown in spring, or as soon as ripe, either in a well-prepared bed outdoors, or in pots or trays of compost in a cold frame or cool greenhouse. The cultivars will not come true from seed. *Vegetative* – Divide and replant the rhizomes in autumn or early to mid-spring, either replanting straight away, or potting up the sections. Take root cuttings from mid-autumn to early winter and insert in trays or pots of gritty compost in a cold frame. The young plants can be potted up or planted in a nursery bed in late spring or early summer, and should be ready for planting in the permanent position in autumn.

Availability: Plants are widely available from many herbaceous nurseries and some garden centres; seeds sometimes crop up in seed-exchange lists.

Problems: Leaves become coarse with age. In variegated forms, leaves are susceptible to scorching from wind or too much sun, so a sheltered position in dappled shade is needed for these. Reversion may happen in variegated cultivars, when plain green shoots appear and may eventually overtake the whole plant, unless kept under control. Remove any shoots that revert to green foliage, but take care not to cut into the roots, as this encourages new green growths.

RELATED FORMS AND CULTIVARS

'Betty Bowring' (syn. 'Alba') – white flowers, seedlings come blue-flowered.

'Blaukuppel' – a new form from Germany, vigorous and floriferous.

'Dawson's White' (syn. 'Variegata') – irregularly cream and green leaves, some broadly margined in creamy-white, some almost completely white; colouring more pronounced in shade; brought from a garden in Holland by Douglas Dawson prior to 1969.

Brunnera macrophylla 'Dawson's White'. (Howard Rice)

'Gordano Gold' – leaves emerge green but acquire patches of golden-yellow. From Mr & Mrs Wills of Walton on Gordano, near Bristol, UK.

'Hadspen Cream' ♀ raised by Eric Smith – the leaves broadly bordered with cream.

'Jack Frost' – a new cultivar with almost completely silver, green-veined leaves, discovered in Walters Gardens (Michigan, USA), and is believed to have originated from *B. macrophylla* 'Langtrees'. The nursery's catalogue describes it as 'a frosty silver overlay with light green venations' and 'leaves resemble cracked porcelain'. This plant has won the First Prize at Plantarium fair in Holland for Best New Perennial of the Year. This is a protected variety, requiring a licence for propagation and distribution.

'Langford Hewitt' – leaves with small patches of golden-yellow, from Bryan Hewitt (London, UK), named after his mother.

'Langtrees' (syn. 'Aluminium Spot') – leaves regularly spotted with silvery-grey; rough, dark green leaves bordered with silver spot. Named after Dr Rogerson's garden in Devon. Comes partly true from seed.

RELATED SPECIES

Brunnera sibirica (Schenk) I.M. Johnston (syn. *Myosotis orientalis* Schenk, *Anchusa neglecta* A. DC.) is similar to *B. macrophylla*, but rather more robust, to 50–60cm (20–24in) high, with slightly larger but fewer flowers, from the wet meadows of Siberian taiga. It is not in general cultivation in the West, but is commonly grown in Russia, and recently brought to UK by the author, who bought several pieces of rhizome from a street stall in Moscow. Will soon be available from Monksilver Nursery.

The genus *Buglossoides* Moench, 1794
Gromwell

NAMING
Etymology: *Buglossoides* – from *Buglossum* (an old synonym for some *Anchusa* species), and Greek suffix *-oides*, resembling.
Synonym: Commonly included within the genus *Lithospermum* L.

GENUS CHARACTERISTICS AND BIOLOGY
Number of species: Approximately 10–15.
Description: Bristly-hairy annual or perennial herbs or dwarf shrubs, similar to *Lithospermum*. Stems upright or prostrate. Leaves entire, hairy. Inflorescence a terminal cyme with bracts. *Calyx* deeply 5–lobed, enlarged in fruit. *Corolla* white, blue or purple, funnel-shaped or cylindrical, with 5 overlapping lobes, and 5 longitudinal bands of hairs inside. Stamens included in the corolla, inserted below the middle of the tube. Style included, simple. Nutlets smooth or roughly warty.
Distribution and habitats: Europe and Asia; a variety of habitats from open to wooded often in mountains.
Life forms: Therophytes, hemicryptophytes, chamaephytes.
Pollination: Bees and flies.
Fruit dispersal: No specialized mechanism.

USES
GARDEN
The species commonest in cultivation, *B. purpurocaerulea*, is excellent for ground cover and informal gardens. The rarer species, *B. calabra* and *B. gastonii*, can perhaps be admitted to a rock garden, if obtainable. The weedy *B. arvensis* is probably only of interest in a collection of native plants.

MEDICINAL
An infusion of leaves of *B. arvensis* is used as a diuretic in India.

CULTIVATION AND PROPAGATION
See under *B. purpurocaerulea*. Annual species are propagated only from seed.

REFERENCES
BRICKELL, 1996; CHOPRA *et al.*, 1986; CLAPHAM *et al.*, 1952; FISH, 1980; GRIFFITH, 1964; HUXLEY, 1992; STACE, 1991; TUTIN *et al.*, 1972; WALTERS, 2000.

Buglossoides purpurocaerulea (L.) I.M. Johnston
Attractive, fast ground-covering perennial for wilder parts of the garden, with creeping stems to 60cm (2ft) long, and shorter, upright flowering stems, bearing velvety-textured flowers that are purple-red in bud but turning intense blue.

'This, for a short time in early summer, puts up 6–7" [inch] sparsely leafy stems which bear terminal twin racemes of most striking velvety, deep, intense blue flowers … The plant is apt to be invasive, as its habit is to make long arching

Buglossoides purpurocaerulea. (George McKay)

sprays, up to a foot or more, which eventually bend over and root at the tips. It is, however, well worthwhile for the deep vividness of its flowers, so long as this propensity is allowed for.' (Anna N. Griffith)

'… can become a nuisance if left to itself, for although the flowers are lovely in a deep shade of blue, they are quite overshadowed by the long leafy stems, which fling themselves out in all directions, settle down and root themselves into new plants, to start the game all over again.' (Margery Fish)

NAMING
Etymology: *purpurocaerulea* – purple-blue, referring to the colour of flowers.
Synonym: *Lithospermum purpurocaeruleum* L.
Common names: *English*: Blue gromwell, Purple gromwell; *French*: Grémil Pourpre-Bleu; *German*: Rotblauer Steinsame, Blauer Steinsame; *Italian*: Erba perla azzurra; *Russian*: vorobeinik purpurovo-goluboi; *Slovak*: kamienka modropurpurová.

DESCRIPTION
General: A sprawling, hairy perennial, arising from thin creeping rhizomes.
Stems: Hairy; sterile shoots long, creeping; flowering shoots shorter, upright, usually unbranched; 15–60cm (6–24in) long.
Leaves: Lanceolate to narrowly elliptical, sharply pointed, 3.5–8cm ($1^3/_8$–$3^1/_4$in) long and 7–15mm ($^5/_8$–$^5/_8$in) wide, tapering gradually to a short stalk; dark green above, light green beneath.
Inflorescence: 2–3 terminal leafy cymes elongating in fruit.
Flowers: *Calyx* 6–8.5mm, lobes linear, sharply pointed, bristly hairy. *Corolla* 14–19mm across, at first reddish-purple, but soon turning bright blue.
Nutlets: Shiny-white, smooth, 3.5–5 × 3–3.5mm.
Chromosomes: 2n=16.

GEOGRAPHY AND ECOLOGY
Distribution: Most of Europe, including Britain, where it is rare (in south Wales and south-west UK).
Habitat: Woodland margins, scrub, in Britain also in hedgerows, to 1600m (5250ft) above sea level, mostly on chalk and limestone.
Flowering: April to June in the wild.
Associated plants: Grows with a wide variety of companions, e.g. in oak (*Quercus petraea*, *Q. pubescens*, *Q. robur*) woodland, with *Lathyrus pannonicus*, *Dictamnus albus*, *Iris graminea*, in Czech Republic, and in Siberian steppe oak woodland, with

Pulmonaria mollis, Lathyrus niger, Polygonatum latifolium and *Vincetoxicum hirundinaria*.

CULTIVATION

Hardiness: Zone 6.

Sun/shade aspect: Sun to shade, dappled shade is best.

Soil requirements: Any well-drained soil that is ideally moisture-retentive; does best on neutral to alkaline soils but will tolerate some acidity.

Garden uses: Good ground cover, especially in a wild garden, woodland garden, under trees and shrubs, or in a mixed border. Not for the well-groomed areas.

Spacing: 60cm (2ft).

Propagation: *Seed* – Mostly unnecessary as vegetative propagation is so easy. Germinates best with a period of warm/cold stratification: sow in spring, in pots/trays of compost at 18–22°C (64–72°F) for 2–4 weeks, move to -4 to +4°C (25–39°F) for 4–6 weeks and finally move to 5–12°C (41–54°F) for germination, or sow as soon as ripe in an outdoor seed bed and allow for the natural stratification to take place. *Vegetative* – By division of rooted stems, or cuttings.

Availability: Available from many herbaceous and wildflower nurseries. Seed can be obtained from a few wildflower merchants and seed-exchange schemes.

Problems: Can be invasive, increasing rapidly by vegetative means where happy, so best confined to the wilder parts of the garden.

RELATED SPECIES

Buglossoides gastonii (Bentham) Johnston (syn. *Lithospermum gastonii* Bentham), from woods and rocky slopes in western Pyrenees (France) is a similar plant, to 30cm (12in) high, which differs by having stalkless leaves, smaller flowers (to 14mm (¹/₂in) across) and yellowish, less-smooth nutlets. This species is included as Rare on the IUNC Red List of Threatened Plants. It is rarely available but worthwhile seeking out from legitimate sources. Another interesting species that is probably not in cultivation is ***Buglossoides calabra*** (Ten.) I.M. Johnston (syn. *Lithospermum calabrum* Ten.), from mountains of southern Italy, with slender procumbent stems and blue flowers (with no purple) 17–20mm (⁵/₈–³/₄in) across.

Corn gromwell, ***Buglossoides arvensis*** (Linnaeus) I.M. Johnston (syn. *Lithospermum arvense* (L.)), is a weedy annual with hairy stems 5–50cm (2–20in) long and small, usually white flowers; found in arable fields and rough ground throughout much of Europe. It needs sun, and may be of interest to a native wild plant collection, or a specialist herb garden; otherwise, it is of little garden merit.

Cerinthe major. (Gordon Smith)

The genus *Cerinthe* Linnaeus, 1753
Honeywort

NAMING

Etymology: *Cerinthe* from Greek *keros,* wax, and *anthos,* a flower, in reference to the wax that bees were supposed to obtain from the flowers.

GENUS CHARACTERISTICS AND BIOLOGY

Number of species: 10 species.

Description: A genus of annual, biennial and a few perennial herbs, often rather fleshy, usually hairless or almost so, sometimes with small white warts tubercles. Leaves alternate, usually glaucous/blue-green. Flowers borne in a terminal cyme, usually branched, with large, often showy bracts that conceal the calyces. *Calyx* lobed to at least half its depth. *Corolla* more or less tubular, usually yellow or sometimes purple or red, 5 lobes are usually shorter than the tube and are erect recurved. Stamens included or slightly protruding from the corolla. Style usually protruding from the corolla. Nutlets fused in 2 separate pairs, smooth, slightly beaked, dark brown or black.

Distribution and habitats: Europe, especially Mediterranean area, Turkey, parts of Siberia. A wide range of habitats, from mountain meadows and woods to arable fields and waste ground. Often on calcareous soils.

Life forms: Hemicryptophytes, or therophytes when not overwintering.

Pollination: Flowers of most species are rich in nectar and are very attractive to bees. Some populations of *C. major* are known to 'trick' the bees into pollinating the flowers while offering a limited reward.

Fruit dispersal: No specialized mechanism.

USES

GARDEN

All species are attractive plants for a mixed or herbaceous border, and also good in containers. They are often rich in nectar, so are good for inclusion in a garden designated for bees. Smaller species can be admitted to a large rock garden, or even treated to a pot culture in the alpine house.

ECONOMIC

In the past, *C. major* was economically important as a nectar source for honeybees.

OTHER

Cerinthes make pretty but short-lasting cut flowers.

CULTIVATION

All *Cerinthe* species are reasonably hardy. The majority need full sun and well-drained soil in order to thrive, though *C. minor* and *C. glabra* can tolerate some shade. Most cerinthes prefer neutral to alkaline soils, but will usually withstand some acidity. A sheltered position is preferable. Overwintering plants should not be kept too wet, otherwise the roots are likely to rot. Remove fading flowerheads to prevent self-seeding.

PROPAGATION

Only from seed. Sow in pots or trays of compost in early spring, or directly into the flowering position outdoors from mid-spring onwards, covering lightly. In mild areas, you can try sowing in early autumn, and the overwintering plants will be larger and earlier-flowering, though dying sooner. Germination takes 1–2 weeks at 15–17°C (59–63°F), and is usually good.

REFERENCES

BECKETT, 1993–94; BLAMEY & GREY-WILSON, 1993; BRICKELL, 1996; DAVIS, 1978; HAY, 1955; HUXLEY, 1992; MANSFIELD, 1945; PHILIPS & RIX, 1999; RICE, 1999; ROBINSON, 1898; TUTIN *et al.*, 1972; WALTERS 2001

Cerinthe glabra Miller

A hairless, ascending or upright perennial or biennial 15–50cm (6–20in) high, grey-green, with oblong to heart-shaped leaves, and yellow tubular flowers with 5 red spots in the throat. For sunny, well-drained positions.

NAMING

Etymology: *glabra* – naked, hairless.
Synonym: *Cerinthe alpina* Kit. in Schultes.
Common names: *English*: Smooth honeywort; *French*: Mélinet Glabre; *German*: Alpen Wachsblume; *Italian*: Erba-vajola alpina; *Slovak*: voskovka holá.

DESCRIPTION

General: A hairless perennial or biennial.
Stems: 15–50cm (6–20in), ascending to upright, usually much-branched.
Leaves: Smooth, grey-green. Basal leaves narrowly elliptic, blade to 12cm (4½in) long and 3.5cm (1⅜in) wide, blunt, narrowing into 4–6cm (1½–2½in) stalk; stem leaves ovate to oblong, to 7cm (2¾in) long and 3cm (1¼in) wide, blunt, stalkless, base heart-shaped, clasping the stem.
Inflorescence: Elongated leafy cymes, bracts usually greenish.
Flowers: *Calyx* divided almost to base, lobes lanceolate, blunt, hairless. *Corolla* pale yellow, with 5 red spots in the throat, 8–14mm long, tubular-bells with rounded, erect or recurved lobes.
Nutlets: Dark brown, about 4 x 3mm.

GEOGRAPHY AND ECOLOGY

Distribution: Pyrenees, Alps, Carpathians, Balkans, Caucasus, Turkey.
Habitat: Meadows and damp woods, roadsides, often in shaded places, generally on calcareous soils, to 2600m (8500ft) in the Mediterranean, to 3300m (10,800ft) in Turkey.
Flowering: May to July in the wild.

CULTIVATION

Hardiness: Zone 5.
Sun/shade aspect: Sun or light shade; tolerates deeper shade in warm climates.
Soil requirements: Well-drained but, ideally, moisture-retentive soil.
Garden uses: As for the genus.
Spacing: 15–20cm (6–8in).
Propagation: As for the genus.
Availability: Plants sometimes available from herbaceous nurseries; seed can be found in some catalogues and the specialist societies' exchange lists.
Problems: Short-lived so must be renewed from seed frequently.

Cerinthe major. (Masha Bennett)

Cerinthe major 'Purpurascens'. (Howard Rice)

Cerinthe major Linnaeus

A rather fleshy, bluish-green, upright or spreading annual 30–60cm (12–24in) high, with spoon-shaped leaves that may have small white warts when young, and tubular, usually yellow flowers, surrounded by large, conspicuous bracts that are often purplish, blue-green or chocolate brown. A nice plant for a sunny, well-drained spot.

NAMING
Etymology: *major* – large.
Common names: *English*: Honeywort, Great(er) honeywort; *French*: Grand Mélinet; *German*: Grosse Wachsblume; *Italian*: Erba-vajola Maggiore; *Spanish*: Ceriflor.

DESCRIPTION
General: A glaucous annual.
Stems: 18–60cm (7–24in), usually 30–40cm (12–16in), upright, simple or branched at the top.
Leaves: Basal leaves oblong or spoon-shaped, broadest above the middle, to 5cm (2in) long and 1.8cm ($^3/_4$in) wide, blunt, base narrowing into a short stalk. Stem leaves oval or oblong-elliptic, to 6.5cm ($2^5/_8$in) long and 2.5cm (1in) wide, stalkless, clasping the stem with a heart-shaped base. All leaves covered with white swellings.
Inflorescence: With large, oval to almost heart-shaped bracts that are grey-green with a reddish tinge or dark purple.

Flowers: *Calyx* divided almost to base, lobes lanceolate-elliptic, pointed, fringed with hairs. *Corolla* yellow with a dark reddish-brown or reddish-purple ring at the base, occasionally all purple, 15–30mm ($^5/_8$–$1^1/_4$in) long, with very short, recurved, oval lobes.
Nutlets: About 4–5mm, dark brown.

GEOGRAPHY AND ECOLOGY
Distribution: Throughout the Mediterranean region, except Balearic Islands or Cyprus.
Habitat: Banks, stony slopes, roadsides, cultivated, fallow and waste places, damp places, from sea level to 100m (330ft).
Flowering: March to June in the wild.
Associated plants: In olive groves of southern Italy, it makes a lovely combination with *Borago officinalis*.

CULTIVATION
Date of introduction: 1596.
Hardiness: Zone 5.
Sun/shade aspect: Full sun.
Soil requirements Any well-drained soil from acid to alkaline.
Garden uses: As for the genus.
Spacing: 15–20cm (6–8in) apart.
Maintenance: The tallest stems may need some support in exposed sites.
Propagation: As for the genus.
Availability: Plants are sometimes available, but easier to find as seed in catalogues and exchange lists.

RELATED FORMS AND CULTIVARS

In recent years, a number of forms of *C. major* have appeared in cultivation, the best known being the rather striking **'Purpurascens'** with purplish flowers and purple bracts. A recent selection, **'Yellow Candy'**, is said to have bright yellow and chocolate-brown flowers enclosed in blue-green bracts; together with another new form, **'Yellow Gem'**, it is unfamiliar to me as yet. *Cerinthe* **'Kiwi Blue'** is probably a hybrid, and is distinguished by its very deep purple-blue bracts.

RELATED SPECIES

Violet honeywort, **Cerinthe retorta** Sibthorp & Smith, from rocky places and roadsides in the Balkan Peninsula and Turkey (Anatolia), is similar, but corolla is smaller, 10–15mm long and 3–5mm wide, curved above, with a constricted throat, pale yellow with a violet apex. Less common in cultivation.

'C. retorta is a beautiful kind, the floral leaves of a rich purple tint, and from among them peep the yellow-purple-tipped flowers in charming contrast.' (William Robinson)

Cerinthe minor Linnaeus

An upright, blue-green annual or biennial to 60cm (2ft) high, with spoon-shaped or oblong leaves covered in small white warts. The honey-scented, pale yellow, tubular flowers often have 5 purple-violet spots towards the base, but are usually mostly hidden by the bracts. For sunny, well-drained situations.

'...Heart-shaped leaves being covered in tiny white warts, whilst the tubular flowers are pale yellow and have purple-brown markings on the tube. They carry a rich honey-like perfume, and are a great attraction for bees.' (...)

NAMING

Etymology: *minor* – small.
Common names: *English:* Lesser honeywort; *Finnish:* Pikkuvahakukka; *French:* Petit Mélinet; *German:* Kleine Wachsblume; *Italian:* Erba-vajola Minore; *Russian:* voskovnik maly, voskotzvetnik maly; *Slovak:* voskovka menéia.

DESCRIPTION

General: A glaucous annual or biennial.
Stems: 15–60cm (6–24in), sometimes to 75cm (30in) high, upright, branched above or simple, rounded or somewhat angled, hairless, glaucous-purple.

Leaves: Basal leaves oblong to spoon-shaped, dying early, to 15cm (6in) long, gradually narrowing towards the stalk, blunt. Lower stem leaves similar shape, narrowing towards the base, clasping the stem, to 8cm (3¼in) long and 2.8cm (1⅛in) wide. Upper stem leaves oval, with heart-shaped base. All leaves blunt, thick, with white warts, glaucous-green often.
Inflorescence: Dense cymes, with oblong-oval bracts, usually greenish with more or less heart-shaped base.
Flowers: *Calyx* deeply divided, more or less to the base, 5–7mm long, outer lobes oblong-ovate, inner lobes lanceolate, pointed, margins fringed with hairs. *Corolla* 10–14mm long, divided, to a third to a half of its depth, into linear-lanceolate, pointed segments, pale yellow, often with 5 purple-violet spots at the base of segments.
Nutlets: Nutlets about 3.5 x 3mm, brown.

GEOGRAPHY AND ECOLOGY

Distribution: Much of Europe, Caucasus, Mediterranean, Asia Minor.
Habitat: Cultivated, fallow and waste places, roadsides, stony slopes, field margins, open woodland, rocky places, alpine pastures, to 2400m (7870ft) above sea level.
Flowering: April to July in the Mediterranean, May to August in Turkey.

CULTIVATION

Date of introduction: 1570.
Hardiness: Zone 5.
Sun/shade aspect: Sun, but will tolerate light shade.
Soil requirements: Well-drained soil.
Garden uses: A useful filler in borders, beds, can be added to rock gardens, containers, and used as a cut flower. Attractive to bees.
Spacing: 20cm (8in).
Propagation: As for the genus.
Availability: Seed is sometimes available at seed merchants, and appears on specialist societies' seed-exchange lists.
Problems: The flowers are mostly hidden by bracts.

RELATED FORMS AND CULTIVARS

Cerinthe minor subsp. auriculata (Ten.) Domac, differs by its perennial habit, rough pedicels (smooth in subsp. *minor*), swollen rather than cylindrical corolla, and blackish, as opposed to brown, nutlets. It occurs in Alps, Apennines, Sicily, the Balkans and Turkey. Probably not in cultivation but worth seeking out because it is perennial.

The genus *Cynoglossum* Linnaeus, 1753
Hound's-tongue

NAMING

Etymology: *Cynoglossum* – from Greek *kunoglosson*, dog's tongue, referring to shape of the leaves.

Synonyms: *Paracynoglossum* Popov; *Pardoglossum* E. Barbier & Mathez. Species belonging to a number of different genera, such as *Eritrichium* Schrad., *Mertensia* Roth and *Lindelofia* Lehm., have been mistakenly assigned to this genus in the past.

GENUS CHARACTERISTICS AND BIOLOGY

Number of species: 50–60.

Description: A genus of perennial, biennial or sometimes annual herbs. Leaves alternate, hairy, basal leaves long-stalked. Flowers borne in an elongating one-sided cyme, usually without bracts. *Calyx* 5–lobed, divided to more than half its depth, enlarged in fruit, with spreading or reflexed lobes. *Corolla* funnel-shaped or cylindrical, with a short tube and widely spreading, broad, overlapping lobes; with 5 scales in the throat that may be pyramidal, oblong or almost crescent-shaped. Stamens included in corolla. Style short. Nutlets 4, ovoid or spherical, covered with hooked spines. 60 species, in temperate and tropical regions of the northern hemisphere.

Distribution and habitats: Cosmopolitan distribution. A wide range of habitats, from woodlands and mountains, to grasslands, wastelands and seaside.

Life forms: Hemicryptophytes (perennials and biennials) and therophytes (annuals).

Pollination: Insects, mainly bees. Some species, like *C. amabile*, are attractive to butterflies.

Fruit dispersal: The burlike nutlets are covered with hooked spines. They attach themselves to fur and clothing, and are dispersed by this means to other suitable habitats.

Animal feeders: Weevil *Ceutorhynchus cruciger* feeds on *C. officinale*.

USES

GARDEN

A number of species are excellent border plants, such as *C. nervosum* and *C. amabile* (the latter is also good for bedding), while most others are more suited to wilder parts of the garden, or a herb collection. Cynoglossums are relatively unpalatable to rabbits and are thus useful in gardens where they are a problem. Some of the more decorative species, such as *C. amabile*, make pretty, though short-lived, cut flowers.

MEDICINAL

The leaves and roots of *C. officinale* are analgesic, antihaemorrhoidal, antispasmodic, astringent, digestive, emollient and slightly narcotic. This species contains allantoin, which has painkilling and healing effects.

The leaves and roots of *C. officinale* have been used extensively as a medicinal herb in the past – externally as a compress for insect bites, minor injuries, leg ulcers and as a suppository for haemorrhoids. It was also taken internally, to treat coughs and diarrhoea, and as a painkiller. It was thought to be effective in treatment of cancer and insomnia. However, like many other members of the borage family, this plant contains pyrrholizidine alkaloids (cynoglossin and consolidin), and may be potentially carcinogenic. It can also cause skin irritation and allergy. May be subject to legal restrictions.

The root of *C. grande* has been used by Native Americans as a dressing on inflamed burns and scalds, and also to treat stomach-aches and venereal diseases.

CULINARY

The young leaves of *C. officinale* are said to be edible, raw or cooked, though both the smell and taste of the plant are unpleasant. It is known to have been toxic to cattle, and may have a potentially carcinogenic effect, so caution must be taken with this plant. The root of *C. grande* is edible when cooked, and was eaten by Native Americans.

CULTIVATION

Cynoglossums are mostly hardy and easy to grow in any well-drained soil, though their preference is for fertile, deep, moisture-retentive soils. Most will be happy in a sunny position, and some, like *C. grande* or *C. officinale*, tolerate light shade.

PROPAGATION

Seed All species can be propagated from seed, and this is the main method. Seeds can be sown in late spring/early summer or, when fresh, in pots or an outdoor seedbed, covering to about 6mm ($^1/_4$in) deep. Annuals and biennials can be sown in their permanent position. There is some evidence that darkness can improve germination.

Vegetative Some perennial species can be carefully divided in early to mid-spring and replanted or potted up. Root cuttings in early winter are also a possibility.

PROBLEMS

Many species are rather weedy, and it may be necessary to remove fading flower heads to prevent self-seeding. Cynoglossums have long taproots and may be difficult to transplant successfully.

REFERENCES

BLAMEY & GREY-WILSON, 1993; BOWN, 1995; CLAPHAM *et al.*, 1952; GENDERS, 1971; HAY, 1955; HUXLEY, 1992; MANSFIELD, 1949; PFAF, 1992–02; POLUNIN, 1980; POLUNIN & STAINTON, 1984; STACE, 1991; THOMAS, 1993; TUTIN *et al.*, 1972; WATERS, 2000

Cynoglossum amabile Stapf & Drummond

A pretty biennial or short-lived perennial, often grown as annual, to 60cm (2ft) high, with a basal rosette of long, felty, grey-green leaves and blue, sweet-smelling flowers. For a sunny, well-drained spot.

'Has a name [amabile] which conveys a perfectly true impression of the plant itself. … The flowers are of a delicate but soft celestial blue … its beauty should be admired under the open sky, and it should be backed with some of the paler Evening Primroses, far away from blatant Anchusas and flaunting Verbenas.' (Thomas Mansfield)

NAMING

Etymology: *amabile* – lovely.
Common names: *English*: Chinese forget-me-not, Chinese hound's-tongue; *Chinese*: dao ti hu; *Finnish*: Tiibetinkoirankiel.

DESCRIPTION

General: A hairy biennial or perennial.
Stems: 45–60cm (18–24in), upright, covered with long hairs.
Leaves: Basal leaves to 20cm (8in) long, oblong to lanceolate or elliptical, stalked. Stem leaves smaller, stalkless. All leaves hairy, grey-green.
Inflorescence: Many terminal and axillary racemes.
Flowers: *Corolla* funnel-shaped, 6–12mm long and 5mm across, blue, sometimes white or pink.
Nutlets: 2–3mm, with distinct margins.
Chromosomes: 2n=24

GEOGRAPHY AND ECOLOGY

Distribution: East Himalayas, west China. Naturalized in several states of the USA.
Habitat: Meadows, forests, roadsides.
Flowering: May–September in the wild.

CULTIVATION

Hardiness: Zone 7.
Sun/shade aspect: Sunny position is required.
Soil requirements: Any well-drained soil, tolerates both acid and alkaline conditions.
Garden uses: Great for mixed borders or informal bedding, also useful in a wildlife garden, as it is attractive to bees and butterflies.
Spacing: 20cm (8in) apart.
Propagation: As for the genus. Can be grown as an annual, sown in early spring under glass, or as a biennial, sown in summer in the open ground.
Availability: Seed is widely available from many seed merchants.

RELATED FORMS AND CULTIVARS

Cynoglossum amabile f. *roseum* is identical to the species except for pink rather than blue flowers. A number of seed strains is available:
'Avalanche' – form with white flowers
'Firmament' – more compact, to 40–45cm (16–18in), with deeper, indigo-blue flowers freely produced. Sown early under glass will flower the same year. *'…is more compact growing and has flowers of deeper blue, and is no better for either.'* (T. Mansfield)
'Mistery Rose' – rose pink flowers, height 60–80cm (24–32in).

Cynoglossum creticum Miller

A softly hairy biennial, with upright, rather robust stems to 60cm (2ft), white-felted lanceolate leaves, and unusual pale blue flowers that are veined in a purple or deep blue shade. An interesting but uncommon plant for a sunny border.

NAMING

Etymology: *creticum* – of Crete, not exactly true, as this species is much more widespread.
Synonym: *Cynoglossum pictum* Aiton.
Common names: *English*: Blue hound's-tongue; *French*: Cynoglosse de Crète; *German*: Kretische hundszunge; *Italian*: Lingua-di-cane a fiori variegati; *Russian*: chernokoren' kritsky; *Spanish*: Cinogloso azul.

DESCRIPTION

General: A hairy biennial
Stems: 30–60cm (1–2ft), angular, upright, branched above, softly white-hairy.
Leaves: Stem leaves oblong to lanceolate, densely hairy both above and underneath, shortly stalked or clasping the stem.

Inflorescence: Branched cymes without bracts, elongating in fruit.
Flowers: *Calyx* lobes 6–8mm, oblong, softly hairy. *Corolla* broadly funnel-shaped, purplish in bud, opening blue, with conspicuous deeper inky-blue or purple net-veins, 7–9mm across.
Nutlets: 5–7mm, with a thickened edge and dense hooked spines of unequal lengths.

GEOGRAPHY AND ECOLOGY
Distribution: Southern Europe, parts of Central Asia.
Habitat: Cultivated, fallow and waste land, olive groves, vineyards and roadsides, moist sites by streams.
Flowering: February to June in the wild, usually summer in cultivation in temperate climates.

CULTIVATION
Hardiness: Zone 7.
Sun/shade aspect: Sun.
Soil requirements: Any well-drained soil but ideally moisture-retentive.
Garden uses: An unusual subject for a mixed border.
Spacing: 20–30cm (8–12in) apart.
Propagation: From seed only, as for the genus.
Availability: Not generally available, but seed may appear on exchange lists.

Cynoglossum creticum. (Masha Bennett)

RELATED SPECIES
Another interesting species is **C. cheirifolium** Linnaeus (syn. *Pardoglossum cheirifolium*), from western Mediterranean and Portugal. It is a white-felty biennial, to 40cm (16in) tall, with lanceolate leaves and small flowers that are variable in colour – from very dark purple to blue or even yellowish with a purple tinge. This plant is worth growing for the foliage alone, but it is generally unavailable, as is **C. columnae** Ten., a pretty annual growing to about 30–45cm (12–18in), with deep blue flowers, native to central and eastern Mediterranean. Both are hardy to Zone 7.

Cynoglossum dioscoridis Vill.
A hairy, upright biennial, to 50cm (20in) tall, with narrowly lanceolate leaves and deep blue flowers. Good plant for a sunny mixed border.

NAMING
Etymology: *dioscoridis* – after Pedanius Dioscoridis (c. AD 40–90), the Greek physician who wrote *De material medica*, the work on substances used in medicine and their properties.
Synonym: *Cynoglossum loreyi* Jordan in Lange.

DESCRIPTION
General: A hairy biennial.
Stems: 15–50cm (6–20in), softly hairy, upright.
Leaves: Stem leaves narrowly lanceolate, more or less clasping the stem, hairy on both sides.
Inflorescence: Cymes without bracts.
Flowers: *Calyx* lobes about 3mm, elliptical to oblong, softly hairy. *Corolla* about 5mm, deep blue, bell-shaped, limb about as long as tube.
Nutlets: 5–6mm in diameter, with a distinct border, outer side densely covered with hooked spines.

GEOGRAPHY AND ECOLOGY
Distribution: Southern and eastern France, north-east Spain.
Habitat: Wood margins, rocky hillsides; calcicole.
Flowering: Summer.

CULTIVATION
Hardiness: Zone 6.
Sun/shade aspect: Sun or very light shade.
Soil requirements: Well-drained soil, preferably neutral to alkaline.
Garden uses: Grow in a mixed border.
Spacing: 25–30cm (10–12in).
Propagation: From seed only, as for the genus.
Availability: Plants available from some nurseries, and seed can be found in some catalogues.

Cynoglossum grande Douglas

An upright perennial, to 90cm (3ft) tall, with oval or elliptical leaves that are almost hairless above, and stems carrying cymes of blue flowers. For sun or light shade in borders or open woodland.

NAMING
Etymology: *grande* – large.
Common names: *English*: Great hound's-tongue, Pacific hound's-tongue, Blue-buttons.

DESCRIPTION
General: A sparsely hairy perennial.
Stems: 30–90cm (1–3ft) high, upright, hairless.
Leaves: Basal leaves to 15cm (6in) long and 10cm (4in) wide, oval or elliptical, sparsely hairy or hairless above, densely hairy beneath, long-stalked. Upper stem leaves to 3cm (1^1/$_4$in), stalkless.
Inflorescence: Loose cymes.
Flowers: *Calyx* lobes 5–7mm, narrowly oblong, densely hairy. *Corolla* 8–12mm, blue or sometimes purple, with cylindrical tube and lobes rounded; scales crescent-shaped.
Nutlets: 5–6mm, flattened globose, with barbed spines.

GEOGRAPHY AND ECOLOGY
Distribution: Western North America.
Habitat: Woods.
Flowering: April to May in the wild.

CULTIVATION
Hardiness: Zone 8.
Sun/shade aspect: Sun or dappled shade.
Soil requirements: Any well-drained soil, tolerates both acidity and alkalinity.
Garden uses: Open woodland garden, herbaceous or mixed border, herb garden, wildflower garden (in western North America).
Spacing: 30cm (12in).
Propagation: As for the genus.
Availability: Occasionally available from nurseries, seed may appear on exchange schemes.

RELATED SPECIES
Wild comfrey, **C. virginianum** Linnaeus, native to upland woodland in eastern USA, grows to about 75cm (2^1/$_2$ft) tall, and bears 20cm (8in) racemes of blue flowers on unbranched stems. Hardy to Zone 4.

Cynoglossum nervosum Benth in C.B. Clarke

An upright, white-hairy perennial about 60cm (2ft) high, with basal rosettes of narrow leaves and leafy stems carrying clusters of deep blue flowers. A very

Cynoglossum nervosum. (Gordon Smith)

attractive plant for sunny mixed and herbaceous borders.

'A gay plant and lovely in contrast to Origanum vulgare *'Aureum'.' (Graham S. Thomas)*

NAMING
Etymology: *nervosum* – veined.
Synonym: *Omphalodes nervosa* Edgew. in C.B. Clarke
Common names: *Dutch*: Himalya-hondstong

DESCRIPTION
General: A hairy perennial.
Stems: To 60cm (2ft) or sometimes 90cm (3ft), upright.
Leaves: Basal leaves to 4cm (1^1/$_2$in) long and 1.5cm (^{5}/$_8$in) wide, oval or oval-oblong, rounded, gradually narrowing into a stalk to 4cm (1^1/$_2$in) long, white-hairy throughout. Lower stem leaves to 12cm (4^1/$_2$in) long and 2cm (^{3}/$_4$in) wide, lanceolate, pointed, long-stalked, upper stem leaves to 7cm (2^3/$_4$in) , broadly lanceolate, stalkless.
Inflorescence: To 10cm (4in) long in fruit, forked.
Flowers: *Calyx* lobes to 4.5mm in fruit, oval-oblong, blunt, covered in long hairs outside, nearly hairless inside. *Corolla* blue, funnel-shaped to 8mm across.
Nutlets: To 4.5 x 3mm, ovoid, with a distinct margin, covered with hooked spines.

GEOGRAPHY AND ECOLOGY
Distribution and Habitat: Mountains, specifically the Himalayas (West Pakistan, Kashmir, north-west India).
Flowering: Early summer in cultivation.

CULTIVATION
Hardiness: Zone 5.
Sun/shade aspect: Sunny position.
Soil requirements: Any well-drained, not-too-rich soil. In soil that is too fertile, the plants acquire an ungainly habit.
Garden uses: Herbaceous and mixed borders, wildlife garden (attracts bees).
Propagation: From seed, as for the genus.
Availability: Commonly available from many herbaceous nurseries and some garden centres, also some seed suppliers.
Spacing: 45–60cm (18–24in).

Cynoglossum officinale Linnaeus
An upright, softly hairy biennial to 60–90cm (2–3ft) tall, with a basal rosette of long, 'dog's-tongue'-shaped leaves, and cymes of small, dull reddish-purple flowers. The leaves, when crushed, have a smell similar to that of mice.

> *'Due to the presence of esters or fatty acids, they emit a fur-like smell, similar to that of the Lizard Orchid, when handled.' (Roy Genders, 1971)*

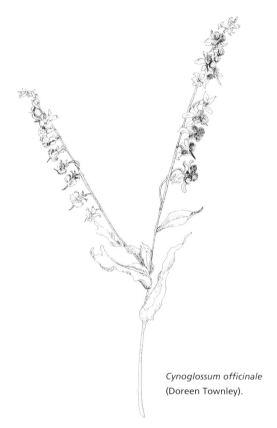

Cynoglossum officinale
(Doreen Townley).

NAMING
Etymology: *officinale* – sold as a herb.
Common names: *English*: Hound's-tongue, Rats-and-mice, Gypsy-flower (North America); *Danish*: Hundetunge; *Dutch*: Veldhondstong; *Finnish*: Koirankieli, Rohtokoirankielet; *French*: Cynoglosse Officinale, Langue De Chien; *German*: Echte Hundszunge; Gebräuchliche Hundszunge; *Italian*: Lingua-di-cane vellutina; *Norwegian*: Hundetunge; *Russian*: Chernokoren lekarstvenny; *Slovak*: psojazyk lekársky; *Spanish*: Cinoglosa; *Swedish*: Hundtunga; *Turkish*: Kopekdili.

DESCRIPTION
General: A hairy biennial, unpleasant-smelling when crushed.
Stems: 30–60cm (1–2ft), sometimes to 90cm (3ft), upright or nearly so, hairy.
Leaves: Basal leaves to about 15cm (6in) long and 2.5cm (1in) wide, oblong, narrowly oblong or lanceolate, usually pointed, on stalks to 15cm (6in) long. Upper stem leaves stalkless, elliptical to lanceolate. All covered with rather silky hairs on both surfaces.
Inflorescence: Cymes usually branched, normally without bracts, lengthening to 10–25cm (4–10in) after flowering.
Flowers: *Calyx* lobes oblong or oval, to 7mm in fruit, densely hairy. *Corolla* about 10mm across, bell-shaped cylindrical, dull red-purple, rarely whitish.
Nutlets: 5–8mm, flattened, oval, surrounded by a thickened border and covered with short barbed spines.
Chromosomes: 2n=24.

GEOGRAPHY AND ECOLOGY
Distribution: Europe, Asia. In the UK widely distributed but local, north to Angus. Naturalized in North America, including most of the USA, and classified as a noxious weed in Colorado, Oregon, South Dakota, Washington and Wyoming, and certain areas of Canada.
Habitat: Grassy places, woodland margins, often on dry soils, on sand, gravel, chalk or limestone, particularly near the sea.
Flowering: June to August in the UK.

CULTIVATION
Hardiness: Zone 6.
Sun/shade aspect: Sun or part shade.
Soil requirements: Any well-drained soil, though preferably neutral to alkaline. Tolerates very dry, gravelly and chalky soils.
Garden uses: Wildflower garden, wildlife garden (bees), herb garden, seaside garden.
Spacing: 30–45cm (12–18in).

Propagation: From seed, as for the genus. Germination can be erratic.
Availability: Available mostly as seed and occasionally as plants from a few wildflower and herb suppliers.

RELATED SPECIES

Green hound's-tongue, **Cynoglossum germanicum** Jacq., found from West Europe east to the Carpathians, Caucasus and Asia Minor, is the only other *Cynoglossum* species native to the UK, where it occurs in woods and on hedgebanks in southern England and the Midlands, and is a rare, protected plant. It is naturalized in eastern North America. It is similar to *C. officinale*, but is green rather than grey, and has smaller flowers, about 5mm across. It has been available from at least one wildflower nursery and one seed merchant in the past.

Cynoglossum wallichii G. Don

An upright, white-hairy biennial, usually grown as annual, about 60cm (2ft) high. It is similar to the more common *C. amabile*, but with smaller flowers that are often of a paler blue shade (plants with darker blue flowers are sometimes considered a separate species, *C. glochidiatum*). For sunny or lightly shaded beds and borders.

> *'Nutlets covered with hooked bristles, which become attached to one's socks and are very difficult to remove.' (Oleg Polunin and Adam Stainton, for* C. glochidiatum*)*

NAMING

Etymology: *wallichii* – after Nathaniel Wallich (Nathan Wolff, 1786–1854), plant collector, born in Copenhagen.
Synonym: *C. glochidiatum* Wallich ex Benth – this is sometimes considered a separate species.
Common names: *English*: Prickly hound's-tongue; *Chinese*: xi nan lu li cao; *Russian*: chernokoren' kryuchkovaty [for *C. glochidiatum*].

DESCRIPTION

General: A hairy biennial.
Stems: Upright, to 60cm (2ft), sometimes larger.
Leaves: Basal leaves to 17cm (6½in) long and 3cm (1¼in) wide, obovate to lanceolate, prominently veined, somewhat pointed to blunt, densely white-hairy throughout, stalked. Stem leaves short-stalked or stalkless.
Inflorescence: Dense, terminal cymes, without bracts.
Flowers: *Calyx* lobes to 2mm, oblong, erect, densely hairy. *Corolla* pale to dark blue, to 4mm, tube to 2mm; scales in throat almost square, hairy.
Nutlets: To 2.5 x 2mm, ovoid, pointed or rounded, with a distinct margin.
Chromosomes: 2n=24.

GEOGRAPHY AND ECOLOGY

Distribution: Central and Southern Asia. Naturalized in the USA (Massachusetts).
Habitat: Cultivated areas, pastures, forests, 1200–4000m (3900–13,100ft) above sea level.
Flowering: May to August in the wild, similar in cultivation in the UK.

CULTIVATION

Hardiness: Zone 6.
Sun/shade aspect: Sun or light shade.
Soil requirements: Well-drained, tolerates both acidity and alkalinity.
Garden uses: Filler in beds and borders.
Spacing: 30–45cm (12–18in).
Propagation: As for the genus. Sow in early spring to grow as an annual, or in late spring–early summer to grow as a biennial, at 17–22°C (63–72°F).
Availability: Seed may be available from merchants.

RELATED SPECIES

Ceylon hound's-tongue, **Cynoglossum zeylanicum** (Vahl) Thunb. in Lehmann (syn. *C. denticulatum* A. DC), is a rather similar plant, 30–90cm (1–3ft) tall, but more branched and with larger, oblong, softly hairy leaves, and bright blue flowers. It is native from Afghanistan to Sri Lanka and in Japan, and is naturalized in the south of the USA. It is hardy to Zone 7.

The genus *Echium*
Linnaeus, 1753
Viper's Bugloss

NAMING

Etymology: *Echium* – name used by Dioscorides, possibly derived from Greek *echis* – viper, the seeds being shaped like a viper's head.
Synonym: *Megacaryon* Boiss.

GENUS CHARACTERISTICS AND BIOLOGY

Number of species: Around 40 species.
Description: A genus of annual, biennial or perennial herbs or shrubs, with tubercle-based bristles and an underlayer of usually short hairs. Inflorescence spike-like or paniculate made up of cymes with bracts, often enlarged in fruit. There may

be thousands of flowers at a time on the woody Canary Island echiums. *Calyx* 5-lobed, divided almost to base. *Corolla* bilaterally symmetrical (zygomorphic), red, blue, purple, yellow or white, funnel-shaped, with a tapering tube and usually oblique, open limb. A ring (so-called annulus) of 10 minute scales or tufts of hairs, or sometimes a collar-like membrane, is present at the base of the corolla. Stamens 5, unequal, inserted below the middle of the corolla. Style protruding; stigma head-like or deeply cleft. Nutlets 4, ovoid to 3-angled, erect, wrinkled.

Distribution and habitats: Europe, Africa, western Asia, Micronesia; in a wide variety of habitats. Canary Island echiums may grow in laurel forests (*E. piniana* and *E. callithyrsum*), pine forest (*E. virescens*), subalpine zones (*E. wildpretii* & *E. auberianum*). Many are rare or endangered in their native habitats, and the Royal Botanic Gardens, Kew is curently running a conservation programme for *E. piniana*. Annual and biennial species mostly in dry places, often as weeds.

Life forms: Therophytes (annuals), hemicryptophytes (biennials and perennials), and nanophanerophytes (shrubs).

Pollination: Bees, flies and Lepidoptera (butterflies and moths). Canary Island echiums are pollinated by honeybees (*Apis mellifera*) and solitary bees *Anthophora*, also the bumblebee *Bombus terrestris canariensis*. Some of these endemic echium species show characteristics that are more typical of bird-pollinated flowers, such as red corollas and dilute nectar. Species like chiffchaff, *Phylloscopus collybita canariensis*, could be among the pollinators. Hawkmoth *Macroglossum stellatarum* and also small bees and wasps visit *E. wildpretii*, but probably do not pollinate. Butterflies attracted to Echiums include pipevine swallowtail (*Battus philenor*) and zebra swallowtail (*Eurytides marcellus*) in the USA.

Fruit dispersal: No specialized fruit dispersal mechanism; the ripe nutlets are probably shaken out by wind onto the ground near the parent plant.

Animal feeders: The following insects are known to feed on *E. vulgare*: caterpillars of Lepidoptera, *Coleophora pennella, Ethmia bipunctella, E. terminella, Tinthia tineiformis, Vanessa cardui* and weevil *Ceutorhynchus geographicus*.

USES

GARDEN
Annual and biennial species are useful fillers for beds and borders; also good for wildlife gardens, as the nectar-rich flowers are very attractive to bees, butterflies and other insects. Shrubby echiums from

Canary Islands are striking specimen plants, under glass or in a mixed border outdoors in sufficiently mild areas.

MEDICINAL
Echium vulgare has been widely used in herbal medicine, as a diuretic, and to treat fevers, headaches and chest conditions, as well as for a variety of skin disorders. In the past it was considered to be a preventive and remedy for viper bites.

CULINARY
Young leaves of *E. vulgare* are edible raw or cooked, and, when cooked, can be used as a spinach substitute. The leaves are somewhat hairy and if you want to use them in salads, they should be chopped finely. WARNING: There is some evidence that echium leaves may be toxic to humans, so do not use them without prior consultation with a doctor.

OTHER USES
A red dye can be obtained from the roots of *E. vulgare*. This plant is an excellent nectar source for bees, and in some countries is grown specially for honey production.

CULTIVATION
CANARY ISLAND *ECHIUM* SPECIES
This is a unique group of plants that require very different conditions from other members of the genus, so the cultivation details are given here separately. Although they can be cultivated in any well-drained soil, with moderate amounts of organic matter, a special mixture is recommended for growing these plants: 40 per cent of humus-rich substrate, 30 per cent grit and 30 per cent crushed lava or pumice. For subalpine species, such as *E. wildpretii, E. simplex* and *E. auberianum*, even less organic matter is advised, so the recommended proportions are 10 per cent humus, 30 per cent sand and 60 per cent crushed lava.

Water the plants regularly and never allow the soil to dry completely. Slow-release fertilizer, such as Osmocote, can be added at half the recommended amount in late spring.

In winter, keep plants in pots at around 10°C (50°F) in a well-lit position. Reduce watering to a minimum, unless grown in a warmer glasshouse where they will need more water. The hardiest Canary *Echium* species, such as *E. wildpretti* and *E. auberianum*, can be tried outdoors in milder parts of Europe, in a sunny spot with some protection from excessive wet, and sheltered from cold winds by a wall or hedge.

Most Canary Island echiums will flower within three to four years from sowing, monocarpic species sometimes sooner. The latter die after flowering and producing seed.

OTHER *ECHIUM* SPECIES
Annual and biennial echiums are mostly very hardy and easy to grow in full sun in any well-drained soil. The soil should not be too rich, or the flowering may be reduced. They tolerate drought and alkalinity well, but dislike root disturbance.

Some echiums can be weedy, so it is a sensible precaution to remove the faded flowerheads to prevent self-seeding.

PROPAGATION
CANARY ISLAND *ECHIUM* SPECIES
Seed It is difficult to separate the four nutlets from each other without damaging them, so they should be sown together. Sow at a minimum temperature of 15°C (59°F) in pots, covering the seeds up to about 5mm (¹/₄in) deep, placing them in a propagator or greenhouse. Keep slightly moist but never too wet. Most will germinate quickly within one or two weeks, but some species, such as *E. webbii* and *E. gentianoides*, may take up to three months. The seedlings of most species need light, except for those native to forests (such as *E. callithyrsum* and *E. piniana*) which may scorch in direct sun and are better in a lightly shaded spot. Transplant seedlings to 20cm/5-litre (8in/1-gallon) pots when the fifth pair of real leaves has appeared – do this very carefully, as these plants deeply resent root disturbance. Transplant to final 10-litre (2-gallon) pots when about 50cm (20in) high, or 30cm (12in) in diameter for monocarpic species.
Vegetative It is possible to propagate woody species by cuttings from one-year-old stems, which is not easy and does not produce flowering plants any faster than by sowing seed.

OTHER *ECHIUM* SPECIES
Seed Biennial echiums can be sown about 6mm (¹/₄in) deep *in situ* in late summer–early autumn for flowering next year, but many will flower in the first season if sown in late winter or early spring in pots or trays at around 15°C (59°F). Prick out into individual small pots as soon as the seedlings are large enough to handle, taking care not to damage the taproot. Germination usually takes about two weeks.

PROBLEMS
In winter, at low light levels and high humidity,

Canary Island echiums may be susceptible to fungal diseases. Their inability to withstand hard frosts also presents a problem when grown outdoors in temperate climates. Red spider mite (*Tetranychus urticae*) and glasshouse whitefly (*Trialeurodes vaporariorum*) may be a problem under glass.

Annual and biennial echiums can become excessively weedy, and some, including *E. vulgare*, possess extremely irritating bristles that may cause a serious rash. They may also cause cattle poisoning.

REFERENCES
BRAMWELL & BRAMWELL, 1974; BRICKELL, 1996; GILLETT & WALTER, 1998; HUXLEY, 1992; KRAEMAR & SCHMITT, 1997; MALKMUS, 2002; PFAF, 1992–02; PHILLIPS & RIX, 1997; TUTIN et al., 1972; WALTERS, 2000

Canary Islands *Echium* species

Echium callithyrsum Webb in Bolle
A loosely branched, woody shrub up to 2.5m (8ft) in height and width, with smooth light brown bark, broadly ovoid-lanceolate, hairy, handsome dark green leaves, with prominent veins, and rather short clusters of red to blue flowers.

NAMING
Etymology: *callithyrsum* – with beautiful bunches, referring to the inflorescenses.

DESCRIPTION
General: A loosely branched shrub.
Stems: Robust, woody, to 2.5m (8ft) high, but often shorter, branched, with smooth light brown bark.
Leaves: Oval to lanceolate, bristly hairy above, covered with simple hairs beneath, prominently veined.
Inflorescence: A rather short, more or less oval, branching thyrse.
Flowers: *Corolla* pale red to deep blue.
Chromosomes: 2n=16.

GEOGRAPHY AND ECOLOGY
Distribution: Gran Canaria (central to north regions, Tenteniguada, El Rincón, rocks below El Juncalillo). Included on the IUNC Red List of Threatened Plants as Vulnerable.
Habitat: Slightly moist, partially shaded spots in lower pine forest and upper laurel forest, at 800–1500m (2600–4900ft).
Flowering: Summer.

Echium candicans. (Gordon Smith)

CULTIVATION

Hardiness: Zone 9.
Sun/shade aspect: Light shade.
Spacing: 1m (3ft).
Availability: Rarely available, will need to be searched for at specialist seed suppliers.
Cultivation, propagation and garden uses: See under the genus description for details.

Echium candicans Linnaeus

A woody biennial shrub to about 2m (6½ft) high, with whitish papery bark, and stout branches with silvery-hairy lanceolate leaves with prominent veins. Flowers are white or blue with white streaks, produced in numerous spikes in spring and early summer. For a warm sunny site.

NAMING

Etymology: *candicans* – becoming white.
Synonym: *Echium fastuosum* Jacq.
Common name: *English*: Pride of Madeira.

DESCRIPTION

General: A hairy, monocarpic shrub.
Stems: 1.8–2.5m (6–8ft) high, white-hairy, with whitish papery bark and stout branches, leafy near apex.
Leaves: Lanceolate to oval-lanceolate, to 25cm (10in) long and 4cm (1½in) wide, with silvery hairs and bristles, and veins prominent beneath.
Inflorescence: Dense, terminal, cylindrical spikelike panicle to 25–30cm (10–12in).
Flowers: *Pedicels* extremely short. *Calyx* 4–5mm, to 7mm in fruit, with lanceolate, pointed lobes. *Corolla*

funnel-shaped, 9–11mm long, white or blue streaked white, red-purple in bud; limb to 2mm, hairy along mid-vern, with a ring of up to 10 tiny conical lobes at the base of the corolla. Stamens protruding from the corolla, with pink, hairless filaments. Style protruding, with the tip split into two.

Nutlets: To 2mm, wrinked, with small elongated projections (papillae).

Chromosomes: 2n=16.

GEOGRAPHY AND ECOLOGY

Distribution: Madeira.

Habitat: Lower parts of the open laurel forest.

Flowering: May to August in the wild.

CULTIVATION

Date of introduction: 1816.

Hardiness: Zone 9.

Spacing: 1m (3ft).

Availability: One of the more commonly available of the Canary Island *Echium* species, plants can be found in a number of specialist nurseries and seeds are sold by some suppliers.

See the details of cultivation, propagation and garden uses under the genus description.

Echium pininana Webb & Berth.

An unbranched, bristly shrub, to 2m (6¹/₂ft) or even higher, with lanceolate or oval leaves in a rosette to 1m (3ft) in diameter, and huge inflorescences of numerous blue flowers. Will die after flowering. Will suit dappled shade on moist soil.

NAMING

Etymology: *pininana* – common name of the plant on Las Palma.

Synonyms: *Echium pinnifolium* invalid; *E. piniana* Pitard & Proust.

Common names: *English*: Giant viper's-bugloss; *Spanish*: Pininana.

DESCRIPTION

General: A monocarpic, bristly shrub.

Stems: To 75cm (2¹/₂ft), unbranched, 3.5cm (1³/₈in) wide, bristly, with leaves at apex.

Leaves: To 50cm (20in) long and 6–10cm (2¹/₂–4in) wide, lanceolate to oval, with large-based, adpressed bristles; veins prominent, especially beneath.

Inflorescence: A pointed, cylindrical panicle of lateral cymes, to 3.5m (11¹/₂ft), sparsely bristly.

Flowers: *Calyx* bell-shaped, lobes to 4mm, oval, blunt, hairy on margins. *Corolla* blue or sometimes

white with bluish streaks, narrowly funnel-shaped, to 13mm, sparsely covered in soft hairs, limb to 2mm. Ring (annulus) at the base of corolla narrow, irregularly lobed, with long hairs. *Stamens* protruding far from the corolla, with hairless filaments. *Style* bristly, with a deeply cleft stigma.

Nutlets: About 2mm, with very short spines.

Chromosomes: 2n=16.

GEOGRAPHY AND ECOLOGY

Distribution: Canary Islands: Las Palma (northern parts at about 600m/1970ft). This plant is extremely rare in its native habitat, and is included in *The IUCN Red List of Threatened Plants* as Vulnerable.

Habitat: Half shaded to open spots in laurel forest in the North.

Flowering: April to May in the wild.

CULTIVATION

Hardiness: Zone 9.

Sun/shade aspect: Needs a half-shaded place and always moist, peaty soil.

Propagation: Usually flowers within three years.

Availability: One of the most commonly grown of all Canary Island echiums, available as plant from many nurseries, and sometimes as seed.

See the details of cultivation, propagation and garden uses under the genus description.

Echium wildpretii H.Pearson in J.D. Hooker f.

A magnificent, half-hardy monocarpic shrub to 2–3m (6¹/₂–10ft) high, or sometimes much shorter, with narrowly lanceolate, silver-hairy leaves arranged in a dense rosette of up to 60–100cm (2–3ft) in diameter. Inflorescence consists of 1–1.5m (3–5ft) long spires and may contain several thousands of funnel-shaped red flowers, blooming from late spring into summer. This plant is a symbol of Tenerife.

NAMING

Etymology: *wildpretii* – after Wolfredo Wildpret de la Torre (b. 1933), Spanish botanist.

Synonyms: *Echium bourgaeanum* Coincy.

Common names: *English*: Tower of Jewels; *Spanish*: Taginaste roja.

DESCRIPTION

General: A hairy, monocarpic shrub.

Stems: To 2–3m (6¹/₂–10ft) high, sometimes much shorter, unbranched.

Leaves: 20–50cm (8–20in) long, linear to oblanceolate, densely long-hairy on both surfaces, stalkless. Veins not prominent below.

Inflorescence: Dense pyramidal panicle 1–1.5m (3–5ft) or sometimes longer, with numerous, many-flowered, lateral cymes.

Flowers: *Calyx* 9–10mm ($^3/_8$–$^7/_{16}$in), with oval to lanceolate lobes, bristly. *Corolla* red, 10–14mm ($^7/_{16}$–$^1/_2$in), funnel-shaped, lobes of limb more or less equal, to 3mm, hairless to densely bristly. Annulus a narrow, sparsely hairy, uneven ring. *Stamens* protruding from the corolla, reddish, hairless. *Style* protruding, bristly, with a forked tip.

Nutlets: To 2mm, conical, blackish, wrinkled.

Chromosomes: 2n=16.

GEOGRAPHY AND ECOLOGY

Distribution: Canary Islands: Tenerife, 1700–2200m (5570–7200ft). Becoming rare and as such is included in the IUNC Red List of Threatened Plants.

Habitat Subalpine semi-desert at about 2100m (6890ft), descending to open pine forest in the south, open slopes and rocky fields.

Flowering: April to July in the wild.

CULTIVATION

Hardiness: Zone 9. In winter keep at 7–10°C (45–50°F). May stand very slight frost for brief periods if the rosette is sheltered from rain.

Availability: Sold by a number of nurseries, particularly specialists in unusual plants.

See also details of cultivation, propagation and garden uses under the genus description.

RELATED FORMS AND CULTIVARS

Very similar is the subspecies **Echium wildpretii subsp. *trichosiphon*** (Svent.) Bramwell, which is very rare in the open areas in pine forests and subalpine regions of Las Palma, at 1600–1800m (5250–5900ft), and is probably not in cultivation. *Echium wildpretii* x *pininana* hybrids are available in the trade; they have rich purple-pink flowers and grow to about 3m (10ft) high.

Other Canary Island species

These are uncommon in cultivation, but are worth trying if obtainable. Many of these species are rare and endangered, so please do check whether the plants or seeds you are acquiring are from a legitimate source.

Echium acanthocarpum Svent.

Description: A much-branched shrub up to 1.8m (6ft). Smooth, densely short-haired, yellowish-green, shiny leaves.

Distribution: La Gomera (Roque Agando, El Cedro). Included in *The IUNC Red List* as Vulnerable.

Habitat: Rock crevices in laurel forest in central parts, 800–1000m (2600–3280ft), very rare.

Echium aculeatum Poir.

Description: Low, compact, branching shrub up to 1.2m (4ft) in height. Narrowly lanceolate leaves, densely pubescent with prominent, very spiny, short hairs on central nerve and margins.

Distribution: Tenerife, La Gomera, El Hierro.

Habitat: Open spots in succulent scrub and pine forest on lava and rocky soils at 300–1000m.

Associated plants: Grows in *Pinus canariensis* forests with plants such as *Cistus monspeliensis*, *Euphorbia obtusifolia* subsp. *regis-jobae*, *Micromeria hyssopifolia*, *Salvia canariensis*.

Echium auberianum Webb & B.

Description: Perennial with basal rosette of up to 30cm (12in) in diameter, densely pubescent, linear to oblanceolate leaves.

Distribution: Tenerife, very rare and endangered species. Listed in the IUNC Red List of Threatened Plants as Endangered.

Habitat: Subalpine zone in pure pumice or crushed lava fields, at 2100–2400m (6890–7870ft).

Echium decaisnei subsp. *decaisnei* Webb

Description: A much-branched, robust shrub 1–2m (3–6$^1/_2$ft) in height. Lanceolate light green leaves with spiny hairs above, on the underside spiny hairs limited to central nerve and margins. Dome-shaped inflorescences of white flowers with a slight pink tinge.

Distribution: Gran Canaria.

Habitat: Upper succulent scrub or lower forest zone, open spots in lava fields, 100–1000m (330–3280ft).

Flowering: February to June in the wild.

Echium decaisnei subsp. *purpuriense* Bramwell

Description: Similar to *E. decaisnei* ssp. *decaisnei* with short-haired, broader leaves.

Distribution: Lanzarote, Fuerteventura. Classified as a Vulnerable subspecies in the IUNC Red List of Threatened Plants.

Habitat: Succulent scrub in open spots in rocky soils, at 100–700m (330–2300ft).

Flowering: February to June in the wild.

Echium gentianoides Webb in Coincy

Common name: Mountain Viper's Bugloss.
Description: A shrub about 70cm (28in), occasionally to 1.5m (5ft) in height, sometimes slightly branched, differing from other echiums by its smooth, glaucous, broadly lanceolate leaves, only slightly hairy underneath; bearing large deep blue flowers. Relatively easy to grow.
Distribution: Las Palma, very rare and endangered species, listed as Vulnerable on the IUCN Red List.
Habitat: Pine forests in lightly shaded, sheltered situations, at 1100–1900m (3600–6230ft).
Flowering: February to April in the wild.

Echium giganteum Linnaeus

Description: A many-branched shrub, 2–3m (6$\frac{1}{2}$–10ft) high, with rather softly-hairy, lanceolate or oblanceolate leaves to 20cm (8in) long and 2cm ($\frac{3}{4}$in) wide, and numerous dome-shaped inflorescences of white flowers. Easier to grow than some other Canary Island echiums.
Distribution: Tenerife. Included on the IUCN Red List as Vulnerable.
Habitat: Open, moist spots in Fayal-Brezal scrub (*Erica arborea*) and succulent scrub at 100–700m (330–2300ft).
Flowering: February to May in the wild.

Echium handiense Sventen

Common name: *Spanish*: Taginaste de Jandia.
Description: Similar to *E. callithyrsum*, but with elliptical leaves (up to 17cm/6$\frac{1}{2}$in long and 4cm/1$\frac{1}{2}$in broad), densely hairy and spiny on both sides.
Distribution: Fuerteventura (southern parts, very rare and endangered). Listed as Vulnerable on the IUCN Red List.
Habitat: Open rocky places or rock crevices, 100–400m (330–1300ft).

Echium hierrense Webb in Bolle

Description: A compact, branching shrub with lanceolate to ovoid, silvery shining, smoothly pubescent leaves (up to 8cm/3$\frac{1}{4}$in long and 1.5cm/$\frac{5}{8}$in wide).
Distribution: El Hierro (Region of el Golfo, from Los Roques de Salmor towards Sabinosa, 400–800m/1300–2600ft). Listed as Vulnerable on the IUCN Red List.
Habitat: Open spots in pine forest descending to zone of Fayal-Brezal (*Erica arborea*), at 400–800m (1300–2600ft).

Echium onosmifolium Webb

Description: A many-branching, compact shrub up to 1m (3ft) in height. Linear to lanceolate, blackish-green leaves (up to 10cm/4in long and 1cm/$\frac{7}{16}$in broad), densely covered with short spiny hairs, and white flowers.
Distribution: Gran Canaria (S and SE regions). A Vulnerable species on the IUCN Red List.
Habitat: Succulent scrub to lower pine forest, in open spots in rocky places and lava soils at 400–1500m (1300–4900ft).
Flowering: February to May in the wild.

Echium simplex De Candolle

Common names: *English*: Pride of Tenerife; *Spanish*: Arrebol.
Description: A biennial or monocarpic perennial, with rosettes up to 1.2m (4ft) in diameter of densely silvery-hairy, elliptical- or oval-lanceolate leaves held on a woody, usually unbranched stem up to 50cm (20in) high, topped by cylindrical inflorescences to 2m (6$\frac{1}{2}$ft) long, of white flowers.
Distribution: Tenerife (north-east parts, very rare). Listed as Vulnerable on the IUCN Red List.
Habitat: Moist places in rocky soils in succulent scrub or open grass places, at 50–350m (160–1150ft).
Flowering: April to May in the wild.

Echium strictum L. fil.

Description: A very variable, lax-branching subscrub, sometimes only short-lived up to 1m (3ft) in height. Broad lanceolate to ovoid-lanceolate, dark green leaves are strongly hairy on both sides. Flowers pink and blue.
Distribution: Gran Canaria, Tenerife, La Gomera, Las Palma, El Hierro.
Habitat: Open to partially shaded, slightly moist spots in rocky soils at 200–700m (650–2300ft).

Echium sventenii Bramwell

Description: Larger in all parts than the similar *E. virescens*, leaf margins are usually revolute.
Distribution: Tenerife (south-east, very rare). Another Vulnerable species on the IUCN Red List.
Habitat: Below pine forest in upper succulent scrub at 350–500m (1150–1650ft).

Echium virescens De Candolle

Description: A branched shrub to 1.5–2m (5–6$\frac{1}{2}$ft) high, with lanceolate, silvery-green softly hairy leaves to 15cm (6in) long, and bearing pink or bluish flowers in long, slender, spikelike inflorescences.

Easier to grow than many other Canary Island echiums.

Distribution: Tenerife (across central mountain ranges). Listed as Rare in the IUCN Red List.
Habitat: Open spots in pine forest.
Flowering: February to June in the wild.

Other Echium species

Echium albicans Lagasca & Rodr.

A striking white-hairy perennial, to 75cm (2½ft) high, with a rosette of narrowly lanceolate, bristly leaves and spikes of pink-red or blue-purple, funnel-shaped flowers. An unusual border plant for full sun and well-drained soil.

NAMING

Etymology: *albicans* – becoming white.
Common names: *English*: White-leaved bugloss; *Spanish*: Viborera.

DESCRIPTION

General: An upright, more or less softly hairy and bristly perennial.
Stems: 20–75cm (8–30in), one to several, upright, densely white-hairy and sparsely bristly.
Leaves: 3.5–7cm (1⅖–2¾in) long and 4–10mm (3/16–7/16in) wide, narrowly lanceolate to narrowly oblong, tapering at base, covered in dense, short whitish hairs and some rather sparse bristles.
Flowers: *Calyx* 10–17mm (7/16–5/8in), with long white hairs. *Corolla* pink-red to blue-purple, funnel-shaped, 16–26mm (5/8–1in). *Stamens* 2 or 3, protruding from the corolla.

GEOGRAPHY AND ECOLOGY

Distribution: Endemic to southern Spain.
Habitat: Rocks and screes in mountains, especially on lime-rich soils.
Flowering: Summer.
Associated plants: In the wild, its companions include such plants as stinking hellebore (*Helleborus foetidus*).

CULTIVATION

Hardiness: Zone 7.
Sun/shade aspect: Full sun.
Soil requirements: Gritty, well-drained soil, ideally from neutral to alkaline.
Garden uses: An impressive border plant.
Spacing: 30–45cm (12–18in).
Propagation: From seed as for the genus.

Availability: Plants and seeds are rarely available, and must be searched for, from specialist suppliers.

RELATED SPECIES

Another perennial bugloss is *Echium angustifolium* Miller (syn. *E. sericeum* Vahl; *E. diffusum* Sibthorp & Smith), from Crete, Greece and Turkey, with upright, white-bristly stems to 40cm (16in), linear to narrowly oblong leaves and, in spring and early summer, bearing red-purple to violet, funnel-shaped flowers to 22mm (7/8in) long, with long-protruding stamens.

Echium boissieri Steudel

An imposing, bristly biennial, 60–250cm (2–8ft) tall, with a basal rosette of linear or lanceolate leaves, and usually a single spike of funnel-shaped, pink flowers with red filaments. Uncommon in gardens, but worth trying in a sunny mixed border.

NAMING

Etymology: *boissieri* – after botanist Pierre Edmond Boissier (1810–55).
Synonym: *Echium pomponium* Boissier.
Common name: *English*: Candle bugloss.

DESCRIPTION

General: A bristly, upright biennial.
Stems: 60–250cm (2–8ft) high, usually unbranched.
Leaves: Basal leaves linear or narrowly lanceolate, white-bristly. Stem leaves narrowly elliptical to lanceolate.
Inflorescence: A dense spike.
Flowers: *Calyx* to 9mm (3/8in) long. *Corolla* rose-pink, narrowly funnel-shaped, to 18mm (3/4in) long. *Stamens* protruding from the corolla, with red filaments.

GEOGRAPHY AND ECOLOGY

Distribution: Spain, Portugal.
Habitat: Mountains, dry places.
Flowering: Summer.

CULTIVATION

Hardiness: Zone 7.
Availability: This plant can occasionally be found in the catalogues of specialist suppliers, available both as plant and seeds. Its cultivation and propagation details are as for other biennial *Echium* species.

Echium plantagineum Linnaeus

A hairy annual or biennial, usually around 30cm (12in) tall, with conspicuously veined basal leaves and rich reddish-purple flowers. Easy to grow in any sunny, well-drained spot.

Echium russicum in north-eastern Turkey. (Erich Pasche)

Naming
Etymology: *plantagineum* – from *Plantago*, probably referring to the shape of the leaves.
Synonyms: *Echium lycopsis* invalid, not L., *E. maritimum* Wildenow.
Common names: *English*: Purple bugloss, Salvation Jane, Paterson's curse (Australia); *German*: Wegerrichblättriger natternkopf; *Spanish*: Sonaja, Viborera; *Turkish*: Engerek otu.

Description
General: A softly hairy annual or biennial.
Stems: 20–60cm (8–24in), upright, from one to many.
Leaves: Basal leaves to 14cm (5½in) long and 4cm (1½in) wide, oval to lanceolate, bristly hairy, with prominent lateral veins. Stem leaves lanceolate or oblong, with a somewhat heart-shaped base, the upper one stalkless and half-clasping.
Inflorescence: A panicle.
Flowers: *Calyx* 1–12mm (1/16–1/2in), to 15mm (5/8in)in fruit. *Corolla* red turning blue-purple, 18–30mm (3/4–1¼in), funnel-shaped, hairy on veins and margins. *Stamens* 2 protruding from the corolla. *Style* with a forked stigma.

Geography and ecology
Distribution: From southern and western Europe (including south-west England, where it is a rare species) to Caucasus; naturalized in Australia where it is a noxious weed, known as 'Paterson's curse'.
Habitat: Dry habitats, often by the sea.
Flowering: Late winter to early summer in the wild.

Cultivation
Hardiness: Zone 7.
Spacing: 20–30cm (8–12in).
Availability: Occasionally seed available from some catalogues.
Problems: A very serious weed in Australia, toxic to cattle, horses and sheep.
Cultivation and propagation as for other biennial echiums.

Related species
Other biennial echiums of interest include **Echium creticum** Linnaeus, from the western Mediterranean area and southern Portugal, to 90cm (3ft), with bristly, narrowly oblanceolate basal leaves and panicles of reddish-purple or carmine-pink flowers that turn blue-purple, 15–40mm(5/8–1½in) long, with 1 or 2 protruding stamens.

Echium russicum J.F.Gmelin
One of the most attractive biennial echiums, to 50cm (20in) tall, with a rosette of narrow bristly leaves and a spike of deep-red flowers. For any well-drained spot, especially a mixed border.

Naming
Etymology: *russicum* – Russian.
Synonyms: *Echium rubrum* Jacq. not Forssk., *E. maculatum* L.
Common names: *Russian*: sinyak krasny; *Slovak*: hadinec cerveny.

Description
General: An upright, bristly biennial.
Stems: To 50cm (20in), usually unbranched, bristly.
Leaves: 5.5–10cm (2¼–4in) long and to 1cm (7/16in) wide, narrowly elliptical to lanceolate, pointed, covered with dense soft bristles.
Inflorescence: To 30cm (12in) long, a spikelike panicle.
Flowers: *Calyx* 5–7mm. *Corolla* dark red to crimson, 9–12mm, narrowly funnel-shaped, softly hairy outside. *Stamens* 4 or 5, protruding from from the corolla; filaments orange below, crimson above. *Style* with a headlike, somewhat two-lobed stigma.
Chromosomes: 2n=24.

GEOGRAPHY AND ECOLOGY
Distribution: Eastern, central and south-east Europe, Western Asia.
Habitat: Meadows, slopes, scrub.
Flowering: Summer.

CULTIVATION
Hardiness: Zone 5.
Sun/shade aspect: Full sun or very light shade.
Soil requirements: Any well-drained soil.
Garden uses: A nice plant for a mixed border, especially when grouped.
Spacing: 30cm (12in).
Propagation: As for the genus.
Availability: Sold by a number of herbaceous plant specialists.

RELATED FORMS AND CULTIVARS
'Burgundy' has long spikes of dark red flowers, but does not seem to be in cultivation at present.

Echium vulgare Linnaeus
A bristly biennial, 30–90cm (1–3ft) high, with rosettes of narrow leaves, and spikes of blue-purple funnel-shaped flowers that are very attractive to bees and other insects. Good for any dry spot in full sun, but can get weedy in ideal conditions.

NAMING
Etymology: *vulgare* – common.
Synonyms: Seed strains offered under the name of *E. plantagineum* are more likely to belong to this species.
Common names: *English*: Viper's bugloss, Blue-weed, Blue-devil; *Chinese*: lan ji; *Danish*: Slangehoved; *Dutch*: Slangekruid; *Finnish*: Neidonkieli, Piennarneidonkieli; *German*: Gewöhnlicher Natternkopf; *Italian*: Viperina Azzurra; *Norwegian*: Ormehuvud; *Russian*: sinyak obyknovenny; *Slovak*: hadinec obycajny; *Spanish*: Viborera; *Swedish*: Blåeld, snokört.

DESCRIPTION
General: An upright, bristly biennial.
Stems: 30–90cm (1–3ft), bristly, usually unbranched.
Leaves: Basal leaves 5–15cm (2–6in) long and 1–2cm ($^{7}/_{16}$–$^{3}/_{4}$in) wide, lanceolate to narrowly elliptical, bristly; upper leaves narrowly lanceolate, stalkless.
Inflorescence: A spike or panicle.
Flowers: *Calyx* 5–7mm, to 10mm in fruit. *Corolla* blue to blue-violet, 10–19mm, broadly funnel-shaped. *Stamens* 4 or 5, long-protruding, with blue-purple filaments. *Style* with a deeply cleft stigma.

Echium vulgare growing with *Verbascum* species.
(Erich Pasche)

Nutlets: About 2.5mm, ovoid, with small swellings.
Chromosomes: 2n=16, 32.

GEOGRAPHY AND ECOLOGY
Distribution: Europe. Classified as a noxious weed in Washington State and areas of Canada and Australia.
Habitat: Calcareous and dry soils, often by the sea, also on walls, in gravel pits.
Flowering: July to October in the wild.
Associated plants: In Kyrgyzstan, grows with orange hawkweed (*Pilosella aurantiacum*), mulleins (*Verbascum* sp.), *Codonopsis clematidea*, tuberous pea (*Lathyrus tuberosus*), and toadflax (*Linaria vulgaris*). In central Asia, where it is grown as a nectar plant for hive bees, vast expanses of blue can be seen, sometimes interspersed with the pink of sainfoin (*Onobrychis viciifolia*).

CULTIVATION
Hardiness: Zone 5.
Sun/shade aspect: Full sun.
Soil requirements: Any well-drained soil, tolerates a high degree of alkalinity, and drought.
Garden uses: Grow in a wildflower garden, mixed borders, on walls, in sand or gravel beds, any dry spots. On strongly alkaline soils may succeed in thin grass.

Excellent bee plant, also attractive to butterflies and
moths. Dwarf seed strains are available for bedding.
Spacing: 20–30cm (8–12in).
Propagation: From seed – sow in late winter or early
spring to flower the same year or in late summer to
autumn *in situ*. Germination usually takes place
within 2–3 weeks at 15°C (59°F).
Availability: Available from many seed suppliers,
especially wildflower specialists, and also as plant
from some nurseries.
Problems: In ideal conditions can be very invasive,
and is listed as a noxious weed in the USA
(Washington State), New South Wales, Western
Australia and Tasmania.

RELATED FORMS AND CULTIVARS
The so-called **Drake's form** originated from Jack
Drake's Inshriach Nursery. There are some seed
strains that are often sold as *E. plantagineum* varieties:
'Blue Bedder' – 30cm (12in) tall, blue flowers.
'Moody Blues' – 30cm (12in) tall, flowers range from
light to deep blue, with some mauves and purples.

RELATED SPECIES
Pale bugloss, **Echium italicum** Linnaeus, from
southern and south-central Europe, is a similar plant,
40–100cm (16–40in) tall, with larger basal leaves, with
narrower and paler blue flowers, which can sometimes
also be yellowish- or pinkish-white, and have long
protruding stamens with transluscent filaments. It is
rather less attractive than *E. vulgare* and so is rarely
grown. It is listed as a noxious weed in Australia.

The genus *Eritrichium* Schrader, 1820
Alpine Forget-me-not

NAMING
Etymology: *Eritrichium* – from Greek *erion*, wool, and
trichos, hair.
Synonyms: *Metaeritrichium* W.T. Wang. Some
species now belonging to *Chionocharis* I.M Johnston,
Cryptantha Lehm., *Trigonotis* Steven, *Plagiobothrys*
Fisher & Meyer and *Mertensia* Roth have been
included with *Eritrichium* in the past. Also, some
plants in this genus were previously assigned to
Myosotis L. and *Omphalodes* Miller. It is sometimes
argued that the correct spelling should be '*Eritrichum*'.

GENUS CHARACTERISTICS AND BIOLOGY
Number of species: There are widely ranging estimates

Eritrichium tianschanicum in Ala Archa, Kyrgyzstan.
(Masha Bennett)

of the number of species within the genus. Both the
AGS Encyclopedia and Mabberley's *The Plant-Book*
quote around 30, the *RHS Dictionary* gives the figure of
70 (which is possibly a mistake), and the European
Garden flora estimates 4 to 10, depending on whether
E. nanum is treated in a wide or narrow sense.
Description: A genus of low, dense, often cushion-
forming and sometimes short-lived perennial herbs.
Stems and leaves hairy. Flowers in simple or branched,
terminal 1-sided cymes. *Calyx* divided into 5 lobes.
Corolla forget-me-not-like, funnel-shaped or with very
short tube and spreading limb; throat-scales present,
often yellow. Stamens included in the corolla-tube.
Fruit of 4 small nutlets, hairy or hairless.
Distribution and habitats: Circumboreal and in the
mountains of central Europe, the Himalaya and western
North America; on mountain rocks, screes, alpine
tundra, some on sea shores.
Life forms: Chamaephytes.
Pollination: Probably by bees.
Fruit dispersal: No specialized mechanism, but the
melting snow probably plays a role in dispersing
E. nanum and other high alpine species.

USES
There is no information on medicinal or economic use
of eritrichiums in the literature. The members of the
genus are known to contain pyrrholizidine alkaloids.

GARDEN
These are the ultimate rock garden plants, but many
are very difficult and the alpine house is usually
considered the only feasible option, though they may
also succeed in tufa in a trough.

CULTIVATION

Although all eritrichiums are completely hardy, the majority are very difficult to grow. They normally need full sun, though in areas with high light levels may also grow in partial shade. A very sharply drained growing medium is essential. The best chance of success is in the alpine house, where they can be tried in clay pots or in tufa. Good air circulation under glass is essential. The mixture recommended by the AGS *Encyclopaedia of Alpines* is 1 part John Innes seed compost to 1 part leaf mould and 1 part gritty sand, ideally mixed with crushed slate or granite. A deep collar of grit will help prevent rotting. If trying eritrichiums outside, trough cultivation is a viable option, but they must be protected from winter wet with a pane of glass. Once the plants are established, it is best to leave them undisturbed, which is especially relevant to *E. nanum*, which deeply resents disturbance.

PROPAGATION

Seed Ideally should be sown as soon as ripe, but mid- to late winter is not too late – this is the usual time of seed distribution for many specialist plant societies, which are the most accessible source of seed for the members of this genus. Sow in pots, in a mixture similar to that for cultivation but with extra grit or tufa dust added. The seeds are fine and must be left uncovered. The pots should be placed in a cold frame, and exposure to frost will assist germination. Alternatively, stratify artificially by placing the pots or trays with seeds in a refrigerator at 4°C (39°F) – three to four weeks should be sufficient for most species, though some, including *E. nanum*, may need up to six weeks of cold treatment or longer. After stratification, move to a cold greenhouse. Seedlings should be pricked out as soon as large enough to handle, with utmost care.

Vegetative Softwood stem cuttings can be taken in summer and inserted in pots of gritty compost. Some species can be divided in mid-spring, but not *E. nanum*, which hates disturbance. Pot up the divisions and grow on in a cold frame or greenhouse.

REFERENCES

BECKETT, 1993–94; BIRD & KELLY, 1992; GRIFFITH, 1964; HEATH, 1964; HUXLEY, 1992; INGWERSEN, 1991; MALYSHEV, 1997; OHWI, 1965; TUTIN *et al.*, 1972; VVEDENSKY, 1962; WALTERS, 2000; WU & RAVEN., 1995.

Eritrichium canum (Bentham) Kitamura

A tufted, short-lived perennial, to about 15cm (6in) high, with rosettes of silvery-grey, hairy leaves and clusters of light blue flowers around 7mm across. A nice rock garden plant, easier to grow than most *Eritrichium* species.

NAMING

Etymology: *canum* – grey.
Synonyms: *Eritrichium rupestre* (Pallas) Bunge; *E. strictum* Decaisne, *E. pectinatum* (Pallas) De Candolle, *Echinospermum canum* Bentham.
Common name: *Chinese:* hui mao chi yuan cao.

DESCRIPTION

General: A tufted perennial, with sterile rosettes and flowering stems, covered in silvery-grey hairs. A very variable species.
Stems: Several to many, variable in height, from low to 15–20cm (6–8in) and occasionally taller, upright to decumbent, often woody at base, unbranched or shortly branched above, densely silvery-hairy.
Leaves: Basal leaves 20–50mm ($^3/_4$–2in) by 2–4mm ($^1/_{16}$–$^3/_{16}$in), linear or narrowly lanceolate, pointed, shortly stalked. Stem leaves smaller, stalkless, numerous; all with dense, adpressed silky-white hairs.
Inflorescence: Groups of 2 or 3 terminal raceme-like cymes, first dense then loose, with linear bracts in the lower part.
Flowers: *Pedicel* erect, 10–15mm ($^7/_{16}$–$^5/_8$in) in fruit, shortly hairy. *Calyx* lobes about 2mm ($^1/_{16}$in) long in fruit, oval, blunt, silky-hairy. *Corolla* to 5–7mm ($^1/_4$–$^3/_8$in) across, light blue, bell-shaped to rotate, lobes rounded, spreading; throat scales broad yellow or orange.
Nutlets: 1.5–2.5mm, with stout hooked bristles on margins, sometimes hairy. In a pyramid, with margins with stout hooked bristles.
Chromosomes: 2n=24.

GEOGRAPHY AND ECOLOGY

Distribution: Afghanistan, northern India, Kashmir, Nepal, Pakistan, Siberia, Mongolia, Tibet, north-east China.
Habitat: Rock ledges, screes, gravelly slopes, sandy river banks, at 2400–5600m (7870–18,350ft).
Flowering: June to August in the wild.

CULTIVATION

Hardiness: Zone 5.
Spacing: 20cm (8in) apart.
Availability: Seed frequently appears in seed-exchange schemes, under a variety of different synonyms. Cultivation, propagation and uses as for the genus.

Eritrichium howardii (A. Gray) Rydberg

An attractive mat-forming perennial to 10cm (4in) tall, with small, bristly narrow leaves and deep blue flowers about 7mm across, in early summer. Easier to grow than the similar *E. nanum*, but still rather unpredictable.

NAMING

Etymology: *howardii* – after Richard Alden Howard (b.1917).
Synonyms: *Cynoglossum howardii* A. Gray; *Omphalodes howardii* A. Gray.
Common names: *English*: Howard's forget-me-not, Howard's eritrichium.

DESCRIPTION

General: A dense, mat-forming perennial, similar to *E. nanum* but less compact and more bristly rather than softly hairy.
Stems: 1–12cm ($^7/_{16}$–5in), grey-bristly.
Leaves: Basal leaves linear-oblanceolate, to 15mm ($^5/_8$in), covered in stiff, not silky, greyish or yellowish hairs.
Flowers: *Corolla* 6–9mm across, darker blue than in *E. nanum*.
Nutlets: 2 × 2mm, broadly pyramid-shaped, oblique, shortly hairy; dorsal surface oval, almost flat.

GEOGRAPHY AND ECOLOGY

Distribution: USA (Montana, Wyoming, Washington).
Habitat: Dry hills, submontane zone.
Flowering: May to July in the wild.

CULTIVATION

Hardiness: Zone 3.
Availability: Uncommon in cultivation, but may appear on seed exchange lists.
Cultivation and propagation as for the genus, but this species is generally a bit easier than *E. nanum*.

Eritrichium nanum (Villars) Schrader in Gaudin

The ultimate alpine plant, a very low cushion-forming perennial with small silvery-hairy leaves and forget-me-not-like, sky-blue flowers with yellow centres. Very hard to grow and flower well.

'… it is a small cushion, as if made of blue-green, scarified velvet, clinging limpet-like to the rock. The stemless flowers, like crystalline forget-me-nots, are of a blue that is uniquely, almost shockingly pure.' (Richard Bird and John Kelly)

'… must be the despair of all alpine gardeners … So easy to collect from its native habitat with all its roots intact but once introduced to gardens in the lowlands it promptly sighs for its vast desolate high screes and departs with a rapidity that will amaze you.' (Royton E. Heath)

NAMING

Etymology: *nanum* – small, dwarf.
Synonym: *Omphalodes nana* A.Gray.
Common names: *English*: King of the Alps, Herald of Heaven; *French*: Eritriche nain; *German*: Himmelsherold; *Italian*: Eritrichio nano; *Spanish*: Rey de los Alpes.

DESCRIPTION

General: A very low, densely tufted, silvery-hairy perennial, with woody rootstock, forming cushions 2–10cm ($^3/_4$–4in) across (rarely much more than 2cm/$^3/_4$in across in cultivation).
Stems: Stems almost absent or hardly taller than the rosettes, occasionally up to about 5cm (2in).
Leaves: Basal leaves oblong-lanceolate to linear, about 10mm ($^7/_{16}$in) long, grey-green, on both surfaces covered with long, silky hairs, that are densest at the leaf tip where they form a tuft. Stem-leaves, if present, are very small.
Inflorescence: Short-stalked, dense racemose cymes of 3–7 flowers, with bracts.
Flowers: *Pedicels* absent. *Calyx* lobes narrowly elliptical, much longer than the tube, very hairy. *Corolla* sky blue opening from pinkish (or rarely white), 6–9mm across, with yellow scales in throat, longer than broad.
Nutlets: Ovoid, 3-angled, 1–2 × 2.5mm, hairless, the margins comblike or sometimes entire.

GEOGRAPHY AND ECOLOGY

Distribution: Alps, Carpathians, Altai Mountains, Himalayas, China.
Habitat: Acid rocks and screes, often on granite, sometimes dolomite, 2500–3600m (8200–11,800ft); in the Himalayas at 3600–5000m (11,800–16,400ft).
Flowering: July to August in the wild.
Associated plants: In the Alps it may be encountered in the company of *Androsace alpina*, *Artemisia umbelliformis*, *Phyteuma hedraianthifolium*, *Gentiana brachyphylla*, moss campion (*Silene acaulis*), cyphel (*Minuartia sedoides*) and *Woodsia alpina*.

CULTIVATION

Hardiness: Zone 6.
Availability: Seed frequently appears in alpine societies' exchange schemes, and sometimes in seed catalogues.
For cultivation and propagation details, see the genus description.

Eritrichium nanum* subsp. *jankae (Simonkai)
Javorka, endemic to the Carpathians and included on
The IUCN Red List of Threatened Plants, differs from
the typical *E. nanum* in being even more densely
white-hairy, with longer hairs, taller stems (usually
5–7.5cm/2–3in) and wider leaves, 3–4.5 mm. A
naturally occurring variant of this plant, **var.
*simonkianum***, endemic to Piatra Craiului Range at
1700m (5570ft), is considered a little easier in
cultivation than the species, as the conditions in its
habitat are somewhat drier and warmer than the
norm for this plant.

Eritrichium villosum (Ledebour) Bunge

A tuft-forming perennial, often only 2–3cm
($^3/_4$–$1^1/_4$in) high and wide, with oval, hairy, grey-
green leaves. Short flower stems each carry small,
rounded, flat, blue flowers in early summer. This
little plant is similar to the famous King-of-the-Alps
(*E. nanum*), of which it is sometimes considered a
subspecies, and differs by often growing rather
taller, having fewer basal leaves and shorter corolla.
It is also a little easier to grow.

> *'Pads of grey-haired leaves and heads of rich
> blue flowers on short stems.' (Will Ingwersen)*

NAMING
Etymology: *villosum* – shaggy-haired, referring to the
leaves.
Synonyms: *Eritrichium nanum* subsp. *villosum*
(Ledebour) Brand; *E. elongatum* (Rydberg) W.Wight; *E.
aretoides elongatum* Rydberg; *Myosotis villosa* Ledebour.
Common names: *English: Pale alpine forget-me-not*
(for *E. elongatum*); *Russian:* nezabudochnik mohnatyi.

DESCRIPTION
General: A densely tufted, mat-forming perennial.
Stems: 1–10cm ($^7/_{16}$–4in) long in flower.
Leaves: Woolly, grey-green; basal closely overlapping,
5–8mm long, oblong or oblanceolate, stem leaves
linear or linear-oblong, loosely silky-hairy.
Flowers: *Corolla* light blue, to 7mm across, crest yellow.
Nutlets: About 2mm long, oblique; dorsal surface
ovate, smooth and shining.
Chromosomes: 2n=24.

GEOGRAPHY AND ECOLOGY
Distribution: Western USA, northern Asia, Arctic
Russia.
Habitat: Hills and mountains, alpine meadows and
Arctic tundra.

Flowering: June to July in the wild.
Associated plants: In northern Tian Shan in
Kazakhstan, can be found with plants such as *Primula
algida, Iris Bloudowii, Fritallaria pallidiflora*.

CULTIVATION
Hardiness: Zone 7.
Availability: Seed sometimes offered for sale or can
be found in exchange schemes, usually under the
name of *E. elongatum*.
Other cultivation and propagation details as for the
genus.

Other *Eritrichium* species of interest

Eritrichium aretioides (Cham.) De Candolle
Synonyms: *E. argenteum* W.F. Wight; *E. nanum
aretioides* A. Gray not Herder; *Omphalodes nana
aretioides* A. Gray; *Oreocarya pulvinata* A. Nels.
(white-flowered form).
Description: A small, cushion- or mat-forming, silky-
hairy perennial, 1–8cm ($^7/_{16}$–$3^1/_4$in) high in flowers
and to 25cm (10in) across, with white- or grey-hairy,
elliptical or oblanceolate leaves 0.3–12mm long,
bearing cymes of 1 to 6 bright blue or occasionally
white flowers 4–6mm across.
Distribution: Arctic Russia, Siberia, Alaska, Colorado
to Utah.
Habitat: Mountains, gravelly places, sandy sea shores
in the Arctic.
Associated plants: In Colorado grows with alpine
bluebells (*Mertensia alpina*), mountain avens (*Dryas
octopetala*), mountain spiderwort (*Lloydia serotina*),
alpine primrose (*Primula angustifolia*) and moss
campion (*Silene acaulis subacaulescens*).

Eritrichium nipponicum Makino
Synonyms: *E. yesoense* Nakai, *E. sachalinense* Popov,
Hackelia nipponica Makino.
Common name: *Japanese:* Miyama-murasaki.
Description: A bristly, tufted to cushion-forming
perennial, with a thick caudex, 10–20cm (4–8in)
ascending stems, a rosette of narrowly lanceolate,
grey-hairy leaves 3–6cm ($1^1/_4$–$2^1/_2$in) long. Blue
flowers 8mm across borne in cymes with a few bracts,
in summer. The plant named *E. sachalinense* in
Russian floras is probably the same species, differing
mainly by smaller flower size.
Distribution: Japan, Sakhalin.
Habitat: Gravelly and sandy slopes.

Eritrichium aretoides in Rocky Mountain National Park, Colorado. (Masha Bennett)

Eritrichium sajanense (Malyschev) Sipl.

Synonym: *E. pauciflorum* subsp. *sajanense* Malyschev.

Description: A cushion-forming perennial, 5–10cm (2–4in) high, with greenish, linear, silky-hairy leaves, and blue flowers 6–10mm across. Ingwersen describes this plant: *'Tufts of silky, silvery hairy leaves grow in rocky crevices and are adorned by typically caerulean blue, golden-eyed flowers.'*

Distribution: Siberia, Mongolia.

Habitat: Rocky crevices, stony slopes, especially on carbonate-rich material.

Eritrichium sericeum (Lehm.) De Candolle

Synonyms: *E. kamtschaticum* Kom.; *Myosotis sericea* Lehm.

Description: A perennial to about 10cm (4in), sometimes 20cm (8in) high, forming small, rather loose mats, with numerous pointed, lanceolate leaves 10mm long and 2–3mm wide, white with felty hairs, bearing single or paired clusters of blue flowers 5–10mm in diameter.

Distribution: Siberia, Kamchatka.

Habitat: Stony slopes and screes, mostly on limestone.

Eritrichium tianschanicum Iljin

Common names: *Russian*: Nezabudochnik tianshansky.

Description: A hairy perennial 5–15cm (2–6in) tall, forming small loose mats, with slender branches, rosettes of densely hairy oblong-lanceolate leaves 1–2cm ($^7/_{16}$–$^3/_4$in) long and up to 6mm ($^1/_4$in) wide, and smaller, lanceolate stem leaves, and cymes of 3 to 10 white or cream, scented flowers, each about 7mm ($^3/_8$in) across.

Distribution: Central Asia (Tian Shan and Pamiro-Alai).

Habitat: Alpine meadows and tundra, stony places.

Associated plants: In the Tien Shan Mountains in Kyrgyzstan, grows beside *Primula turkestanica*, *Hegemone lilacina*, *Viola tianshchanica* and *Leontopodium ochroleucum*.

The genus *Heliotropium* Linnaeus, 1753

NAMING

Etymology: *Heliotropium* – from Greek *helios*, sun, and *trope*, turning; referring to the old belief that the plant's flowers turned with the sun.

Synonyms: *Beruniella* Zakirov & Nabiev; *Bourjotia* Pomel; *Bucanion* Steven; *Cochranea* Miers; *Euploca* Nutt. Some species of genus *Tournefortia* L. have previously been assigned to *Heliotropium*.

Common names: *English*: Heliotrope, Turnsole; *German*: Sonnenwende.

GENUS CHARACTERISTICS AND BIOLOGY

Number of species: According to various estimates, from 100 to 250 species.

Description: A genus of more or less hairy annuals, perennials and shrubs, usually with alternate, entire, stalked leaves. Flowers are variously coloured, often purple or white, usually in terminal or axillary, spike-like or raceme-like scorpioid cymes without bracts. *Corolla* 5-lobed, from white to deep purple, sometimes yellow, with a long slender tube and an abruptly expanded limb, usually with teeth between the lobes. Stamens included in the corolla-tube. Style is also included, usually very short, with a large, disc- or cone-shaped stigma that may be entire or with 2–4 lobes. Nutlets up to 4, variously ornamented.

Distribution and habitats: Mostly warmer parts of the Old World and the New World, in dry, open habitats.

Life forms: Therophytes (annuals), hemicryptophytes, chamaephytes and phanerophytes.

Pollination: A variety of insects, including bees, butterflies and moths, are especially attracted to flowers of *H. arborescens*.
Fruit dispersal: No specialized mechanism.

USES

GARDEN

Heliotropes are commonly used as dot plants for bedding, or in containers. In Latin America and other warm areas, *H. arborescens* is sometimes grown as a low hedge, but it is not hardy enough for this purpose in temperate climates. It also makes an admirable small standard shrub. It is of value in a wildlife garden, where the nectar-rich flowers will attract butterflies, bees and other insects. As it attracts night-flying insects, *H. arborescens* is a plant to be included in a 'bat-friendly' garden. The less common tender species, such as *H. amplexicaule* and *H. convolvulaceum*, are not as suitable for formal bedding as *H. arborescens*, and are best added to a sunny mixed border, or used in containers. The weedy-looking *H. europaeum* is probably only of interest in a specialist native plant collection.

The flowers of showy species and hybrids can be used for cutting and fresh flower arrangements, and last better if at first the cut stems are plunged deep in water and kept in a cool dark place for several hours.

MEDICINAL

Heliotropium arborescens has been used to make a herbal tea for fever. In India, *H. ellipticum* is used to treat boils, ulcers and the bites of snakes and dogs.

ECONOMIC

An essential oil obtained from the flowers of *H. arborescens* is used in perfumery.

CULTIVATION

Most cultivated heliotropes are tender, and should either be raised from seed every year as annuals, or overwintered under glass. When grown outdoors, they require full sun and well-drained, moderately fertile soil. Tender species including *H. arborescens* and its hybrids should be overwintered at minimum night temperature of 5–7°C (41–45°F). Under glass, good ventilation is required when temperatures rise above 13°C (55°F).

To obtain bushy habit in *H. arborescens*, shoot tips are pinched out, and straggly plants can be cut hard back in spring and will re-sprout. For more cultivation details, see the individual species descriptions.

Heliotropium arborescens 'Marine'. (Howard Rice)

PROPAGATION

Seed Heliotrope seeds germinate without any special treatment. For tender species and varieties, they should be sown in late winter or early spring and germinated in a heated propagator. See more details under individual species descriptions.
Vegetative *Heliotropium arborescens* cultivars are normally propagated by cuttings – see details under this species.

PROBLEMS

None of the popular species or cultivars are hardy in temperate climates, so they have to be raised from seed every year or overwintered in a heated glasshouse. Those species that are hardy are mostly weedy plants, of little interest to the average gardener. Some species, such as *H. amplexicaule*, have naturalized and become serious weeds in dry regions, including the United States, with the problem exacerbated by the fact that they are poisonous to stock. A degree of success has been achieved in using leaf-feeding beetles, *Deuterocampta* and *Longitarsus*, as biocontrol agents.

REFERENCES

BRICKELL, 1996; GENDERS, 1994; HAY, 1955; HUXLEY, 1992; MANSFIELD, 1955; PHILLIPS & RIX, 1998; SINGH & KACHROO, 1976; TUTIN *et al.*, 1972; WILKINSON BARASH, 1993; WALTERS, 2000.

Heliotropium amplexicaule Vahl

A tender, woody-based perennial, to 60cm (2ft) tall, with wavy-margined elliptical leaves that are often slightly clasping. It bears purple flowers, similar to the famous 'Cherry pie' (*H. arborescens*), but unfortunately unscented. An uncommon but interesting plant for a sunny border.

NAMING

Etymology: *amplexicaule* – clasping the stem, referring to the bases of leaves.
Synonyms: *Cochranea anchusifolia* (Poiret) Gurke; *Heliotropium anchusifolium* Poiret
Common names: *English*: Clasping heliotrope, Violet heliotrope, Summer heliotrope.

DESCRIPTION

General: A more or less sprawling, woody-based perennial.
Stems: To 50–60cm (20–24in) long, somewhat branched, woody at base, hairy and glandular above.
Leaves: To 9cm (3^1/$_2$in) long and 2.5cm (1in) wide, oblong to oblanceolate, usually pointed, with a wedge-shaped base and wavy margins, either stalkless and slightly clasping the stem or short-stalked.
Inflorescence: Dense axillary or terminal cymes, to 8cm (3^1/$_4$in) long, on stalk to 10cm (4in) long.
Flowers: *Calyx* lobes 2–5mm, linear-lanceolate, often glandular-hairy. *Corolla* purple, blue or sometimes white, funnel-shaped, unscented; tube to 5mm, limb 4–8mm in diameter. *Stamens* consist of sessile, apiculate anthers. *Style* with a sessile, conical or hemispherical stigma.
Nutlets: Usually only 2 per flower, to 3 x 2.5mm, somewhat compressed, warty, hairless.

GEOGRAPHY AND ECOLOGY

Distribution: South America (endemic to Argentina); naturalized in California, Hawaii, and southern Italy. Listed as a noxious weed in Australia.
Habitat: Scrub, grassy and dry places.
Flowering: Summer.

CULTIVATION

Hardiness: Zone 10; in colder areas must be overwintered in a heated glasshouse.
Sun/shade aspect: Full sun.
Soil requirements: Well-drained, fairly fertile soil.
Garden uses: Grow in a mixed border or in containers. Less suitable for formal bedding than *H. arborescens* and its hybrids.
Spacing: About 45cm (18in).
Propagation: From seed, as for *H. arborescens*.
Availability: Although often cultivated in southern Europe, this is rarely grown in north-western Europe, but can be found in a few nurseries specializing in unusual plants.

Heliotropium arborescens Linnaeus

A tender small shrub, frequently grown as a half-hardy annual, to around 60cm (2ft) high or more, with prominently veined, oval or lanceolate leaves, and usually purple flowers possessing a delightful scent variously described as that of vanilla, almonds or, most famously, cherry pie. Extremely popular in Victorian times, this plant deserves the resurgence in popularity it has been enjoying recently.

> 'The rich odor, reminiscent of violets with a vanilla overtone, is one of the quintessential smells of summer. ... If the fragrance is not enough to endear this plant to evening gardeners, the white-flowered variety is visible at night and the more common purple variety fluoresces in the late-afternoon light.' (Cathy Wilkinson Barash, 1993)

NAMING

Etymology: *arborescens* – treelike, referring to the plant's woody habit.
Synonyms: *Heliotropium corymbosum* Ruiz & Pav.; *H. peruvianum* Linnaeus.
Common names: *English*: Heliotrope, Cherry Pie.

DESCRIPTION

General: A branched shrub, often grown as a half-hardy annual, usually around 60cm (2ft) high but in warmer climates reaching 1.5–2m (5–6^1/$_2$ft) tall.
Stems: Branched, woody.
Leaves: To 8cm (3^1/$_4$in) long and 4cm (1^1/$_2$in) wide, oval to oblong-lanceolate, pointed, with a stalk to 1.5cm (^{5}/$_8$in) long.
Inflorescence: Consisting of short, hairy scorpioid cymes.
Flowers: *Calyx* lobes to 3.5mm, narrowly awl-shaped. *Corolla* deep purple, violet, lavender or white, sweetly fragrant, with hairy tube to 7mm long, and limb to 5mm across with rounded lobes that may be hairy at first. *Stamens* with elongated anthers.
Nutlets: 4, ellipsoid.

GEOGRAPHY AND ECOLOGY

Distribution: South America (endemic to Peru).
Habitat: Scrub.
Flowering: In the wild may flower at most times of the year; in cultivation in temperate climates, normally summer and autumn.

Heliotropium arborescens 'Chatsworth'. (Howard Rice)

Heliotropium arborescens 'White Lady'. (Howard Rice)

CULTIVATION

Hardiness: Zone 10.

Sun/shade aspect: Plenty of light is essential, but under glass the plants should be shaded from the hottest sun at midday.

Soil requirements: Well-drained, moderately fertile soil that is nevertheless reasonably moisture-retentive. If the soil is too rich in nitrogen, it will encourage production of foliage at the expense of flowers. Tolerates both acidity and alkalinity.

Garden uses: Most common use is as a dot plant in bedding schemes, but should be planted in any sunny beds or borders for evening scent, or in a wildlife garden to attract butterflies and bees. In the country of its origin and in warmer places elsewhere, such as parts of southern Ireland, it can be grown as an attractive hedge to 1.5m (5ft) tall. Also good as a cut flower.

Spacing: Depends on cultivar – allow sufficient space for spread of the plant to equal its anticipated height.

Propagation: Sow from late winter to early spring, at 15–20°C (59–68°F), covering the seed lightly. Germination should occur in 2–4 weeks. When the seedlings are large enough to handle, prick the seedlings out into individual pots, gradually harden off and plant out after the last expected frosts.

Cultivars are best propagated from green or semi-ripe cuttings, 5–7cm (2–2³⁄₄in) long, taken in summer (with a heel where possible), and inserted in a propagator with bottom heat of 16–21°C (61–70°F). Pot up and grow on the young plants under glass before planting out in late spring.

Availability: The true species is rarely available, but a number of named hybrids are easily obtainable, either as seed from many seed merchants, or as plants from nurseries specializing in tender and half-hardy plants.

Problems: Tender. Does not transplant well, so if you wish to keep stock plants for vegetative propagation, it is better to grow them permanently in pots.

RELATED FORMS AND CULTIVARS

The following are considered to be hybrids derived from *H. arborescens* and *H. corymbosum* (though often these two are thought to be synonymous), and are usually listed as cultivars of *H. arborescens*.

'Chatsworth' ♔ – vigorous, upright growth, strongly scented, with bright, deep-purple flowers. The RHS Award of Garden Merit plant.

'Dame Alicia de Hales' – pink flowers

'Florence Nightingale' – flowers pale mauve.

'Gatton Park' – 45cm (18in), compact plant, purple-mauve flowers.

'Grandiflorum' – flowers larger than usual.

'Iowa' – compact, with light violet-blue, fragrant flowers; dark green leaves; habit bushy, dense and upright.

'Lemoine' – to 60cm (2ft), flowers deep purple, fragrant.

'**Lord Roberts**' – habit small, flowers soft violet, scented.

'**Marine**' – flowers fragrant, deep violet-blue in clusters up to 15cm (6in) wide; compact, bushy habit, to 45cm (18in).

'**President Garfield**' – bushy habit, very floriferous, small sweetly scented flowers; old variety.

'**Princess Marina**' ♡ – compact, with deep violet-blue, highly scented flowers. The RHS Award of Garden Merit plant.

'**Regal Dwarf**' – very compact, with large heads of dark blue, fragrant flowers.

'**Spectabilis**' see 'W.H. Lowther'

'**The Speaker**' – lavender flowers, scented, similar to 'W.H. Lowther'.

'**W.H. Lowther**' (syn. 'Spectabilis') – flowers pale violet, scented.

'**White Lady**' – dwarf, to 30cm (12in), pure white flowers, tinged pink in bud. Victorian variety, richly perfumed.

'**White Queen**' – white flowers.

Heliotropium convolvulaceum (Nutt.) A.Gray

An unusual, showy annual to 40cm (16in) tall, with oval or lanceolate, hairy leaves, and funnel-shaped fragrant white flowers about 20mm (³/₄in) across, that resemble a small convolvulus (bindweed) and open at night. For a sunny, well-drained spot.

> '... opens its pure white flowers only in the cool of the evening when its delicious perfume with spicy undertones blends to perfection with the scent of Hesperis matronalis and Oenothera.' (Roy Genders)

NAMING
Etymology: *convolvulaceum*, resembling *Convolvulus* (bindweed), referring to the shape of the flower.
Synonym: *Euploca convolvulacea* Nutt.
Common names: *English:* Wide-flower heliotrope, Bindweed heliotrope, Sweet-scented heliotrope.

Description
General: A more or less upright, branching, hairy annual.
Stems: 10–40cm (4–16in), branching, with ascending, elongating branches.
Leaves: To 4cm (1½in) long and 1.5cm (⅝in) wide, hairy, oval to lanceolate, pointed, on stalks to 8mm long.
Inflorescence: Flowers solitary in leaf axils.
Flowers: *Pedicels* to 5mm in fruit. *Calyx* to 8mm in fruit, lobes awl-shaped to narrowly lanceolate. *Corolla*

white with yellow throat, tube to 1mm, hairy outside, limb 15–25mm across, funnel-shaped. *Stamens* with minute filaments, and anthers to 2.5mm. *Style* to 4mm.
Nutlets: To 4mm, hairy, laterally compressed.

GEOGRAPHY AND ECOLOGY
Distribution: Washington to California, south to northern Mexico.
Habitat: Dunes, other sandy places in deserts and arid grasslands.
Flowering: March to October in the wild.

CULTIVATION
Hardiness: Probably Zone 8.
Sun/shade aspect: Full sun is probably required.
Soil requirements: Well-drained soil.
Garden uses: Would make a nice addition to a wildflower border, or in a special evening garden, where it will attract night-flying moths.
Spacing: 30cm (12in).
Propagation: From seed sown in late winter, as for *H. arborescens*. In warm areas, can probably be sown *in situ* outdoors, in spring.
Availability: Generally unavailable, but worth seeking out.

Heliotropium europaeum Linnaeus

A variable, weedy annual, 5–40cm (2–16in) tall, covered in dense, soft white hairs, with oval or elliptical leaves and small white flowers in curved cymes. Only of interest to a specialist wildflower collection.

NAMING
Etymology: *europaeum* – European.
Common names: *English:* Heliotrope, European heliotrope, European turnsole; *Arabic:* Karee, ramrm; *Finnish:* Rikkaheliotrooppi; *French:* Héliotrope d'Europe; *German:* Sonnenwende; *Italian:* Eliotropio Selvatico, Erba Porraia; *Spanish:* Verrucaria; *Turkish:* Akrepotu, Bambulotu.

DESCRIPTION
General: A variable, hairy annual.
Stems: 5–40cm (2–16in) high, upright to ascending, usually branched below, softly hairy.
Leaves: To 6.5cm (2⅝in) long and 3.5cm (1⅜in) wide, oval to elliptical, cuneate to almost rounded at the base, with a rounded or somewhat pointed tip, softly white-hairy throughout, with a stalk up to 4cm (1½in).
Inflorescence: To 8cm (3¼in) long, mostly terminal.
Flowers: *Calyx* lobes to 3mm, narrowly oblong or triangular, densely white-hairy. *Corolla* usually white,

unscented, 2–4.2mm, hairy outside. *Stamens* with anthers 0.7–1mm. *Style* short or absent.
Nutlets: 4, to 2.5mm, oval, hairless to densely hairy, finely wrinkled or warty.
Chromosomes: 2n=24, 32, 48

GEOGRAPHY AND ECOLOGY
Distribution: Europe to central Asia.
Habitat: Open, dry places.
Flowering: Spring–summer.

CULTIVATION
Hardiness: Zone 6.
Sun/shade aspect: Full sun.
Soil requirements: Any well-drained soil; tolerant of drought.
Garden uses: A weedy little plant, this is likely to be of interest only to a specialist collector of native plants.
Spacing: 20cm (8in).
Propagation: No information in literature, but the seeds should germinate without any special treatment if sown in spring.

The genus *Lindelofia* Lehmann, 1850

NAMING
Etymology: *Lindelofia* – named after Friedrich von Lindelof of Dormstadt, nineteenth-century botanist.
Synonyms: *Adelocaryum* Brand; some species have previously been assigned to *Cynoglossum* L., *Solenanthus* Ledeb. and *Omphalodes* Miller.

GENUS CHARACTERISTICS AND BIOLOGY
Number of species: 11 or 12.
Description: A genus of perennial herbs with stout rootstocks, and upright or ascending, usually softly hairy stems. Leaves are more or less long-stalked, the basal ones (and sometimes stems leaves) tapered gradually at the base. Flowers are borne in a terminal panicle of cymes, without bracts. The calyx is divided almost to the base, its lobes are lanceolate to linear. The corolla is usually blue or purple, bell-shaped, funnel-shaped or cylindrical, with the tube being longer than the calyx and at least as long as the limb. Scales in corolla throat are oblong, and may have appendages near the apex. Stamens protruding from the corolla tube, with anthers extending at least partly above the base of the scales. The style is also protruding. Nutlets are attached to

Lindelofia stylosa in Tien Shan Mountains, Kyrgyzstan. (Erich Pasche)

the receptacle by apical scars, and their sides and dorsal surfaces are covered with barbed spines or bristles (glochids).
Distribution and habitats: Southern central Asia (Tien Shan, Pamir Alai), Himalaya, Pakistan, Afghanistan, China; in the mountains.
Life form: Hemicryptophytes.
Pollination: Bees.
Fruit dispersal: The barbed nutlets are probably designed to attach to animal fur.

USES
No known medicinal or culinary uses.

GARDEN
Several species are highly attractive perennials for a herbaceous or mixed border, with some being suitable for a large rock garden.

CULTIVATION
Most species will need a position in full sun when grown in a temperate climate, though in warm areas they should fare well in light shade. Any well-drained soil that is at the same time moisture-retentive and reasonably fertile is suitable. Lindelofias require some shelter because they cannot withstand the prolonged effect of drying winds. Apply a layer of mulch in autumn, such as bark, cocoa shell, or gravel, which will help to protect the crowns from frost.

PROPAGATION

Seed Seed can be sown either as soon as ripe (usually in late summer), or in early spring. 2 to 4 weeks of warmth followed by 4 to 6 weeks of cold treatment (4°C/39°F) is said to improve germination. The young plants should flower in their second year.
Vegetative Divide in autumn or spring, replanting the portions in their permanent positions, or potting them up. Alternatively, take root cuttings in winter.

REFERENCES

BRICKELL, 1996; HUXLEY, 1992; POLUNIN & STAINTON, 1984; PHILLIPS & RIX, 1994; THOMAS, 1993; VVEDENSKY, 1962; WALTERS, 2000; WU & RAVEN, 1995.

Lindelofia anchusoides (Lindley) Lehmann

A more or less upright, hairy perennial, about 50cm (20in) tall, with narrowly lanceolate, greyish leaves and bright blue, funnel-shaped flowers. A good herbaceous plant for sunny borders.

NAMING

Etymology: *anchusoides* – Anchusa-like.
Synonym: *Adelocarym anchusoides* (Lindley) Brand.

DESCRIPTION

General: A hairy perennial arising from a stout rootstock.
Stems: To about 50cm (20in), or sometimes to 90cm (3ft), softly hairy, upright or sometimes decumbent, simple or branched above.
Leaves: Basal leaves 14–40cm (5^1/$_2$–16in) long (including the stalk) and 1.7–7cm (^{5}/$_8$–2^3/$_4$in) wide, lanceolate, pointed, with adpressed bristly hairs above and sometimes beneath, gradually narrowing into a stalk. Stem leaves to 13cm (5in) long and 2cm (^{3}/$_4$in) wide, lanceolate to narrowly lanceolate, stalkless or short-stalked, but not clasping, unlike in *L. longiflora*.
Inflorescence: Terminal and axillary, few-flowered.
Flowers: *Pedicels* to 20mm (^{3}/$_4$in) in fruit, densely hairy. *Calyx* lobes oblong, blunt, densely woolly-hairy, to 6mm (^{1}/$_4$in) long and 2mm (^{1}/$_{16}$in) wide in fruit. *Corolla* blue, sometimes pink or purple, bell-shaped, about 12mm (^{1}/$_2$in)long and to 10mm (^{7}/$_{16}$in)across, with nearly rounded lobes; scales in throat are wedge-shaped, with two short appendages just below the tip. *Stamens* included in the corolla, with arrow-shaped anthers just reaching beyond the bases of scales. *Style* protruding, to 9mm (^{3}/$_8$in)long, threadlike, with a headlike stigma.
Nutlets: To 5 × 4mm, oval to rounded or triangular, covered with dense, hooked bristles to 2mm, finely hairy.

GEOGRAPHY AND ECOLOGY

Distribution: Afganistan, Iran, central Asia and the Himalayas (Himachal Pradesh).
Habitat: In the Himalayas, on stony slopes, at 2100–3600m (6890–11,800ft).
Flowering: June to August.

CULTIVATION

Hardiness: Zone 7.
Sun/shade aspect: Full sun.
Soil requirements: Any well-drained soil.
Garden uses: Well suited to a sunny herbaceous or mixed border. Though originating from mountainous regions, this plant grows far too big for any but the largest of rock gardens.
Spacing: 45cm (18in).
Propagation: As for the genus.
Availability: Plants are sometimes sold by specialist herbaceous nurseries.

Lindelofia longiflora (Bentham) Baillon

A very attractive perennial about 45cm (18in) high, with long, narrow, hairy leaves, and clusters of gentian- or purple-blue flowers to 15mm (^{5}/$_8$in) across, in summer. An excellent border plant for a sunny situation.

'A plant that one cannot pass by without admiring its flower colour.' (Graham S. Thomas)

NAMING

Etymology: *longiflora* – with long flowers.
Synonyms: *Lindelofia spectabilis* Lehmann; *Omphalodes longiflora* (Bentham) De Candolle; *Cynoglossum longiflorum* Bentham.

DESCRIPTION

General: A more or less upright perennial.
Stems: Solitary, or several, 15–60cm (6–24in) high, unbranched, covered with bristly white hairs.
Leaves: Basal leaves lanceolate, pointed to 30cm (12in) long (including the stalks) and 1–3.5cm (^{1}/$_2$–1^3/$_8$in) wide, bristly-hairy, gradually narrowing into a stalk. Stem leaves stalkless, the upper clasping the stem, 4–8cm (1^1/$_2$–3^1/$_4$in) long, oval to oblong-lanceolate, rounded or heart-shaped at the base.
Inflorescence: Usually terminal, to 15cm (6in) long.
Flowers: *Pedicels* to 12mm (^{1}/$_2$in) in fruit, hairy. *Calyx* lobes oblong-elliptical, about 5mm(^{1}/$_4$in), to 10mm (^{7}/$_{16}$in) in fruit, hairy, especially on the margins. *Corolla* deep blue to purple, cylindrical to funnel-shaped, tube to 13mm (^{1}/$_2$in), lobes to 6mm, (^{1}/$_4$in) oval, spreading; throat scales are 'keyhole'-

shaped, notched at the tip, without appendages, unlike in *L. anchusoides*. *Stamens* with filaments to 2mm and anthers to 2mm, the latter protruding above the bases of the scales. *Style* to 15mm, threadlike with a headlike stigma.

Nutlets: To 4 x 2.5mm, oval, with hooked bristles along the margin.

GEOGRAPHY AND ECOLOGY

Distribution: Pakistan, north-west India, Kashmir, western Nepal (western Himalayas).
Habitat: In the Himalayas, found on open slopes at 300–3600m (1000–11,800ft).
Flowering: June to August in the wild.

CULTIVATION

Date of introduction: 1839.
Hardiness: Zone 6.
Sun/shade aspect: Sun.
Soil requirements: Any well-drained soil.
Garden uses: Suitable for herbaceous and mixed border, and also a large rock garden.
Spacing: 45cm (18in).
Propagation: As for the genus.
Availability: Plants are often sold by specialist herbaceous nurseries, and the seed can be found on seed-exchange lists.

Related forms and cultivars

A naturally occurring variant, **L. longiflora var. falconeri** (C.B. Clarke) Brand, is said to be distinguished by its narrowly lanceolate leaves and a longer corolla tube than in the typical *L. longiflora var. longiflora*; and **var. levingii** (C.B. Clarke) Brand has oval or elliptical leaves, and the flowers of a particularly bright blue colour. It is not clear which of these variants are in cultivation.

Lindelofia stylosa (Karelin & Kirilov.) Brand

An unusual perennial, of variable height, usually up to 60cm (2ft), with hairy, oblong or lanceolate leaves, and nodding clusters of tubular, dark purple or wine-coloured flowers with prominent, long styles. Not as showy as some other lindelofias, but an interesting plant worthy of inclusion in a mixed or herbaceous border.

NAMING

Etymology: *stylosa* – with conspicuous styles.
Synonym: *Lindelofia angustifolia* Brand.
Common name: *Chinese:* chang zhu liu licao.

DESCRIPTION

General: A more or less upright perennial with stout roots to 2cm ($^3/_4$in) in diameter.
Stems: Variable in height, 20–100cm (8–40in), hairy, usually branched above.
Leaves: Basal leaves elliptical – or narrowly oblong, or lanceolate, 8–35cm ($3^1/_4$–14in) long, shortly hairy, gradually narrowing into a stalk. Stem leaves narrowly lanceolate or nearly linear, lower ones stalked, upper sessile or nearly so.
Inflorescence: 3–7cm ($1^1/_4$in–$2^3/_4$in) long, to 20cm (8in) in fruit, densely hairy.
Flowers: *Calyx* lobes 5–8mm, slightly unequal, awl-shaped, densely hairy. *Corolla* dark purple or purplish-red, 8–11mm long, with a straight tube and narrowly obovate lobes 3–4mm, erect or spreading. *Stamens* with threadlike filaments about 2mm, and narrowly oblong anthers about 3mm. *Style* 12–15mm, protruding from the corolla, usually slightly curved, with a headlike stigma.
Nutlets: About 6mm, forming a pyramid, with dense hooked bristles particularly on the margins.

GEOGRAPHY AND ECOLOGY

Distribution: China, Mongolia, Central Asia, Pakistan, Afghanistan, northern India and Kashmir.
Habitat: A wide variety of habitats, including meadows, forests, steppes, stony slopes at 1200–4700m (3900–15,400ft).
Flowering: June to August in the wild in the Himalayas.
Associated plants: In the Tien Shan mountains in Kyrgyzstan, grows with plants such as *Arnebia euchroma*, *Codonopsis clematidea*, *Potentilla sersoviana* and *Angelica brevicaulis*.

CULTIVATION

Date of introduction: 1999.
Hardiness: Zone 7.
Sun/shade aspect: Sun or very light shade.
Soil requirements: Any well-drained soil; tolerant of drought.
Garden uses: An unusual plant for a mixed or herbaceous border.
Spacing: 30cm (12in).
Propagation: As for the genus.
Availability: Not generally available, but the seed may appear on societies' exchange lists.

The genus *Lithodora*
Grisebach, 1844
Shrubby Gromwell

NAMING
Etymology: *Lithodora* – from *lithos*, meaning 'stone', referring to hard nutlets.
Synonyms: Previously included within the genus *Lithospermum* L.

GENUS CHARACTERISTICS AND BIOLOGY
Number of species: 7.
Description: A genus of dwarf, evergreen or semi-evergreen shrubs. Leaves alternate, entire, linear to elliptical or obovate, often bristly-hairy, with curved-in or inrolled margins. Flowers borne in terminal, loose, leafy cymes of 1 to 10. Calyx 5-lobed more or less to the base, slightly enlarging in fruit, with narrow, unequal lobes. Corolla blue, purple or white, funnel-shaped or salver-shaped (with a long slim tube and abruptly flattened limb); throat without scales or invaginations in the throat, and lacks a basal ring of hairs; sometimes glandular or long-hairy outside. Stamens usually included in the corolla, with thread-like filaments and oblong anthers with a blunt or notched tip. Style simple or branched above, with 2 stigmas, usually included in the corolla. Nutlets usually solitary, sometimes in twos, ovoid, constricted above the base.
Distribution and habitats: Mainly in southern Europe, also Turkey and Algeria; in warm, well-drained habitats, often on alkaline soils, with the exception of *L. diffusa*, which has a strong preference for acid soil.
Life form: Chamaephytes.
Pollination: Bees.
Fruit dispersal: No specialized mechanism.

USES
No known medicinal or economic uses.

GARDEN
The majority are good rock garden plants, though not totally hardy in cold areas. Some species, such as *L. diffusa* and its cultivars, are suitable for groundcover, others, like good colour forms of *L. zahnii*, are worthy of alpine house cultivation.

CULTIVATION
Most species are hardy to Zone 7 or 8, but in colder areas protection will be necessary. All require full sun, and both the habit and flowering will be spoilt if shaded.

Lithodora diffusa 'Inverleith' (Doreen Townley).

Lithodora zahnii in cultivation. (Masha Bennett)

The majority of lithodoras are tolerant of lime, and some, like *L. oleifolia*, thrive in alkaline conditions, but *L. diffusa* and its cultivars need acid soil. Plants may become straggly with age and should be trimmed lightly after flowering to maintain a compact habit.

PROPAGATION
Seed Seed is tricky to collect as there is usually only one nutlet per flower. Sow in early or mid-spring, or in autumn, in a cold frame, in gritty compost, covering lightly.
Vegetative Take semi-ripe cuttings of side shoots, ideally with a heel, from mid- to late summer and root in gritty compost in a cold frame. Softwood cuttings taken in early summer require bottom heat in a propagator. Layering is easy, and plants often layer themselves naturally.

PROBLEMS
Lithodora species are not fully hardy, and have to be grown under cover in areas colder than Zone 7 or 8. They are not always easy to grow unless their demands are fully satisfied.

REFERENCES
BEAN, 1981; BECKETT, 1993–94; BRICKELL, 1996; BLAMEY & GREY-WILSON, 1993; GREY-WILSON, 1989; HUXLEY, 1992; INGWERSEN, 1991; POLUNIN, 1980; THOMAS, 1990; TUTIN *et al.*, 1972; WALTERS, 2000.

Lithodora diffusa (Lagaska) I.M. Johnston
A low-growing evergreen shrub, 20–30cm (8–12in) tall, with small dark green leaves and vivid gentian-blue flowers, quite beautiful when well grown but tending to get straggly with age. The most popular *Lithodora*, for sunny positions on acid soil.

NAMING
Etymology: *diffusa* – spreading.
Synonyms: *Lithodora prostrata* (Loiseleur) Grisebach; *Lithospermum diffusum* Lagaska.
Common names: *English*: Shrubby gromwell, Scrambling gromwell.

DESCRIPTION
General: A mat-forming, dwarf evergreen shrub.
Stems: To 60cm (2ft) long, bristly-hairy, spreading or sometimes scrambling.
Leaves: Hairy on both surfaces, linear to oblong or elliptical, 7–38mm long and 1–8mm wide, flat or inflexed at the margin.
Inflorescence: Terminal leafy cymes.
Flowers: *Calyx* 5–10mm long. *Corolla* up to 21mm, funnel-shaped or salver-shaped, hairy outside; throat with a wide ring of very dense, white, soft, long hairs.
Nutlets: Usually one or two, to 4 x 2mm, pale brown to greyish, smooth, oblong.
Chromosomes: 2n=24.

Lithodora hispidula in southern Turkey. (Erich Pasche)

GEOGRAPHY AND ECOLOGY

Distribution: South-west Europe (Spain, Portugal, France), north to north-west France.
Habitat: *Pinus* woods, scrub, hedges, and maritime sands, usually on acid soil.
Flowering: Late spring–summer.
Associated plants: In Picos de Europa grows with a wide variety of plants, including *Gentiana acaulis*.

CULTIVATION

Hardiness: Zone 7; the hardiest species.
Sun/shade aspect: Full sun is preferable in temperate climates, though it will withstand light shade, especially in the warmer areas.
Soil requirements: Very well-drained but moisture-retentive, acid soil – unlike other members of the genus, this species does not tolerate alkalinity.
Garden uses: In a rock garden, as a groundcover in warm, sunny spots, banks.
Spacing: 60cm (2ft) apart.
Propagation: As for the genus; cultivars should only be propagated vegetatively.
Availability: The species itself is rarely grown, but in recent years an increasing number of cultivars has become available from nurseries and garden centres.

RELATED FORMS AND CULTIVARS

'Alba' – white-flowered, with pale green leaves; often the flowers become bluish, this effect more pronounced with age.
'Cambridge Blue' – light blue flowers, low-growing.
'Compacta' – compact habit.
'Grace Ward' ♀ – low, trailing habit, to 15cm (6in) high; narrow dark green leaves, azure-blue flowers. Similar to the most popular 'Heavenly Blue', the flowers more luminous in colour. The RHS Award of Garden Merit plant.
'Heavenly Blue' ♀ – deep azure-blue flowers, rather low-growing; very similar to 'Grace Ward', but the colour of flowers is somewhat matt in comparison. The RHS Award of Garden Merit plant.
'Inverleith' – a low-growing form.
'Picos' – a form introduced from the Picos de Europa by Ron McBeath; it is very low-growing and dense, creating neat carpets, but less floriferous than other forms.
'Star' – white flowers with a contrasting, star-shaped blue centre – a reverse effect of that in the better-known *Omphalodes cappadocica* 'Starry Eyes', and more sharply defined. Protected by the European Plant Breeder's Rights. A sport of *Lithodora diffusa* 'Heavenly Blue'.

Lithodora hispidula (Sibthorp & Smith) Grisebach

A dwarf evergreen shrublet to 35cm (14in) high, with branched stems bearing small leathery leaves and blue or violet-blue flowers. Rare in cultivation, but worth trying in a warm spot in a rock garden.

NAMING

Etymology: *hispidula* – with bristly hairs.
Synonym: *Lithospermum hispidulum* Sibthorp & Smith.
Common names: *German*: Borstiger steinsame; *Turkish*: Yanar döner çalisi.

DESCRIPTION

General: A dwarf evergreen shrub to about 35cm (14in) high.
Stems: Much-branched, with short, stiff branches covered in grey-white, fine bristly hairs.
Leaves: Up to 15mm ($^5/_8$in)long and 4.5mm ($^3/_{16}$in)wide, obovate to oblanceolate or oblong, dark green, leathery, more or less flat, bristly-hairy on both sides.
Inflorescence: Cymes with 1 to 4 flowers.
Flowers: *Calyx* about 7mm, lobes with bristly hairs. *Corolla* blue or blue-violet, sometimes reddish or

whitish, hairless, with tube about 12mm long and limb 10mm across, with short rounded lobes.
Nutlets: Ovoid, 3-angled, white, with minute warts.

GEOGRAPHY AND ECOLOGY
Distribution: Eastern Mediterranean (Crete, Karpathos, Turkey, Cyprus, Syria).
Habitat: Rock crevices, dry banks.
Flowering: February to June in the wild, summer in cultivation in temperate climates.

CULTIVATION
Hardiness: Zone 8.
Sun/shade aspect: Needs full sun.
Soil requirements: Well-drained, gritty soil.
Garden uses: Rock garden or dry bank.
Spacing: 30–40cm (12–16in) apart.
Propagation: As for the genus.
Availability: Available from few specialist alpine nurseries.

RELATED SPECIES
Lithodora fruticosa (Linnaeus) Grisebach (syn. *Lithospermum fruticosum* L.) is a 30cm (12in) tall shrublet from southern Europe, with narrow leaves and blue flowers, hardy to Zone 8.

Lithodora oleifolia (Lapeyr.) Grisebach ♔
A dwarf evergreen shrub, with a semi-upright, suckering habit, usually about 20cm (8in) high, with smallish, obovate, dull green leaves and small clusters of blue flowers opening from pink buds. The RHS Award of Garden Merit plant. A nice small shrub for a sunny, well-drained position.

> *'A wanderer, emitting from underground stems tufts of rounded, greyish hairy leaves and cymes of flowers, often pink in bud but opening light blue.' (Will Ingwersen)*

NAMING
Etymology: *oleifolia* – with leaves like in *Olea* (olive).
Synonym: *Lithospermum oleifolium* Laperyr.

DESCRIPTION
General: A dwarf, suckering evergreen shrub.
Stems: 10–45cm (4–18in), slender, ascending, loosely branched, often leafless at the base.
Leaves: To 4cm (1½in) long and 1.5cm (⅝in) wide, obovate to oblong or spoon-shaped, blunt, short-stalked, dull green, with sparse stiff hairs above, and silky-hairy underneath. Leaves mostly crowded at tips of non-flowering branches, or at the base of flowering shoots.

Inflorescence: Loose terminal cymes of 3 to 7 flowers.
Flowers: *Calyx* 5–8mm, with soft white hairs, lobes linear, to 1mm wide. *Corolla* pale pink in bud becoming sky blue, softly hairy outside and hairless inside; tube 8–12mm long, limb 6–9mm across, with rounded lobes.
Nutlets: About 3 × 2mm, broadly ovoid, grey-white, smooth, with a short beak.

GEOGRAPHY AND ECOLOGY
Distribution: Eastern Pyrenees (Spain).
Habitat: Rocky places, to 1100m (3600ft).
Flowering: May to July in the wild.

CULTIVATION
Hardiness: Zone 7.
Sun/shade aspect: Needs full sun.
Soil requirements: Very well-drained soil, ideally from neutral to alkaline.
Garden uses: Rock garden.
Spacing: 30–45cm (12–18in) apart.
Propagation: As for the genus.
Availability: Available from a number of specialist alpine nurseries, and the seed may sometimes appear in specialist societies' exchange schemes.

Lithodora rosmarinifolia (Tenore) I.M. Johnston.
A half-hardy, dome- or cushion-shaped evergreen shrublet, 30–60cm (1–2ft) high, with dark green leaves resembling those of rosemary, and blue to white flowers in summer. For a very warm and sunny position outdoors, or the alpine house.

NAMING
Etymology: *rosmarinifolia* – rosemary-leaved.
Synonym: *Lithospermum rosmarinifolium* Tenore.
Common name: *English:* Rosemary-leaved gromwell.

DESCRIPTION
General: A cushion-forming, dwarf, evergreen shrub, grey-hairy when young.
Stems: Upright to spreading, densely branching, 30–60cm (1–2ft) high, with grey hairs above.
Leaves: 2.5–6cm (1–2½in) long and to 1cm (⁷⁄₁₆in) wide, lanceolate to linear, pointed, rigid, dark green, densely grey-bristly beneath, margins usually inrolled.
Inflorescence: Loose terminal cymes.
Flowers: *Calyx* about 6mm (¼in) long, with white bristly hairs. *Corolla* blue, lilac or white, tube about 12mm (½in) long, the limb 15–17mm (⅝in) across with oblong, rounded lobes; throat hairless and slightly glandular, the tube hairy on the outside.
Nutlets: Whitish, smooth.

GEOGRAPHY AND ECOLOGY

Distribution: Southern Italy, Sicily, north-eastern Algeria.

Habitat: Rock crevices.

Flowering: January to May in the wild, in cultivation in temperate areas mostly in summer.

Associated plants: In the wild, it may be found growing with *Convolvulus cneorum*, *Santolina neapolitana*, *Seseli polyphyllum*, *Campanula fragilis*, *Centaurea cineraria*, *Asperula crassifolia*, *Euphorbia dendroides*.

CULTIVATION

Hardiness: Zone 8; grow under cover where temperatures fall below −5°C (23°F).

Sun/shade aspect: Needs as much sun and warmth as possible.

Soil requirements: Gritty, perfectly drained soil or compost, which is at the same time moisture-retentive.

Garden uses: Rock garden in a perfectly drained, sunny crevice, or the alpine house.

Spacing: 45cm (18in).

Propagation: As for the genus.

Availability: Sold by a few specialist alpine nurseries, but generally uncommon.

Lithodora zahnii. (Masha Bennett)

Lithodora zahnii (Heldr. in Halácsy) Halácsy I.M. Johnston

A dwarf evergreen shrublet, to 40–60cm (16–24in), with narrow leathery leaves, and salver-shaped flowers that vary in colour from whitish to pale blue. A very attractive plant for a sheltered spot in a rock garden, or for pots under glass.

> *'At its best it is a wonderful shrub, but the flower colour varies and quite often they are as much white as blue and rather wishy-washy.' (Jack Elliott)*

NAMING

Etymology: *zahnii* – after Zahn, a fern expert who collected plants in South America.

Synonym: *Lithospermum zahnii* Heldr. in Halácsy.

Common name: *English*: Greek gromwell.

DESCRIPTION

General: A dense, dwarf evergreen shrub.

Stems: 40–60cm (16–24in) high, much branched, with erect or ascending branches, silvery-hairy when young, becoming leafless and woody as they age.

Leaves: 2–4cm ($^3/_4$–$1^1/_2$in) long and 2–4mm wide, leathery, linear to linear-oblong, blunt, greenish or greyish, stiffly hairy above, and softly grey-hairy underneath, strongly rolled under margins.

Inflorescence: Flowers are solitary or in cymes of 2–3, aggregated into larger heads.

Flowers: *Calyx* 7–11mm ($^3/_8$–$^7/_{16}$in) long, with linear lobes, softly white-hairy. *Corolla* white or pale blue, salver-shaped, tube to 10mm ($^7/_{16}$in) long, limb 13–15mm ($^1/_2$–$^5/_8$in) across, abruptly expanding with oval, spreading lobes. *Stamens* with anthers to 2.5mm.

Nutlets: Smooth.

GEOGRAPHY AND ECOLOGY

Distribution: Endemic to southern Greece (south-east of Kalamai). Included as Rare on *The IUCN Red List of Threatened Plants*.

Habitat: Cliffs and rocks; endangered in its habitat.

Flowering: May to June in the wild; in cultivation may flower intermittently until mid-autumn.

CULTIVATION

Hardiness: Zone 7 or 8. Should be able to tolerate temperatures falling down to between −5°C (23°F) and −10°C (14°F), for brief periods of time.

Sun/shade aspect: Full sun.

Soil requirements: Well-drained gritty soil.

Garden uses: Rock garden; small specimens of good colour forms also make admirable pot plants for the alpine house.

Spacing: 45cm (18in).

Propagation: As for the genus.

Availability: Sold by a number of alpine and rock plant nurseries.

The genus *Mertensia* Roth, 1797
Bluebells, Languid-ladies

NAMING
Etymology: *Mertensia* – after Franz Karl Mertens (1764–1831), German botanist, professor at Bremen.
Synonyms: *Steenhammera* Rchb.; *Pneumaria* Hill.; *Mertensianthe* M. Popov; *Casselia* Dumort. A number of species are sometimes separated into the genus *Pseudomertensia* Riedl. Some species have previously been included under *Eritrichium* Schrad., *Pulmonaria* L. and *Lithospermum* L.

GENUS CHARACTERISTICS AND BIOLOGY
Number of species: From 40 to 50, according to various estimates.
Description: A genus of perennial herbs, hairy or hairless. Stems from one to many from each root, decumbent or upright. Leaves entire, linear to heart-shaped, alternate. Flowers borne in loose or dense modified scorpioid cyme, without bracts. Calyx 5–parted, sometimes bell-shaped, often enlarged in fruit. Corolla in various shades of blue or purple, often from pink buds, occasionally white, tubular, funnel- or bell-shaped, with the tube at least as long as the calyx, throat often with scales, limb with 5 lobes, spreading or nearly erect. Nutlets are flattened, strongly 3–angled, fleshy, smooth.
Distribution and habitats: Europe, Asia, North America; one species (*M. pulmonarioides*) reaching Central America; in a wide variety of habitats, from high mountains (*M. alpina*) and streamsides (*M. sibirica*), to seashores (*M. maritima*) and woodlands (*M. pulmonarioides*).
Life forms: Geophytes and hemicryptophytes.
Pollination: Visited mostly by bees, such as bumblebees – *Bombus* species, and solitary bees *Andrena*, *Halictus*, *Osmia* and *Prosopis*. Long-tongued insects, like bumblebees, obtain the nectar legitimately, but may also 'rob' the flowers by piercing holes in the corolla, reaching the nectar without pollinating the flower. Some *Mertensia* species, such as *M. pulmonarioides*, are also attractive to hummingbirds. A few species, like *M. maritima*, are self-pollinating.
Fruit dispersal: Most species do not have any specialized mechanism, though the nutlets of species growing on streambanks (such as *M. ciliata*) and on floodplains (like *M. pulmonarioides*) are probably carried by water. Those of maritime species, such as *M. maritima* and *M. simplicissima*, are dispersed by sea currents.

Mertensia ciliata habitat: Fall River, Colorado.
(Masha Bennett)

Animal feeders: A wide variety of mammals is known to consume *M. ciliata*: deer, bear and domestic sheep feed on the shoots; pikas harvest, dry and store it for winter, and elk (wapiti) not only feed on the plants but seek out large stands for use as bedding and birthing places (Dorward & Swanson, 1993).

USES
GARDEN
Due to the wide range of habits and requirements, different mertensias can be grown in a wide variety of garden habitats: mixed herbaceous borders (*M. pulmonarioides*, *M. paniculata*); stream-side (*M. ciliata*, *M. franciscana*, *M. pulmonarioides*); woodland garden (*M. sibirica*); rock garden or alpine house (*M. alpina*, *M. viridis*, *M. primuloides*, *M. moltkioides*, *M. simplicissima*, *M maritima*); sand or gravel bed (*M. maritima*, *M. simplicissima*).

The leaves of *Mertensia maritima* can be eaten raw or cooked, and are said to taste of oysters, hence the plant's common name – oysterplant. The root of this species is eaten by the Inuit of Alaska, and the flowers are also edible.

CULTIVATION
Most mertensias are perfectly hardy and can be grown in Zone 4 or even 3, though some, for instance the Himalayan species, are hardy only to Zone 6. The majority of species are happy in a degree of shade, apart from *M. maritima*, *M. simplicissima* and *M. alpina*, which require full sun in order to thrive. Soil requirements differ between the species. Those native to wooded habitats, like *M. paniculata*, prefer moist but well-drained, rich humusy soil, ideally with plenty of leaf mould mixed in. Those primarily found on stream banks, such as *M. ciliata*, are happiest in permanently moist but not stagnant sites. Some species, for instance *M. alpina*, require perfectly drained, gritty soil or compost mix, and like dryish conditions after flowering, when they enter a dormant period. Mertensias of seashores, like *M. maritima* and *M. simplicissima*, will grow in almost pure grit or gravel, but will need sufficient depth for their long roots to penetrate.

PROPAGATION
Seed Seeds should be sown as soon as ripe into pots or trays of good-quality compost, and germinated in a cold frame. Seed of many species loses viability in storage, and so ideally should be sown fresh.

About four weeks warm followed by eight weeks of cold stratification may be helpful, especially with older seeds. At least some species show hard coat dormancy, and scarifying or carefully removing the seed coat should enhance germination.

In experiments with *M. ciliata* seeds, testing treatments aiming to overcome dormancy, Pelton (1961) found the most effective method to be the complete removal of the thick nutlet coat, combined with breaking the inner coat (90 per cent germination). Soaking in concentrated sulphuric acid achieved 77 per cent germination, and shaving one corner of the seedcoat with a razor, 52 per cent germination. Cold moist stratification at 5°C (41°F) for up to 346 days resulted in only around 7 per cent germination. It is possible that, in the wild, seeds germinate only after spending two years in the soil for the hard-coat dormancy to be sufficiently overcome by means of fungal and bacterial activity.

The germinated seedlings should be protected from direct sunlight. Once large enough to handle, pot up the seedlings individually, and grow on in a cold frame.

Mertensia alpina. (Masha Bennett)

Vegetative Division should be carried out during early spring or autumn, replanting the portions immediately in their flowering positions, or potting the sections up. Division is not easy and needs to be done carefully, as plants generally resent being disturbed. Species with fleshy roots can be propagated by taking root cuttings in winter.

PROBLEMS
Mertensias may suffer from a variety of ailments in cultivation, including powdery mildew, leaf spot caused by fungus *Septoria poseyi*, and rust. Slugs are a menace to many species, especially those with hairless leaves. Many species are difficult to keep, and die out soon after planting.

REFERENCES
BECKETT, 1993–94; BRICKELL, 1996; CLEMENTS & LONG, 1923; HITCHCOCK *et al.*, 1959; HUXLEY, 1992; OHWI, 1965; PELTON, 1961; POLUNIN & STAINTON, 1994; RYDBERG, 1954; SHISHKIN & BOBROV, 1969–77; WEBBER & WITTMANN, 1996; WILLIAMS, 1937.

Mertensia alpina (Torr.) G. Don

A delightful little plant, usually about 15cm (6in) tall, with dark green oblong or lanceolate leaves, and clusters of fragrant, widely funnel-shaped, deep-blue flowers in summer. It is not an easy plant to grow, but well worth trying in a rock garden, scree or alpine house.

NAMING
Etymology: *alpina* – alpine.
Synonyms: *Mertensia tweedyi* Rydb.; *M. obtusiloba* Rydb.; *M. brevistyla obtusiloba* A. Nels.; *Pulmonaria alpina* Torr.; *Cerinthodes alpinum* Kuntze.
Common names: *English*: Alpine mertensia, Alpine chiming bells, Alpine bluebells.

DESCRIPTION
General: A tufted perennial arising from a branching caudex.
Stems: 2.5–30cm (1–12in) tall, hairless, loosely upright, ascending or nearly prostrate, one to numerous.
Leaves: Basal leaves 1–7cm ($^7/_{16}$–2$^3/_4$in) long and 5–30mm ($^1/_4$–1$^1/_4$in)wide, oblong to narrowly lanceolate, with a winged stalk. Stem leaves stalkless or nearly so, 1–6cm ($^7/_{16}$–2$^1/_2$in) long and 3–20mm ($^1/_8$–$^3/_4$in)wide, broadly to narrowly lanceolate or elliptical, gradually reducing in size upwards. All leaves bristly-hairy above, hairless beneath, dark green.
Inflorescence: Becoming slightly paniculate with age.
Flowers: *Pedicels* 1–10mm long, bristly-hairy or hairless. *Calyx* 2–5mm long, divided almost to the base, often slightly accrescent; lobes oblong or narrowly lanceolate, pointed or blunt, with hairy margins. *Corolla* dark blue, 7–11mm long; tube 3–6mm long and 2mm wide, hairless within; limb expanded, 2–6mm long and 5–11mm wide; scales conspicuous, often almost closing the throat. *Stamens* with anthers 0.9–1.3mm long and shorter filaments. *Style* 2–5mm long.
Nutlets: About 2mm long, wrinkled.

GEOGRAPHY AND ECOLOGY
Distribution: North America: Rocky Mountains (Montana, Wyoming, Idaho, Colorado, Utah, New Mexico).
Habitat: Open gravelly slopes and meadows at high altitudes in the mountains, alpine tundra, often above treeline.
Flowering: July to August in the wild.
Associated plants: On Pike's Peak, the author observed it growing with such plants as *Androsace chamaejasme carinata*, *Dryas octopetala*, *Eritrichium aretioides*, *Geum rossi turbinatum*, *Lloydia serotina*,

Mertensia alpina on Pike's Peak, Colorado. (Masha Bennett)

Mertensia ciliata in Colorado. (Masha Bennett)

Primula angustifolia, *Rhodiola integrifolia* and *Silene acaulis subacaulescens*, at 3300–3900m (10,800–12,800ft).

CULTIVATION
Hardiness: Zone 3.
Sun/shade aspect: Full sun.
Soil requirements: Gritty, well-drained soil that is also moisture-retentive. Keep dryish during the dormant period after flowering.
Garden uses: Rock garden, scree or alpine house.
Spacing: 20cm (8in).

Propagation: From seed, as for the genus. Cold stratification is probably necessary.
Availability: Occasionally offered by some alpine specialists, and the seed frequently crops up on the alpine and rock garden societies' exchange lists.
Problems: Not easy to grow, as it tends to disappear during the dormant period.

Mertensia ciliata (James) G. Don

An elegant, upright perennial, 30–60cm (1–2ft) tall, with oval, glaucous-green leaves, and clusters of tubular-bell-shaped, sky-blue flowers in late spring and early summer. A lovely plant for any moist situation, best by waterside or in dappled shade.

> *'Pink buds open to light blue tubular flowers over glaucous leaves; a charming picture.'*
> *(Graham S. Thomas)*

NAMING

Etymology: *ciliata* – ciliate, referring to hairy leaf margins and calyces of this plant.
Synonyms: *Mertensia polyphylla* Greene; *M. punctata* Greene; *M. pallida* Rydberg (a white-flowered form); *M. incongruens* Macbr. & Pays.; *Pulmonaria ciliata* Torr.
Common names: *English*: Tall chiming bells, Tall fringed bluebells, Mountain bluebells.

DESCRIPTION

General: A perennial arising from a branched, woody caudex.
Stems: 10–120cm (4–48in), upright or ascending, usually numerous, pale and somewhat bluish below, hairless.
Leaves: Hairless or often bristly-hairy, ciliate on the margins; somewhat glaucous; more or less prominently veined. Basal leaves (sometimes absent), 4–15cm (1½–6in) long and 3–10cm (1¼–4in) wide, long-stalked, oval, oblong or lanceolate with a somewhat heart-shaped base, blunt or somewhat pointed. Stem leaves lanceolate, narrowly elliptical or oval, 2–15cm (¾–6in) long and 1–5cm (⁷/₁₆–2in) wide, the uppermost stalkless.
Inflorescence: Terminal dense cymes in young plants; in larger plants, a branched, open panicle.
Flowers: *Pedicels* 1–10mm long, hairless, papillose or rarely with a few short bristly hairs. *Calyx* 1–3mm long, divided nearly or quite to the base, lobes with rounded or almost pointed tip, ciliate on margins, bristly inside. *Corolla* blue opening from pink buds, 10–17mm long; tube 6–8mm, interior hairless or with a ring of crisped hairs below the middle; limb 4–10mm long, moderately expanded; scales conspicuous, hairless or softly hairy. *Stamens* with broad filaments 1.5–3mm long, and narrower anthers 1.2–2.5mm long. *Style* about as long as corolla, or shortly protruding from it.
Nutlets: 3.5 × 2.3mm, wrinkled.
Chromosomes: 2n=12, 24.

GEOGRAPHY AND ECOLOGY

Distribution: USA (Wyoming, Montana, Colorado, Utah, Nevada, Idaho, Oregon, California, northern New Mexico), Mexico.
Habitat: Moist habitats in mountains – stream banks, lake shores, wet meadows, wet cliffs, spruce-fir forest, krummholtz zone, aspen forest; most frequently at 3000–3860m (9850–12,660ft).
Flowering: June to August in the wild.
Associated plants: The following plants were found by the author, growing with *M. ciliata* in one of the sites in Colorado Rockies: marsh-marigold (*Caltha polysepala*), avalanche lily (*Erythronium grandiflorum*), *Primula parryi*, *Ranunculus alsimaefolius* and globeflower (*Trollius albiflorus*).

CULTIVATION

Hardiness: Zone 4.
Sun/shade aspect: Dappled shade is best, and sun is acceptable as long as the soil is sufficiently moist.
Soil requirements: Moist soil, ideally slightly acid to neutral. Will tolerate drier soil, especially with some shade, but is likely to lose much of its appeal in dry conditions.
Garden uses: In the wild, this likes nothing better than standing with its feet in running water – so perfect for stream-sides and pool-sides (though stagnant moisture is not as good). Will succeed in drier border.
Spacing: 30cm (12in).
Propagation: Division, root cuttings, seed. This plant is quite easy to propagate by division in autumn or spring, and also by root cuttings in winter. The genus description advises on overcoming the hard-coat dormancy in this species.
Availability: This is one of the few more commonly available representatives in the genus, sold by a number of herbaceous specialists. The seed is frequently seen on the seed lists of specialist societies.
Problems: Does not achieve its full potential in drier soil of a border.

RELATED FORMS AND CULTIVARS

'Blue Drop', a selection by Alan Bloom, does not appear to be in cultivation at present. Plants with flowers remaining pink (rather than turning blue) at maturity, and white-flowered forms are also known for this species.

Mertensia echioides (Bentham) Bentham & Hooker

A mat-forming, softly hairy perennial 15–30cm (6–12in) high, with oblong or lanceolate leaves and many-flowered clusters of blue flowers around 12mm (½in) long, from late spring to midsummer, and sometimes again in early autumn. A nice plant for groundcover under trees and shrubs, rock garden or border.

NAMING

Etymology: *echioides* – possibly referring to another plant of borage family, *Arnebia pulchra* (syn. *Echioides longiflorum*).
Synonyms: *Pseudomertensia echioides* (Benth.) Ried.; *Lithospermum echioides* Benth.

DESCRIPTION

General: A softly hairy perennial with creeping rhizomes.
Stems: 15–30cm (6–12in), unbranched, upright or sometimes decumbent, softly hairy.
Leaves: Basal leaves to 9cm (3½in) long and 2.5cm (1in) wide, lanceolate or oval to elliptical-oblong, rather blunt or pointed, narrowing to a long stalk. Stem leaves nearly stalkless. All leaves hairy.
Inflorescence: Dense, many-flowered cyme, to 13cm (5in) long.
Flowers: *Pedicels* to 5mm. *Calyx* to 4mm long, with linear, hairy lobes. *Corolla* deep blue to purple-blue, to 4mm long, funnel-shaped or nearly cylindrical, with ascending, and not spreading lobes. Scales in corolla throat inconspicuous or absent. *Stamens* have linear filaments and oblong anthers to 2mm. *Style* to 15mm, protruding from the corolla, with a headlike stigma.
Nutlets: About 2–2.5mm, oval-triangular, usually white, shiny.

GEOGRAPHY AND ECOLOGY

Distribution: Pakistan, Iran, Kashmir, Tibet.
Habitat: Rocks, banks in mountains, 2700–3600m (8850–11,800ft).
Flowering: June to August in the Himalayas, in cultivation usually early to midsummer, sometimes producing a second flush of flowers in early autumn.

CULTIVATION

Hardiness: Zone 6.

Sun/shade aspect: Light shade is best.
Soil requirements: Well-drained but moisture-retentive, humus-rich soil.
Garden uses: Makes nice groundcover in dappled shade of trees and shrubs, or can be grown in a rock garden, peat bed, alpine house, or at the front of a herbaceous/mixed border.
Spacing: 30cm (12in).
Propagation: Can be divided easily, or grown from seed.
Availability: Sold by a few herbaceous and alpine specialists, and the seed appears regularly on exchange lists.

Mertensia lanceolata (Pursh) A. De Candolle

A pretty, variable perennial 20–45cm (8–18in) tall, with leaves that can be hairless and glaucous or bristly and greyish-green, to 14cm (5½in) long, and clusters of tubular or bell-shaped flowers of pure pale sky-blue, sometimes almost white or deeper blue. For a larger rock garden or front of the border, this is a charming plant that deserves to be more widely grown.

> *'My favorite is our native* M. lanceolata, *a dwarf dryland species found in the foothills of the Rockies. Seeing the gorgeous sky-blue flowers is one of the great signs of spring around here. Even better, it's become a weed in our rock garden.'* (Bob Nold)

NAMING

Etymology: *lanceolata* – lance-shaped, referring to leaves.
Synonyms: *Mertensia marginata* G. Don; *M. linearis* Greene; *M. papillosa* Greene; *M. lanceolata* var. *aptera*, Ckll. in Macbr.; *Pulmonaria lanceolata* Pursh.
Common names: *English:* Lanceleaf mertensia, Lanceleaf chiming bells, Foothill mertensia, Prairie bluebells.

DESCRIPTION

General: A very variable, hairy or hairless perennial.
Stems: 10–45cm (4–18in) high, upright or ascending, simple or branch, usually several from each rootstock.
Leaves: Hairless beneath, either hairless or shortly bristly/pimply above. Basal leaves stalked, 3–14cm (1¼–5½in) long and 1.2–4cm (½–1½in) wide, oval-lanceolate; with rather prominent parallel veins, stalk longer or shorter than the blade. Stem leaves linear to broadly lanceolate or oblong-elliptical, pointed or blunt, 1.5–10cm (⅝–4in) long, 2–30mm (1⁄16–1¼in) wide, stalkless often semi-clasping.

Inflorescence: Usually dense at first but becoming loosely paniculate in age.

Flowers: *Pedicel* hairless to bristly, 1–20mm long. *Calyx* 2–9mm long, the lobes 1–6mm long, lanceolate to triangular, pointed or blunt, hairy on margins and sometimes on the outside. *Corolla* tube 3–6.5mm long, interior usually with a dense ring of hairs near the base; limb 2.5–9mm long, usually bell-shaped; throat scales more or less conspicuous, hairless or softly hairy. *Stamens* with filaments 2–4mm long and anthers 1.5–2mm long. *Style* shorter or longer than the corolla.

Nutlets: 2.5–3mm long, mostly evenly wrinkled on the backs and laterally, black.

GEOGRAPHY AND ECOLOGY

Distribution: North Dakota, South Dakota, Nebraska, Saskatchewan, Montana, Wyoming, Colorado, Idaho, New Mexico.

Habitat: Very varied, including hillsides, prairies, forest, woodland edges, stream banks, grassy and rocky places.

Flowering: April to August in the wild.

Associated plants: I have observed *M. lanceolata* growing in association with a great variety of plants in different habitats in Colorado, some of which are listed below:

> Estes Park, 2250m (7380ft), grassy waste ground: *Echinocactus viridiflorus, Eriogonum umbellatum, Erodium cicutarium, Erysimum cheiranthoides* subsp. *altum, Euphorbia* sp., *Oenothera coronopifolia, Cryptantha virgata, Penstemon virgatus* subsp. *asa-grayi, Potentilla argentea, Rosa woodsii* and *Verbascum* sp.
> Moraine Park, 2400m (7870ft), forest edges and meadow: *Antennaria* sp., *Bistorta bistortoides, Erigeron* sp., *Geranium richardsonii, Cryptantha virgata, Penstemon virens, Potentilla floribunda, Purshia tridentata* and *Pulsatilla patens* subsp. *multifida*.
> Rocky Mountain National Park, 2400m (7870ft), beside a stream, in grass: *Astragalus drummondii, Dodecatheon pulchellum, Iris missouriensis, Lathyrus leucanthus, Sisyrinchium montanum, Thermopsis divaricarpa* and *Vicia americana*.

CULTIVATION

Hardiness: Zone 4.

Sun/shade aspect: Full sun to moderately deep shade.

Soil requirements: Prefers light, well-drained soil, but seems to adapt easily to a wide variety of conditions.

Garden uses: Rock garden, raised beds, front of the border, light shade under trees and shrubs.

Spacing: 20–30cm (8–12in).

Propagation: From seed, as for the genus.

Availability: Plants are rarely available from nurseries, but the seeds sometimes appear in societies' seed-exchange lists.

RELATED FORMS AND CULTIVARS

This is a very variable species, with several naturally occurring forms, such as **M. lanceolata var. secundorum** Ckll., with the leaves bristly on both sides. Plants with white and nearly-white flowers are also common in the wild.

Mertensia maritima (Linnaeus) S.F. Gray

A beautiful plant of northern seashores, forming a mat of prostrate stems to over 1m (3ft) in diameter, with bluish leaves and turquoise-blue flowers opening from pink buds. Not easy to please, it can be tried in a rock garden or sand bed in full sun.

> *'It forms mats of ovate rather fleshy blue-grey leaves, and in July flings procumbent stems bearing large, flat flowers of opalescent shades of turquoise opening from buds of soft pink and mauve. A well-flowered, healthy plant forms a beautiful picture.'* (Anna N. Griffith)

NAMING

Etymology: *maritima* – growing by the sea.

Synonyms: *Mertensia parviflora* G. Don; *Pulmonaria maritima* L.; *P. parviflora* Langsdorff; *Pneumaria maritima* Hill; *Lithospermum maritimum* Lehm.; *Casselia maritima* Dumort.; *C. parviflora* Dumort.; *Steenhammera maritima* Reichb.; *Hippoglossum maritimum* Hartm.; *Cerinthodes maritimum* O.Kuntze.

Common names: *English*: Oyster plant, Sea lungwort, Oysterleaf, Seaside smooth-gromwell, Northern shorewort, Sea bugloss, Light wort, Ice plant; *Danish*: Hestetunge; *Finnish*: Merihalikka; *French*: pulmonaire de Virginie, sanguine de mer; *German*: Seestrandlungenkraut; *Greenlandic*: Sioqqat naanii; *Icelandic*: Blálilja; *Norwegian*: Østersurt; *Russian*: mertenziya morskaya; *Swedish*: Ostronört, fjärva.

DESCRIPTION

General: A hairless, glaucous, fleshy perennial, branched and spreading, forming large, loose mats to over 1m (3ft) in diameter, with unusual cable-like roots that allow maximum anchorage of the plant in sand.

Stems: To 60cm (2ft) long, leafy, trailing, glabrous and glaucous with purplish tinge.

Leaves: 2–10cm (1–4in) long and about 5cm (2in) wide, oval to spoon-shaped, blunt or pointed, sparsely to densely papillose/pimply above, the lower stalked, the upper stalkless; leaves gradually reduced in size towards the end of the stems.

Inflorescence: Loose, often branched, with leaflike bracts.

Flowers: *Pedicels* slender, 2–10mm, elongating and often recurved in fruit. *Calyx* 2–6mm long, with oval lobes, divided to near the base; calyx much enlarged in fruit, lobes becoming broad and rounded. *Corolla* first pink, becoming turquoise blue, rarely white, tube 6–7mm long, with distinct notched scales in throat; limb about 6mm across, bell- or funnel-shaped, divided to about one-third into oval-triangular lobes. *Stamens* with filaments rather narrower and longer than the anthers. *Style* 2–5mm long, included in corolla.

Nutlets: 3–6mm, smooth, flattened, black when ripe, outer coat fleshy, becoming papery.

Chromosomes: 2n=24.

GEOGRAPHY AND ECOLOGY
Distribution: Coasts of most of northern Europe, excluding the Baltic (southwards to 54° in Britain and Ireland); Greenland; in North America, the coast of Alaska, south to British Columbia, east to Massachusetts; only Anadyr in Asia.

Habitat: Sands, shingle and gravel, by the sea; often just above high-water mark – growing closer to water than most other seaside plants.

Flowering: Mainly June to August in Scotland.

Associated plants: In Scotland, on shingle beaches grows with sea purslane (*Honckenia peploides*), sea campion (*Silene uniflora*), silverweed (*Potentilla anserina*) and common sorrel (*Rumex acetosa*).

CULTIVATION
Hardiness: Zone 3.

Sun/shade aspect: Full sun is essential.

Soil requirements: Needs gravelly or sandy soil in full sun. Try in a deep bed of almost pure coarse sand or grit, with some organic matter mixed in, such as leaf mould.

Garden uses: Grow in a wildflower garden, especially by the sea, in a shingle or sand bed, in a rock garden.

Spacing: Spreads over 90cm (3ft) if growing well, but may not survive until reaching this size.

Propagation: Seed germinates easily, if sown in pots of sandy compost when ripe, or in spring. Seed does not seem to lose viability as rapidly as some other mertensias. In the wild, seed is dispersed by sea currents and can survive for periods of time in sea water.

Availability: Sold by a number of nurseries, including some wildflower suppliers and herbaceous specialists.

Seed regularly listed in societies' seed-exchange schemes.

Problems: *Mertensia maritima* is not easy to grow, and often dies without an obvious reason. You could try growing the closely related and similar-looking *M. simplicissima* instead, which usually does better in cultivation.

RELATED FORMS AND CULTIVARS
There is a naturally occurring form, **M. maritima f. albiflora** Fern., with corolla starting whitish, rather than pink becoming blue, which occurs in Alaska, Quebec and Nova Scotia. It is probably less attractive than the usual form, and does not appear to be cultivated.

Mertensia paniculata (Aiton) G. Don
An elegant perennial, around 60cm (2ft) high, with greyish-green, lanceolate leaves, and nodding pale blue flowers opening from pink buds. An excellent plant for waterside or under shrubs and trees.

'Though the narrow bell flowers are small, their profusion – in branching nodding sprays – and cool Delft blue makes a delightful picture amongst the profuse grey-green leaves.'
(Graham S. Thomas)

NAMING
Etymology: *paniculata* – bearing panicles.

Synonym: *Pulmonaria paniculata* Aiton.

Common names: *English:* Tall mountain mertensia, Tall bluebells.

DESCRIPTION
General: A variable perennial, arising from a stout rhizome or caudex.

Stems: Very variable, from 10–90cm (4–36in), normally 30–75cm (1–2½ft), hairless or hairy, unusually numerous.

Leaves: Basal leaves 5–20cm (2–4in) long and 2.5–14cm (1–5¾in) wide, elliptical lanceolate to oval, or almost heart-shaped; roughly hairy above and below, with a stalk 10–25cm (4–10in) long. Stem-leaves 5–18cm (2–7in) long and 1–8cm (²/₅–3in) wide, oval or lanceolate, pointed, short-stalked or stalkless.

Inflorescence: Branched, open panicles.

Flowers: *Pedicels* 1–30mm, bristly or hairy, usually reflexed in fruit. *Calyx* 2–7mm long, divided to below the middle or to the base, bristly. *Corolla* first pink, gradually turning blue, sometimes white, tube 4.5–7mm long, sometimes hairy within, with

Mertensia primuloides in Cruickshank Botanic Garden, Aberdeen. (George McKay)

conspicuous scales in throat; lobes 6–9mm, spreading. *Stamens* with broad, conspicuous filaments, 1.5–3.5mm long and anthers 2.2–3.4mm long. *Style* elongated, often slightly protruding.
Nutlets: Wrinkled.

GEOGRAPHY AND ECOLOGY
Distribution: Northern parts of North America (Alaska, Quebec, Ontario, Iowa, Idaho, Oregon, Washington).
Habitat: Stream banks, wet meadows, damp thickets and wet cliffs, from the foothills to high elevations in the mountains.
Flowering: May to August in the wild; in cultivation may produce a second flush of flowers if the flowered stems are cut back.
Associated plants: On Iron Mountain in Oregon, grows in moister meadows and fields with plants such as *Aquilegia formosa*, *Polemonium carneum* and *Lilium washingtonianum*. In British Columbia, it can be found in spruce (*Picea*) forests with *Lathyrus*

ochroleucus, *Galium boreale*, *Shepherdia canadensis*, *Geocaulon lividum*, *Mitella nuda* and *Vaccinium membranaceum*.

CULTIVATION
Hardiness: Zone 4.
Sun/shade aspect: Light to moderately deep shade.
Soil requirements: Moist, humus-rich, well-drained soil.
Garden uses: Woodland garden, stream bank, poolside, mixed or herbaceous border.
Spacing: 60cm (2ft) apart.
Propagation: By seed or division, as for the genus.
Availability: Sold by some herbaceous plant specialists, and the seed is sometimes available on exchange lists.

RELATED FORMS AND CULTIVARS
Mertensia paniculata var. borealis (Macbride) Williams, differs by having hairless leaves (at least above) that are sometimes glaucous, and occurs from Montana and British Columbia to Oregon and Washington.

Mertensia primuloides (Decaisne) C.B. Clarke
A small tufted perennial to 15cm (6in) high, with hairy, lanceolate leaves, and dense clusters of funnel-shaped flowers that are usually indigo-blue. A nice little plant for semi-shade in leafy soil.

NAMING
Etymology: *primuloides* – *Primula*-like, referring to the shape of the flowers.
Synonyms: *Pseudomertensia primuloides* (Decaisne) Riedl; *Eritrichium primuloides* Decaisne.

DESCRIPTION
General: A small tufted perennial.
Stems: To 15cm (6in), bristly, densely leafy below.
Leaves: Basal leaves to 7cm (2³⁄₄in) long and 1cm (⁷⁄₁₆in) wide, lanceolate to oblong or narrowly lanceolate, pointed or blunt.
Inflorescence: Solitary, short few-flowered.
Flowers: *Pedicels* to 35mm(1³⁄₈in). *Calyx* to 4mm (³⁄₁₆in) long, lobes oblong to narrowly lanceolate, hairy. *Corolla* deep blue, or can be white or yellow, with tube about 13mm long, and limb about 10mm (⁷⁄₁₆in)in diameter; with conspicuous scales. *Stamens* with very short filaments.
Nutlets: To 3mm, brown, oval.

GEOGRAPHY AND ECOLOGY
Distribution: Afghanistan, Pakistan, Tibet.

Mertensia pulmonarioides. (George McKay)

Habitat: Woods.
Flowering: Summer.

CULTIVATION
Hardiness: Zone 5.
Sun/shade aspect: Semi-shade is preferable.
Soil requirements: Moist but well-drained, light soil, ideally enriched with leaf mould.
Garden uses: Rock garden, peat-bed or alpine house.
Spacing: 15cm (6in).
Propagation: From seed that benefits from about eight' weeks cold stratification (see genus description).
Availability: Sold by a couple of alpine nurseries, and the seed is likely to appear on exchange lists.

Mertensia pulmonarioides Roth
A lovely upright perennial, usually growing 30–60cm (1–2ft) high, with grey-green lanceolate to oval leaves, and drooping clusters of purple-blue, funnel-shaped flowers in spring. Dies down completely by midsummer. A great plant for any lightly shaded spot on moist soil.

'… at best its foliage on first appearance is of the most lovely amethyst shades of green, and from its tall crosiers dangle comfrey-like flowers of luminous sky-blue, set against the pink and mauve shades of the unopened buds. The whole plant has a stately and architectural beauty.' (Anna N. Griffith)

NAMING
Etymology: *pulmonarioides* – *Pulmonaria*-like.
Synonyms: *Mertensia virginica* (L.) Pers. in Link; *Pulmonaria virginica* L.
Common names: *English*: Virginia bluebells, Virginian cowslip, Roanoke-bells.

DESCRIPTION
General: A hairless perennial with thick, fleshy rhizomes.
Stems: 10–70cm (4–28in) high, upright.
Leaves: Basal leaves 4–20cm (1$^1/_2$–8in) long and 2–12cm ($^3/_4$–5in) wide, elliptic to oval, hairless, sometimes slightly pimply above, long-stalked. Stem leaves 4–12cm (1$^1/_2$–5in) long and 2–9cm ($^3/_4$–3$^1/_2$in) wide, short-stalked or stalkless, upper leaves almost clasping.
Inflorescence: Axillary scorpioid cymes.
Flowers: *Pedicels* 3–10mm long. *Calyx* 2–10mm

long, lobes enlarged in fruit. *Corolla* blue, white or pinkish, tube 11–21mm long, with a dense ring of hairs at the base, and small scales in throat; lobes 7–13mm, bell-shaped, scales usually very small. *Stamens* with filaments 4–8mm long and anthers 1.2–2mm. *Style* 5–10mm.
Nutlets: To 3mm, wrinkled.

GEOGRAPHY AND ECOLOGY
Distribution: North America (from New York to South Carolina, and west to Minnesota, Kansas and Alabama).
Habitat: Stream banks, low moist woods, wet meadows and floodplains.
Flowering: Spring.
Associated plants: Grows with other woodland plants, including spring beauty (*Claytonia virginica*), a number of *Trillium* species, including *T. pusillum*, *T. sessile* and *T. recurvatum*, plus bleeding-heart (*Dicentra*), toad-lily (*Tricyrtis*), wild ginger (*Asarum*) and violets (*Viola*), under shrubs such as redbud (*Cercis*), serviceberry (*Amelanchier*) and dogwood (*Cornus*).

CULTIVATION
Date of introduction: 1600s.
Hardiness: Zone 3.
Sun/shade aspect: Sun to shade, best in dappled shade.
Soil requirements: Any moist, humus-rich soil, ideally from slightly acid to neutral.
Garden uses: *Mertensia pulmonarioides* makes a perfect woodland garden plant, also good in mixed borders and on stream-banks, and also grows well on the pool-side. The plant is also good in a wildlife garden, as the flowers are attractive to insects and hummingbirds.
Spacing: 30cm (12in).
Propagation: Can be grown from seed that benefits from cold stratification (see under the genus description), and self-sows if happy. Division of rhizomes is the easiest method.
Availability: The commonest of all mertensias, sold by numerous nurseries and some garden centres; the seed is quite easy to come by, on most seed-exchange lists, and in the catalogues of some commercial suppliers.
Problems: Dies down early, leaving gaps in borders.

RELATED FORMS AND CULTIVARS
Two colour variants are known: **'Alba'** – a form that has pure white flowers, and **'Rubra'** – a form that has deep pink flowers; both are rare in cultivation.

Mertensia sibirica (Linnaeus) G. Don
An upright perennial, around 30–40cm (12–16in) high, with blue-green, hairless, elliptical leaves and clusters of more or less tubular, small blue flowers in spring and early summer. A lovely plant for a shady spot in a border or by waterside.

NAMING
Etymology: *sibirica* – of Siberia.
Synonyms: *Mertensia pterocarpa* (Turczaninow) Tatewaki & Ohwi; *Pulmonaria sibirica* Linnaeus; *Lithospermum sibiricum* Lehm.; *Steenhammera sibirica* Turcz.
Common names: *Chinese*: da ye bin zi cao; *Russian*: mertensia sibirskaya.

DESCRIPTION
General: A hairless perennial with a rhizome about 15mm ($^5/_8$in) thick.
Stems: 30–45cm (12–18in), sometimes to 70cm (28in), upright, usually single, branched at the top.
Leaves: Fleshy, hairless, the lower to 10–20cm (4–8in), oval with a heart-shaped base, but usually withered at flowering time, stalk to 25cm (10in). Upper leaves heart-shaped, rounded, or broadly elliptical. Stem-leaves oval, pointed, stalkless, 3–7cm ($1^1/_4$–$2^3/_4$in) long.
Inflorescence: Of two loose cymes.
Flowers: *Pedicels* 15–25mm ($^5/_8$–1in) long. *Calyx* hairless, 4–5mm ($^3/_{16}$–$^1/_4$in) long, divided nearly to base into oblong-oval or lanceolate lobes. *Corolla* blue or purple-blue, tube 4–5mm ($^3/_{16}$–$^1/_4$in) long, limb 3–5mm ($^1/_8$–$^1/_4$in), with rounded lobes; throat scales conspicuous. *Stamens* with filaments 1.5–2mm. *Style* protruding about 4–5mm from the corolla.
Nutlets: 4–5mm, smooth, pale.

GEOGRAPHY AND ECOLOGY
Distribution: Endemic to Siberia and China (West Sichuan).
Habitat: Riverbanks, on shingle or sand in Siberia, alpine meadows in China.
Flowering: Spring and early summer.

CULTIVATION
Hardiness: Zone 3.
Sun/shade aspect: Lightly shaded position is best.
Soil requirements: In spite of rather different conditions in its native habitat, this plant thrives in cultivation in any well-drained but moist, humus-rich soil.
Garden uses: Grow in a mixed border, in dappled shade under trees and shrubs, on a streambank.
Spacing: 30cm (12in).
Propagation: From seed or by division, as for the genus.
Availability: Supplied by many herbaceous nurseries, and the seed frequently appears on exchange lists.

RELATED FORMS AND CULTIVARS
There is a rare white-flowered form, **'Alba'**.

RELATED SPECIES
Japanese bluebells, or Ezo-ruri-so, **Mertensia yezoensis** Tatewaki & Ohwi (syn. *M. rivularis* var. *japonica* Takeda), *M. pterocarpa* var. *yezoensis* Tatew. & Ohwi), is a similar plant, native to Japan (Hokkaido) and the Russian Far East (southern Kuriles).

Mertensia simplicissima G. Don
A beautiful glaucous perennial, with long trailing stems to 1m (3ft) long, clothed in elliptical fleshy leaves and, in summer, bearing small clusters of turquoise-blue, tubular flowers opening from pink buds. An excellent plant for a rock garden or a gravel bed in sun, easier to grow than the closely related *M. maritima*.

'This gorgeous glaucous-leaf, sprawling bluebell is caviar for slugs.' (Rebecca Day-Skowron)

NAMING
Etymology: *simplicissima* – most simple, although this reference to simplicity is not clear.
Synonyms: *Mertensia maritima* subsp. *asiatica* Takeda; *M. asiatica* (Takeda) Macbr.
Nomenclature note: This plant is sometimes considered identical to *M. maritima* L.
Common name: *Japanese:* Hama-benkei-so.

DESCRIPTION
General: A fleshy, tufted, glaucous perennial.
Stems: To 1m (3ft) long but often shorter, fleshy, leafy, prostrate and trailing along the ground.
Leaves: Basal and lower leaves 3–8cm (1¼–3¼in) long, 2–6cm (¾–2½in) wide, long-stalked, oblong, elliptical, obovate to broadly so, slightly fleshy, hairless on both sides or with sparse short hairs. Upper leaves almost stalkless.
Inflorescence: Hairless cymes with bracts.
Flowers: *Calyx* hairless, pointed, as long as the corolla tube. *Corolla* blue, opening from pink buds, 8–12mm (⅜–½in) long, with a flared limb.
Nutlets: Fleshy.

GEOGRAPHY AND ECOLOGY
Distribution: Japan (Hokkaido, Honshu), Korea, Russian Far East (Sakhalin, Kuriles), Aleutians.
Habitat: Sandy seashores.
Flowering: May to September in the wild.

CULTIVATION
Hardiness: Zone 6.
Sun/shade aspect: Full sun.
Soil requirements: As for *M. maritima*.
Garden uses: As for *M. maritima*.
Spacing: 60cm (2ft).
Propagation: Seed and division.
Availability: Sold by many nurseries and even some garden centres, and seed is available from most societies' lists and some seed merchants.
Problems: Though easier to grow than *M. maritima*, this plant is sometimes so floriferous that it 'flowers itself to death', so it may be necessary to replace it every couple of years or so. Possibly removing the flowered stems to prevent seed-setting will help the plant conserve its resources and prolong its lifespan. As with *M. maritima*, slugs can be a major problem.

Mertensia viridis (A. Nels) A. Nels
A hairy perennial around 30cm (12in) high, with bright green lanceolate leaves that are hairy above, and cymes of bright blue funnel-shaped flowers in late spring and summer. Not the easiest to grow, but a very attractive plant for a rock garden or alpine house.

NAMING
Etymology: *viridis* – green, referring to the bright green colour of leaves, as distinct from the closely related *M. lanceolata* which often has glaucous leaves.

Mertensia viridis var. *cana*. (Masha Bennett)

Mertensia viridis var. *cana* on Hoosier Pass, Colorado.
(Masha Bennett)

Synonyms: *Mertensia lanceolata* var. *viridis* A. Nels;
M. ovata Rydb.; *M. lineariloba* Rydb.; *M. parryi*
Rydb.; *M. perplexa* Rydb., 1904; *M. viridula* Rydb.
Common name: *English:* Greenleaf mertensia, Green
Mertensia.

DESCRIPTION

General: A variably hairy perennial arising from a
fairly stout root.
Stems: 5–40cm (2–16in) tall, upright or ascending,
hairless, ranging from one to many from each
rootstock.
Leaves: Basal leaves lanceolate to oval, 2–10cm (1–4in)
long and 1–4cm (⁷/₁₆–1¹/₂in) wide, stalked, often with
prominent lateral veins. Stem leaves stalkless or nearly
so, lanceolate to broadly oval, 2–7cm (³/₄–2³/₄in) long
and 7–25mm (³/₈–1in) wide, with indistinct veins. All
hairless below, bristly above.
Inflorescence: Dense scorpioid cymes, with flowers
held upright or nodding.
Flowers: *Pedicels* bristly or hairless, 1–10mm long.
Calyx 2–6mm long, the lobes divided almost to the base,
narrowly lanceolate to narrowly oval-lanceolate, pointed
or blunt, with a fringe of hair on margins. *Corolla*
bright, deep blue, tube 3–9mm long, usually with a ring
of crisped hairs inside, limb 4–9mm long (mostly about
5mm), moderately expanded; scales conspicuous,
hairless to densely softly hairy. *Stamens* with filaments
1–3.5mm, anthers 1–2.5mm long (usually about
1.5–2mm). *Style* variable, may be short or long.
Nutlets: 2–3mm, wrinkled.

GEOGRAPHY AND ECOLOGY
Distribution: North America (Montana, Idaho,

Colorado and Utah, northern New Mexico, south-
eastern Oregon, Nevada, California).
Habitat: Mountains: alpine meadows and tundra,
rocky places, hillsides, up to above timber line.
Flowering: May to August in the wild.
Associated plants: On Trail Ridge in Colorado, at
3600m (11,800ft), in rocky alpine tundra, grows
with *Achillea lanulosa*, *Androsace septentrionalis*,
Astragalus sp., *Draba* sp., *Eritrichium aretioides*,
Lloydia serotina, *Phlox pulvinata*, *Polemonium
viscosum*, *Ranunculus adoneus*, dwarf *Salix* sp.,
Saxifraga sp. and *Sedum lanceolatum*.

CULTIVATION
Hardiness: Zone 3.
Sun/shade aspect: Full sun.
Soil requirements: Sharply drained, gritty soil with
adequate humus.
Garden uses: Rock garden, scree or alpine house.
Spacing: 20cm (8in).
Propagation: From seed, as for the genus.
Availability: Occasionally available from alpine
specialists, or look for its seed on societies' exchange
lists.

RELATED FORMS AMD CULTIVARS
Mertensia viridis* var. *cana (Rydberg) Williams, is a
naturally occuring variation, distinguished by having
smaller leaves that are densely grey-hairy on both
surfaces. The plants encountered by the author in the
wild also had a distinct purple-blue flower colour as
opposed to the normal bright blue in *M. viridis*.
On Hoosier Pass, Colorado, in the open fir/spruce
woodland, *M. viridis* var. *cana* is found growing with
Achillea lanulosa, *Aletes acaulis*, *Lewisia pygmaea*,
Polemonium delicatum, *Viola labradorica*, *Trollius
albiflorus* and *Vaccinium myrtillus* ssp. *oreophilum*.
Mertensia bakeri Greene, resembles *M. viridis* and
is sometimes considered as its variant. It is
distributed from Colorado and Utah to New Mexico.

Other Mertensia species of interest

These mertensias are not often found in cultivation,
but may appear on seed-exchange lists. Any other
Mertensia species, not listed here, are also well worth
trying.

Mertensia elongata (Decaisne) Bentham & Hooker
Synonym: *Lithospermum elongatum* Decaisne.

Description: A hairy perennial to 25cm (10in), with oblong or elliptical-lanceolate leaves, to 7cm (2³/₄in) long, and cymes of blue to deep purple-blue flowers in summer.
Distribution: Pakistan, India (Kashmir).
Habitat: Mountain woodlands.

Mertensia franciscana Heller

Synonyms: *M. pratensis* Heller; *M. alba* Rydb, 1904; *M. grandis* Wooton & Standley; *M. pratensis f. alba* Macbr.
Common name: Franciscan bluebells.
Description: A perennial rather similar to *M. ciliata*, to 1m (3ft) tall, with upright or ascending stems, oblong to elliptical leaves 6–20cm (2¹/₂–8in) long, bristly-hairy above, hairy or hairless below, and drooping clusters of tubular, blue or sometimes white flowers, about 15mm (⁵/₈in) long, in mid- to late summer.
Distribution: Southern USA (southern Colorado and Utah, New Mexico, Arizona, eastern Nevada, California).
Habitat: Varied, along rivers, aspen groves, spruce forests, woodlands, wet meadows, dry rocky summits.
Associated plants: In the Sacramento Mountains, it grows in ponderosa (*Pinus ponderosa*) and Douglas fir (*Pseudotsuga menziesii*) forests with *Mimulus glabratus, Dodecatheon pulchellum, Galium trifolium, Euphorbia brachycera* and *Cirsium vinaceum*.

Mertensia longiflora Greene

Synonyms: *M. pulchella* Piper; *M. horneri* Piper.
Description: A perennial, 5–30cm (2–12in) tall, arising from a thick rootstock, with elliptical, rounded leaves (basal ones are rarely present in plants at flowering stage), bearing dense clusters of tubular bright blue flowers 15–25mm (⁵/₈–1in)long, from mid-spring into summer.
Distribution: British Columbia to central Oregon, to north-eastern California, east to Montana and to Idaho.
Habitat: Open or lightly shaded places in the plains and foothills, often with sagebrush or with ponderosa pine, up to 1500m (4900ft). Zone 3.
Associated plants: On Moscow Mountain, Washington State, it grows on grassy slopes with *Geranium viscosissimum, Vicia villosa, Fritillaria pudica, Draba verna* and *Lithospermum ruderale*.

Mertensia moltkioides C.B. Clarke

Synonyms: *Pseudomertensia moltkioides (*Royle) *Kazmi; Myosotis moltkioides* Benth.

Description: A rather tufted plant, 2–10cm (1–4in) tall, with lax rosettes of hairy, elliptical or spoon-shaped leaves, outer stalked, inner ones stalkless, and terminal clusters of long tubular flowers about 2cm (³/₄in) long, with reddish-purple tube and blue, spreading, rounded lobes, with a yellow or white eye, borne from midsummer to early autumn.
Distribution: Pakistan to Kashmir.
Habitat: Rocks, open slopes, coniferous forests, at 3600–4500m (11,800–14,750ft). Zone 6.

Mertensia nemorosa (A. DC.) I.M. Johnston

Synonyms: *Pseudomertensia nemorosa* (DC.) R.R. Stewart & Kazmi; *Eritrichium nemorosum* (A. DC).
Description: A tufted perennial 10–12cm (4–4¹/₂in) high, with elliptical leaves 3–4cm (1¹/₄–1¹/₂in) long with a stalk of similar length, and loose clusters of blue, funnel-shaped flowers in late spring.
Distribution: Afghanistan to India (Kashmir).
Habitat: Forests and scrub, 1500–2400m (4900–7870ft). Zone 6.

Mertensia oblongifolia (Nutt.) Don

Description: An upright perennial, 10–40cm (4–16in) tall, with a rosette of hairless or hairy, deep green, elliptical to oval-lanceolate leaves, 2–15cm (³/₄–6in) long, and blue funnel-shaped flowers, 1–2cm (⁷/₁₆–³/₄in) long, with a slightly expanded limb, borne from mid-spring into summer.
Distribution: North America (Montana, Nevada, Washington, northern Wyoming and Utah, British Columbia).
Habitat: Open hill slopes, meadows, in sagebrush, from the plains to alpine zones in mountains. Zone 4.

The genus *Moltkia* Lehmann, 1817

Naming
Etymology: *Moltkia* – after Count Joachim Gadske Moltke (1746–1818), Danish statesman.
Synonyms: Previously included within the genus *Lithospermum* L., and some species have been assigned to *Onosma* L., *Echium* L. and even *Pulmonaria* L. in the past. *Moltkia doerfleri* is sometimes separated into a different genus, *Paramoltkia* Greuter.

GENUS CHARACTERISTICS AND BIOLOGY

Number of species: From 4 to 8 species, according to various estimates.

Description: A genus of perennial herbs or dwarf shrubs. Stems herbaceous or woody, densely hairy, bristly. Leaves alternate. Flowers in short, terminal cymes with bracts. *Calyx* deeply 5–lobed. *Corolla* blue, purple or yellow, funnel-shaped, without scales or annulus. Stamens usually protruding from the corolla throats, filaments inserted at or above middle of tube; anthers oblong or lanceolate. Style protruding, with entire or notched stigma. Nutlets usually smooth, or finely warty, usually with a prominent keel.

Distribution and habitats: Europe from northern Italy to northern Greece, and south-west Asia; mostly in rocky or dry places, usually in mountains.

Life forms: Chamaephytes (shrubby forms), hemicryptophytes or geophytes (herbaceous forms).

Pollination: By various bees.

Fruit dispersal: No specialized mechanism.

USES

No culinary or medicinal uses of *Moltkia* species are known.

GARDEN

Most species are excellent plants for a rock garden, scree or raised bed, apart from *M. doerfleri* which is too tall for these garden habitats and would be more suited to a herbaceous or mixed border. Shorter forms can also be tried in pots in the alpine house.

CULTIVATION

All species and forms grown require a sunny position, and a spot as warm as possible in a perfectly drained soil or compost that ideally should be moisture-retentive, and with alkaline to neutral pH. The best planting time is probably late spring, to allow plants to establish before the onset of harsh weather. If grown in an alpine house, keep just moist in winter, begin to water moderately from spring until the flowers are over and follow by a few weeks of a dryish regime. Plants grown under glass can be repotted annually in spring.

PROPAGATION

Seed Sow seeds from early to mid-spring or when ripe, in pots or trays of sharply drained compost, top-dressing lightly with grit and placing containers a cold frame. There is some indication that seed loses viability rather quickly in storage, so sowing fresh is probably the best option. Young plants should be potted up and overwintered under cover.

Vegetative Softwood or semi-ripe cuttings of woody species can be taken from early to mid-summer and inserted in perfectly drained compost in a heated propagator at about 15°C (59°F). Trailing stems of dwarf shrubs will layer themselves naturally, or can be assisted by weighing the stems down to the ground in spring. Herbaceous forms, such as *M. doerfleri*, can be increased by division of rootstock.

REFERENCES

BECKETT, 1993–94; BLAMEY & GREY-WILSON, 1993; BRICKELL, 1996; GRIFFITH, 1964; HEATH, 1964; HUXLEY, 1992; INGWERSEN 1991; MABBERLEY, 1997; THOMAS, 1992 ; TUTIN *et al.*, 1972; WALTERS, 2001

Moltkia doerfleri Wettstein

An unusual herbaceous perennial, to 50cm (20in) high, with horizontal rhizomes giving rise to upright stems clothed in lanceolate leaves, and deep purple or violet, tubular flowers in late spring and summer. A nice plant for a sunny, well-drained border.

NAMING

Etymology: *doerfleri* – after Ignaz Doerfler (1866–1950).

Synonyms: *Lithospermum doerfleri* Anon.; *Paramoltkia doerfleri* (Wettst.) Greuter & Burdet.

DESCRIPTION

General: A perennial herb with thick horizontal rhizomes.

Stems: 30–50cm (12–20in), upright, unbranched, hairy.

Leaves: Lanceolate, pointed, sparsely bristly and, on the margins, the lowest leaves are not well developed.

Inflorescence: Terminal cymes, with bracts.

Flowers: *Calyx* about 10mm, lobes falling off individually before the calyx-base. *Corolla* deep purple, 19–25mm, gradually widening above. *Stamens* not protruding from the corolla; with filaments up to 2mm and anthers 2.5–3mm long.

Nutlets: About 4mm.

GEOGRAPHY AND ECOLOGY

Distribution: Endemic to north-eastern Albania.

Habitat: Mountains, in rock crevices. Rare in its habitat, included on *The IUCN Red List of Threatened Plants*.

Flowering: May to June in the wild.

CULTIVATION

Hardiness: Zone 6.

Moltkia doerfleri in cultivation. (Masha Bennett)

Sun/shade aspect: Full sun.
Soil requirements: Any well-drained soil.
Garden uses: Interesting plant for a herbaceous or mixed border; though a true mountain plant, it is too large for an average rock garden.
Spacing: 45cm (18in) apart.
Propagation: As for the genus.
Availability: Uncommon, but can be found in a few specialist herbaceous and rock plant nurseries.
Problems: Rhizomes may be somewhat invasive in ideal conditions.

RELATED SPECIES
Moltkia coerulea (Willldenow) Lehmann (syn. *Onosma coeruleum* Willdenow) is another rarely seen member of the genus, native to stony slopes of semi-deserts in Asia Minor. It is a woody-based herbaceous perennial to 10–30cm (4–12in) high, with lanceolate grey-green leaves, and cylindrical, bright blue flowers to 19mm (^{3}/$_4$in) long, borne in spring.

Moltkia petraea (Trattinick) Grisebach
A small semi-evergreen shrublet, 15–30cm (6–12in) tall, with hairy, greyish-green leaves, resembling those of lavender, and dense clusters of tubular flowers in early summer, pink in bud and opening deep violet-blue, with long-protruding stamens. An excellent rock garden plant, for a hot, sunny situation.

> *'A dense, woody shrublet, with pale green leaves ... and profuse sprays of pinkish-blue flowers, deepening to blue-purple.'* (A.N. Griffith)

Moltkia petraea, south of Jablanca in the former Yugoslavia. (Erich Pasche)

NAMING
Etymology: *petraea* – of rocks, referring to the plant's habitat.
Synonyms: *Echium petraeum* Trattinick; *Lithospermum × froebeli* Sundermann; *L. petraeum* (Trattinick) de Candolle.

DESCRIPTION
General: A densely branched, semi-evergreen dwarf shrub, 20–40cm (8–16in) high.
Stems: Rigid, slender, upright, densely whitish-bristly.
Leaves: 1–5cm (^{7}/$_{16}$–2in) long and 1–6mm (^{1}/$_{16}$–^{1}/$_4$in) across, linear to oblong-lanceolate, blunt or somewhat pointed, sparsely bristly and green above, very densely bristly and whitish beneath; margin more or less curved downwards and inwards.
Inflorescence: Cymes short, forming dense heads, with bracts.
Flowers: *Calyx* about 4mm, with persistent lobes. *Corolla* deep violet-blue, opening from pink buds, 6–10mm long, cylindrical to obconical. *Stamens* strongly protruding from the corolla, with filaments to 8mm long and anthers 2mm.
Nutlets: 2.5–3mm.

GEOGRAPHY AND ECOLOGY
Distribution: Balkan Peninsula, the former Yugoslavia through Albania to central Greece.
Habitat: Rock crevices in mountains.
Flowering: Summer.

CULTIVATION
Date of introduction: 1840.
Hardiness: Zone 7.
Sun/shade aspect: Full sun.
Soil requirements: Perfectly drained soil.
Garden uses: Rock garden, scree, troughs, raised beds, or pot in the alpine house.
Spacing: 30cm (12in).
Propagation: As for the genus.
Availability: Sold by some specialist alpine and rock plant nurseries. Seed sometimes appears on the exchange lists of specialist societies.

Moltkia suffruticosa (Linnaeus) Brand
A dwarf, cushion- or mat-forming deciduous shrublet, to 30cm (12in) high, with woody stems densely covered with dark green, narrow leaves, and dense clusters of tubular, bright blue to purple-blue flowers in summer.

> *'Tufts of long, narrow, grass-like leaves, and 6–9" [inch] sprays of drooping tubular bells of azure blue.'* (Anna N. Griffith)

NAMING
Etymology: *suffruticosa* – sub-shrubby.
Synonyms: *Lithospermum graminifolium* Viviani; *Moltkia graminifolia* (Viviani) Nyman; *Pulmonaria suffruticosa* Linnaeus.

DESCRIPTION
General: A mat-forming, dwarf shrub with trailing stock, producing short, non-flowering, densely leafy stems and taller flowering stems.
Stems: Non-flowering stems short, densely leafy, and from their tips arise slender, upright, bristly flowering stems 5–25cm (2–10in) high.
Leaves: 5–15cm long and 1–4mm ($^1/_{16}$–$^3/_{16}$in) wide, linear, pointed, sparsely bristly and green above, very densely bristly and whitish beneath; leaf margin may be flat or curved downwards and inwards.
Inflorescence: Cymes short and dense, grouped in corymbs.
Flowers: *Calyx* 5–6mm ($^1/_4$in), bristly, with persistent lobes. *Corolla* blue 13–17mm, ($^1/_2$–$^5/_8$in) gradually widening above Stamens protruding from the corolla, filaments about as long as anthers, which are yellow and about 3mm ($^1/_8$in).

Nutlets: About 3mm, ovoid, beaked.
Chromosomes: 2n=16.

GEOGRAPHY AND ECOLOGY
Distribution: Northern Italy (south-eastern Alps, northern Apennines).
Habitat: Rocky and stony places in mountains, usually on limestone, to 1200m (3900ft).
Flowering: July to September in the wild.

CULTIVATION
Date of introduction: 1888.
Hardiness: Zone 7.
Sun/shade aspect: Full sun.
Soil requirements: Well-drained, ideally from neutral to alkaline. Will tolerate strong alkalinity.
Garden uses: Rock garden.
Spacing: 30cm (12in).
Propagation: Cuttings, division, seed.
Availability: Sold by some alpine and rock plant specialists. Seed can sometimes be found on seed-exchange lists of specialist societies.

RELATED FORMS
Moltkia × *intermedia* (Froebel) Ingram (syn. *Lithospermum* × *intermedium* Froebel) is a hybrid between *M. suffruticosa* and *M. petraea*. It is rather similar to *M. suffruticosa*, but is evergreen, usually 20–30cm (8–12in) high, with leaves to 10cm (4in) long, and funnel shaped, clear bright blue flowers in early summer. Leaves more than 4 mm wide. This is an RHS Award of Garden Merit plant. *Moltkia* × *intermedia* 'Froebeli' is a compact selection, around 15cm (6in) high, with bright azure blue flowers.

The genus *Myosotidium* J.D. Hooker, 1859
Chatham Island Forget-me-not

NAMING
Etymology: *Myosotidium* – from *Myosotis*, forget-me-not, referring to the resemblance in flower.
Synonyms: In the past, this has been included under *Myosotis* L. and *Cynoglossum* L.

GENUS CHARACTERISTICS AND BIOLOGY
Number of species: This is a monotypic genus, containing only one species.
Description: See the description of *M. hortensia*.
Distribution and habitat: Chatham Islands (New Zealand); only on the sea shore.

Life form: Chamaephyte.
Pollination: Most likely by Lepidoptera (moths).
Fruit dispersal: Probably by sea currents.

USES
No medicinal or culinary uses are known.

GARDEN
A sumptuous pot plant for a cool greenhouse or conservatory, but in mild, coastal districts, can be grown outdoors in a sheltered herbaceous or mixed border, in a woodland garden or peat bed.

OTHER
The cut flowers can be used in fresh arrangements, though they are not particularly long-lasting.

CULTIVATION
Knowledge of this unusual plant's natural habitat allows conclusions to be drawn about the best conditions for cultivation. Temperature there rarely falls below 0°C (32°F), and while in cultivation *M. hortensia* can tolerate brief periods of frost down to -5–-10°C (14–23°F). If harder frosts are expected, the plant must be protected with bracken, straw, evergreen branches or polythene. Alternatively, it could be grown under cover in a cool greenhouse or conservatory. The sky over Chatham Islands is often overcast, and in cultivation *M. hortensia* usually needs at least some shade in order to thrive, though sometimes it is successful in the open. It grows on the sea shoreline, where under the sand and shells there is a rich layer of decaying organic matter, into which the plant's roots are penetrating; and humus-rich, continually moist, but freely draining soil is essential in cultivation. Being a seaside plant, the Chatham Island forget-me-not is tolerant of coastal conditions, including salt spray, but does not like cold winds.

Mulching with seaweed and fish-derived fertilizer throughout summer provides important micronutrients to plants and helps to keep them in good condition.

Newly planted specimens must be watered thoroughly and regularly throughout their first season, and as needed afterwards.

PROPAGATION
Seed To ensure success, seed should be sown fresh. Nutlets ripen 1–2 months after flowering and, if collected and sown straight away, germination should occur quite quickly (2–3 weeks). If only stored dry seed is available, it may help to soak it, or treat it with gibberilic acid before sowing. The seedlings are susceptible to crown rot if kept too moist. Pot the young plants up in a good-quality compost, and grow on until their second year, when they can be planted out.
Vegetative It is possible to divide large plants carefully in spring, but this is not easy, as they strongly resent disturbance. Pot up the divisions and grow on before planting out.

PROBLEMS
Apart from marginal hardiness and general fickleness, the main problem with *M. hortensia* is that slugs are very fond of the succulent leaves and, in a badly affected garden, may need to be raised off the ground in pots to prevent severe damage. It can also be affected by cucumber mosaic virus.

REFERENCES
ALLAN, 1961; BRICKELL, 1996; HUXLEY, 1992; PHILLIPS & RIX, 1997; THOMAS, 1993; WALTERS, 2001.

Myosotidium hortensia (Decaisne) Baillon
A stout, striking, clump-forming perennial, usually about 45cm (18in) high, with a mound of large, deeply veined, succulent leaves, and heavy clusters of blue forget-me-not-like flowers. Hardy only in milder coastal areas, this is a connoisseur's plant, in need of a sheltered, lightly shaded spot on humus-rich soil.

> *'Imagine a Bergenia with leaves deeply veined like those of Hosta sieboldiana or Viburnum davidii, and flowers of forget-me-not blue, and I need say no more, except to apologize for tantalizing you: it is only hardy in salubrious southern maritime gardens and needs to be fed on seaweed.'* (Graham S. Thomas)

> *'A snob plant of the highest order.'* (Jane Powers)

NAMING
Etymology: *hortensia* – of gardens, from where it was first described.
Synonyms: *Myosotis hortensia* Decaisne; *Cynoglossum nobile* J.D. Hooker; *Myosotidium nobile* (J.D. Hooker) J.D. Hooker.
Common names: *English*: Chatham Island forget-me-not; Giant forget-me-not.

DESCRIPTION
General: A somewhat succulent, evergreen perennial, with stems arising from a stout, cylindrical rootstock.
Stems: 30–60cm (1–2ft) high, occasionally more, hairless apart from a sparse down at the top.

Myosotidium hortensia in a botanic garden. (Gordon Smith)

Leaves: Basal leaves numerous, 15–32cm (6in–12½in) long including the stout stalks; blade broadly oval with a heart-shaped base, thick, fleshy, deeply veined. Upper leaf surface hairless, shiny, lower surface with evenly scattered minute hairs. Stem leaves alternate, upper stalkless.

Inflorescence: Terminal, dense corymb of spirally coiled cymes, 10–15cm (4–6in) diameter.

Flowers: *Pedicels* about 10mm (⁷/₁₆in) long. *Calyx* lobes blunt, oval, to 4mm (³/₁₆in), with stiff hairs. *Corolla* dark to pale blue, about 12–15mm (½–⁵/₈in) across, with a short tube and spreading limb; lobes rounded, white-margined. Scales in the corolla throat yellow. *Stamens* short, not protruding from the corolla. *Style* short, thick with a headlike stigma.

Nutlets: Broadly oval, flattened, with unevenly toothed wing, about 6mm in diameter.

GEOGRAPHY AND ECOLOGY

Distribution: Endemic to the Chatham Islands, New Zealand.

Habitat: Coastal rocky places and sandy beaches. Endangered in the wild due to overgrazing by sheep and cattle.

Myosotidium hortensia (Doreen Townley).

Flowering: September to October in the wild (in cultivation in the UK late spring–early summer).
Associated plants: Grows in company with other coastal plants such as other Chatham Island endemic *Geranium traversi*, and also *Urtica australis*, *Embergeria grandifolia*.

CULTIVATION
Date of introduction: 1859.
Hardiness: Zone 9. In the UK hardy in mild coastal districts in the south and west, otherwise needs protection in winter.
Spacing: 60cm (2ft) apart, though when flourishing, the plants may spread a lot wider.
For other details on cultivation, propagation and garden uses, see the genus description.
Availability: Sold by many suppliers of herbaceous perennials and unusual plants.

The genus *Myosotis* Linnaeus, 1753
Forget-me-nots

NAMING
Etymology: From the Greek *mus*, mouse, and *ous* or *otos*, ear, referring to shape of the leaves.
Synonyms: Plants of many other genera have been previously included under *Myosotis* – for example, *Myosotidium* J.D. Hooker, *Eritrichium* Schrad. and *Brunnera* Steven.

GENUS CHARACTERISTICS AND BIOLOGY
Number of species: According to various estimates, from 50 to about 100, including over 30 species in New Zealand.
Description: A genus of hairy annual, biennial or perennial herbs. Leaves alternate, entire and softly hairy. Flowers usually in paired scorpioid cymes, often without bracts, sometimes with bracts at base, or with leaflike bracts subtending each flower. *Calyx* 5–lobed, often enlarged in fruit. *Corolla* mostly blue opening from pink buds, sometimes white, yellow or purple, wheel-shaped, 5–lobed, usually with a short tube and flat or slightly concave lobes; corolla-throat with 5 white or yellow scales. Stamens 5, usually included in the corolla, filaments inserted near middle of corolla-tube. Style included, stigma headlike. Nutlets usually 4, dark brown to black, hairless, smooth and shiny, ovoid and usually compressed.
Distribution and habitats: Native to the temperate regions of Europe and Asia, and mountains in the

Myosotis sylvatica 'Gold 'n' Sapphires. (Dan Heims)

tropics, including Africa and North America; many endemic species in New Zealand. *Myosotis* can be found in a wide variety of habitats, from arable fields (*M. arvensis*) and woodland (*M. sylvatica*) to high mountains (*M. alpestris, M. olympica*), wetlands (*M. scorpioides*) and seashores (*M. rakiura*).
Life forms: Therophytes (annuals), hemicryptophytes (most perennials) and a few helophytes (*M. scorpioides*).
Pollination: *Myosotis* are attractive to a great variety of insects, including honeybees (*Apis mellifera*); solitary bees such as *Andrena, Osmia*; leaf-cutter bees (*Megachile*); butterflies, including blues (*Lycaena* and *Polyommatus*) and whites (*Pieris*); hoverflies (Syrphidae); true flies (Muscidae); and beetles.
Fruit dispersal: No specialized mechanism, but in *M. scorpioides* water probably plays an important part.
Animal feeders: Forget-me-nots support a variety of insects – for instance, *M. scorpioides* is the only foodplant of the hemipteran bug *Monanthia humili* and one of the foodplants of *Cymus glandicolor*; *M. arvensis* supports the hemipteran bug *Sehirus luctuosus*. Finches feed on the ripe forget-me-not seed.

USES
GARDEN
Forget-me-nots can be used in a variety of garden settings, such as a woodland garden (*M. sylvatica*), water garden (*M. scorpioides, M. rehsteineri*), rock garden (*M. alpestris* and most New Zealand species,

Myosotis alpestris in the Tien Shan Mountains.
(Masha Bennett)

such as *M. explanata*), alpine house (New Zealand forget-me-nots), bedding and containers (*M. sylvatica* cultivars) and wildflower gardens (any native species).

MEDICINAL
Forget-me-nots are of limited importance in herbal medicine, but *M. alpestris* lotion is considered a good remedy for many eye diseases, and the powder made from the dried plant can be applied externally to wounds. *Myosotis scorpioides* has been used to treat whooping cough and bronchitis.

OTHER
Forget-me-nots are charming in flower arrangements, but usually do not last more than a week. *Myosotis sylvatica* cultivars are the most suitable cutting material. Before arranging, condition the cut stems by standing them in a bucket of tepid water, placed in a cool, dark spot for several hours.

The borage family is rarely honoured by appearing in works of art, but Van Gogh depicted forget-me-nots in his painting *Vase with Myosotis and Peonies* (1886).

CULTIVATION
The majority of European and Asian species are hardy to at least Zone 6, but New Zealand and Australian

species are marginally hardy (Zone 8). *Myosotis azorica* and *M. welwitchii* are not frost-hardy and are rarely grown in areas colder than Zone 9. The more tender species can be protected from the worst of the winter weather in a cool glasshouse, or alpine house.

Most forget-me-nots enjoy a fairly sunny situation, the ones tolerant of shade include *M. scorpioides* and *M. sylvatica*. New Zealand forget-me-nots, although requiring plenty of light, must be protected from direct sunshine during the hottest time of the day.

Due to the great diversity of habitats that forget-me-nots inhabit in the wild, it is not possible to recommend soil conditions suitable for all. New Zealand and Australian species, including *M. explanata*, *M. australis* and *M. pulvinaris*, require gritty, perfectly drained soil or compost mix, which at the same time should be water-retentive – adding plenty of grit and sterilized leaf mould will help achieve these properties. *Myosotis alpestris* also thrives in such conditions, although is not as particular about the substrate.

Myosotis scorpioides and *M. rehsteineri* require wet soil: the former can grow on mud or in shallow water. Ordinary, reasonably drained soil is suitable for *M. sylvatica* and its numerous strains.

PROPAGATION
Seed Most forget-me-nots come easily from seed. Use ordinary multi-purpose or seed compost for annuals and larger perennials, and a gritty mixture for smaller alpine species, including the New Zealand forms; cover the seed lightly, and germinate in a garden frame or a cool greenhouse. *Myosotis sylvatica*, *M. arvensis* and *M. scorpioides* can be sown *in situ*, in mid- to late spring, or as soon as ripe. Bedding strains of *M. sylvatica* are normally sown in summer in trays, with eventual pricking out into small pots. Germination usually occurs within 2–4 weeks at about 20°C (68°F), and in at least some species is more successful when the sown seed is kept in the dark.
Vegetative Perennials can be divided in early to mid-spring, replanting them in their permanent position or, for the more difficult alpine forms, potted up. Soft cuttings of perennial species can be taken during summer, rooting them in a gritty mixture, in a cold frame, greenhouse or propagator.

PROBLEMS
In dry conditions, many forget-me-nots are very susceptible to powdery mildew. Aphids can also be a problem, especially on the small New Zealand species, which may be weakened and killed by heavy infestations. Many *Myosotis* are short-lived and require frequent renewal, while some, like *M. arvensis*, are very weedy.

REFERENCES

ALLAN, 1961; BECKETT, 1994; BRICKELL, 1996; CLAPHAM *et al.*, 1952; GENDERS, 1977; GRIFFITH, 1964; HUXLEY, 1992; PHILIPSON & HEARN, 1965; PRESTON & CROFT, 1997; RICE, 1999; STACE, 1991; SWINDELLS & MASON, 1992; TUTIN *et al.*, 1972; WALTERS, 2001.

Myosotis alpestris F.W. Schmidt

A delightful arctic-alpine plant with typical sky-blue, fragrant forget-me-not flowers carried above hairy leaves at a height up to 20cm (8in). For a rock garden or scree. This is a rare British native.

'As the daylight begins to fade, the flowers become deliciously fragrant though it is difficult to detect any perfume at all during daylight.'
(Roy Genders)

NAMING

Etymology: *alpestris* – alpine.
Synonyms: *Myosotis sylvatica* subsp. *alpestris* (F.W. Schmidt) Gams.; *M. rupicola* Smith is a name applied to dwarf forms.
Nomenclature note: Some spring-bedding cultivars are occasionally listed for this species, but these should be under *M. sylvestris*.
Common names: *English*: Alpine forget-me-not; *Chinese*: wu wang cao; *French*: Myosotis alpestre; *Gaelic*: Lus Midhe Ailpeach; *German*: Gebirgs-Vergißmeinnicht; *Italian*: Nontiscordardime alpino; *Polish*: niezapominajka alpejska; *Russian*: Nezabudka alpiiskaya; *Slovak*: nezábudka alpínska; *Spanish*: Nomeolvides alpina.

DESCRIPTION

General: A softly hairy, clump- to mat-forming perennial with short rhizomes.
Stems: 5–20cm (2–8in), occasionally to 35cm (14in), simple or branched, stiff, usually roughly hairy.
Leaves: Basal leaves oblong-lanceolate to spoon-shaped, more or less pointed, long-stalked, about 8cm (3^1/$_4$in) long and 1.5cm (^{5}/$_8$in) wide; stem leaves stalkless. All sparsely to densely hairy, sometimes hairless below.
Inflorescence: Rather short cymes without bracts.
Flowers: *Pedicels* 1–2mm long, in fruit ascending and to 5mm long. *Calyx* bell-shaped, silvery-hairy; lobes narrowly lanceolate, 3–5mm long, to 7mm in fruit. *Corolla* bright blue, tube about 2mm long, limb 4–9mm across, spreading, with rounded lobes, and yellow eye.
Nutlets: 2–2.5 × 1.4–1.7mm, roundish-ovoid, keeled on one surface, dark brown or black, with distinct grooves.
Chromosomes: 2n=24.

GEOGRAPHY AND ECOLOGY

Distribution: Europe, Asia and North America, alpine, subalpine and arctic regions.
Habitat: Basic mountain rocks, alpine meadows, stony places, open forest, 700–3000m (2300–9850ft). A rare British native, in parts of Cumbria and on Ben Lawers, Perth.
Flowering: June to September in the wild, usually spring to early summer in cultivation.
Associated plants: In the central Highlands of Scotland, grows with cyphel (*Minuartia sedoides*), rock whitlow-grass (*Draba norvegica*) and alpine cinquefoil (*Potentilla crantzii*); in N Yorkshire and Westmorland, with blue moor-grass (*Sesleria albicans*), sheep's fescue (*Festuca ovina*), wild thyme (*Thymus polytrichus* subsp. *britannicus*) and harebell (*Campanula rotundifolia*).

CULTIVATION

Hardiness: Zone 4.
Sun/shade aspect: Full sun is ideal; though it can grow in dappled shade, this may spoil the compact habit.
Soil requirements: Gritty, well-drained soil, ideally with some leaf mould mixed in.
Garden uses: Rock garden, scree, wildflower garden.
Spacing: 20cm (8in) apart, or less for the more compact forms.
Propagation: Seed ripens about two months after flowering in the wild (August to September).
Availability: The species occasionally sold by nurseries. Often listed on seed-exchange lists of specialist societies, with varied origins.
Problems: Can be short-lived; propagate frequently to replace older plants.

RELATED FORMS AND CULTIVARS

'Aurea' – has yellow leaves.
'Myosotis rupicola' is the name applied to very compact, dense forms, 2.5–5cm (1–2in) high, with small flowers, but is not distinct from *M. alpestris*.
'Ruth Fischer' – to 20cm (8in) high, has larger leaves which are curled and crumpled and larger, light blue flowers.

RELATED SPECIES

Closely related species include **Myosotis alpina** Lapeyr, (syn. *Myosotis pyrenaica* Pourret), a mat-forming perennial, with stems up to 12cm (4^1/$_2$in), and with deep blue flowers 8mm (^{3}/$_8$in) across, from rocks, screes and mountain pastures of Pyrenees. Asian forget-me-not **M. asiatica** (Vestergr.) Shischkin & Sergievsk (syn. *M. alpestris* subsp. *asiatica* Vesterg in Hulten), to 10cm (4in) high, flowers bright/deep blue to 8mm (^{3}/$_8$in) diameter, almost glabrous, from Arctic and subarctic Russia. This has been the official state flower of Alaska since 1908. *Myosotis suaveolens* Waldst. & Kit. in Willd. (syn. *M. alpestris* subsp. *suaveolens* A. Strid), from the alpine habitats of the Balkans, grows up to 40cm (16in), with deep blue flowers. This is not generally available, but the seed may appear on the exchange lists of specialist societies, and are worth trying.

Myosotis arvensis (Linnaeus) Hill

A variable, weedy annual to 30cm (12in) high or taller, with sprays of tiny, blue forget-me-not flowers. Not out of place in a 'cornfield patch' or a sunny spot in a wildlife garden; otherwise a weed which may be tolerated.

NAMING
Etymology: *arvensis* – of cultivated fields.
Synonyms: *Myosotis intermedia* Link; *M. scorpiodes* var. *arvensis* Linnaeus
Common names: *English*: Field Forget-me-not, Common forget-me-not; Rough forget-me-not (USA); *Danish*: Mark-forglemmigej; *Dutch*: Akkervergeet mij nietje; *Finnish*: Peltolemmikki; *French*: Myosotis des champs; *German*: Acker-Vergißmeinnicht; *Icelandic*: Gleym-mér-ei, Kallarauga; *Italian*: Nontiscordardime Minore; *Norwegian*: Åkerminneblom; *Russian*: nezabudka polevaya; *Slovak*: nezábudka roæná; *Swedish*: åkerförgätmigej.

DESCRIPTION
General: A very variable, upright, softly hairy annual or biennial. The habit greatly depends on the environmental conditions.
Stems: 15–30cm (6–12in), to 50cm (20in), robust, hairy, usually branched at the base.
Leaves: Basal leaves to 8cm (3($^1/_4$in) long and 1.5cm ($^5/_8$in) wide, oblanceolate, scarcely stalked, in a rosette. Stem leaves oblong-lanceolate, pointed, stalkless, all hairy on both surfaces.
Inflorescence: Cymes without bracts, loose, elongated after flowering.
Flowers: *Pedicels* directed upwards in fruit, the lowest to 10mm, shorter above. *Calyx* to 7mm in fruit, with many short, hooked hairs, and narrow triangular lobes, two-thirds to three-quarters of the length of calyx. *Corolla* bright blue, sometimes purple, with tube about 2mm long, limb 3–5mm across, concave/saucer-shaped. *Style* very short.
Nutlets: To 2.5 × 1mm, ovoid, pointed, keeled on one face, dark greenish- to brownish-black, shining.
Chromosomes: 2n=24, 48, 25.

GEOGRAPHY AND ECOLOGY
Distribution: Europe, western Asia, Siberia, north-east Africa; naturalized in North America.
Habitat: Dry places, arable soils, disturbed ground, roadsides, sand-dunes; to 2000m (6550ft).
Flowering: April to October.
Associated plants: Grows in disturbed ground with other ruderals such as Buxbaum's speedwell (*Veronica persica*), corn poppy (*Papaver rhoeas*) and scarlet pimpernel (*Anagallis arvensis*).

CULTIVATION
Hardiness: Zone 6.
Sun/shade aspect: Sun.
Soil requirements: Any well-drained soil, from moderately acid to slightly alkaline. Tolerant of drought.
Garden uses: Grow in a 'cornfield patch', with other annual weeds, or let it seed to fill bare patches. Will thrive by the sea. Useful in a wildlife garden because the birds like the seeds.
Spacing: 10–20cm (4–8in).
Maintenance: Dead-head or pull out to prevent seeding, unless you want to leave the seeds for the birds (and for the new generation of little blue-flowered weedlings).
Propagation: There is rarely, if ever, a need to propagate *M. arvensis,* as it perpetuates itself happily without any outside assistance. Seed is set from late spring onwards, and in the wild usually germinates in autumn, so this is probably the best time to sow, covering the seed to about 2mm ($^1/_{16}$in) deep. The germination rate is said to increase after dry storage.
Availability: One or two wildflower nurseries may stock this, but it is not really worth buying as a plant. Seed is available from a number of seed merchants, especially those specializing in wildflowers.
Problems: A weedy little plant, though can be tolerated in a wild garden, but seeding far too freely for neatly groomed spaces. Prone to powdery mildew.

RELATED SPECIES
Early forget-me-not, *Myosotis ramosissima* Rochel, is a tiny annual (often only 5cm/2in high, though sometimes taller) with bright blue flowers just 2–3mm ($^1/_8$in) across, of similar habitats in Europe, south-west Asia and North Africa. Yellow-and-blue or Changing forget-me-not, *M. discolor* Pers., is a small, slender annual with similarly minute flowers that first open cream or pale yellow, turning pink, violet or blue, and is widespread throughout most of Europe (not Britain). Both are puny little weeds, but may be of interest for a wildflower collection.

Myosotis australis R. Brown

A short-lived perennial forget-me-not, to around 15cm (6in) tall or sometimes more, with one or few rosettes of hairy, spoon-shaped or elliptical, greenish-brown leaves, and clusters of custard-yellow or white flowers to 5mm ($^1/_4$in) across. Easier to grow than some other New Zealand *Myosotis*.

NAMING
Etymology: *australis* – southern.
Common name: *English:* Australian forget-me-not.

DESCRIPTION
General: A tufted annual, biennial or short-lived perennial.
Stems: To 15cm (6in), sometimes to 30cm (12in), branched near the base, with upright or ascending branches.
Leaves: Basal leaves usually in a solitary rosette, spoon-shaped or elliptical, blunt, 2–6cm ($^3/_4$–2$^1/_2$in) long and 4–12mm ($^3/_{16}$–$^1/_2$in) wide, with a stalk to 6cm (2$^1/_2$in) long; upper stem-leaves stalkless, mostly 1–1.5cm ($^7/_{16}$–$^5/_8$in) long, rather pointed. All leaves greenish-brown, finely and densely hairy above, sparser below.
Inflorescence: Terminal cymes, usually simple and without bracts, to 20cm (8in) long.
Flowers: *Pedicels* very short. *Calyx* about 4mm long, lobes over half the length of calyx, sparsely hairy. *Corolla* yellow or white, tube about 4–5mm long, widest at top, limb 3–5mm across, with rounded, concave lobes. *Stamens* with very short filaments attached below scales, anthers barely reaching scales.
Nutlets: More or less pointed and slightly keeled.

GEOGRAPHY AND ECOLOGY
Distribution: Australia, Tasmania, New Zealand (North and South Island).
Habitat: Mountains, in stony, rocky or shingly places, sometimes in tussock grassland.
Flowering: Spring to summer in the wild; usually early to midsummer in cultivation.
Associated plants: In Australia, may be found near *Swainsona* spp., *Velleia paradoxa*, *Wahlenbergia* spp.

CULTIVATION
Hardiness: Zone 8.
Sun/shade aspect: Sun with shade during midday.
Soil requirements: Gritty, moisture-retentive soil.
Garden uses: Rock garden, raised bed, scree, alpine house.
Spacing: 20cm (8in).
Propagation: Relatively easy from seed and cuttings.
Availability: Rarely available, but look out for seed on the exchange lists of specialist societies.

Myosotis azorica H.C. Watson in Hook

A half-hardy, short-lived perennial forget-me-not, around 30cm (12in) high, carrying clusters of intensely blue flowers opening from striking violet-purple buds. Uncommon and would normally need protection of the alpine house, but worth trying. This plant is rare and protected in the wild.

> *'A striking little plant with large, deep blue flowers from reddish-purple buds.'*
> (Anna N. Griffith)

NAMING
Etymology: *azorica* – of Azores.
Common name: *English*: Azores forget-me-not.

DESCRIPTION
General: A short-lived, hairy perennial.
Stems: 20–30cm (8–12in), sometimes twice as high, much-branched, the leafy parts densely hairy.
Leaves: Basal leaves to 10cm (4in) long and 2cm ($^3/_4$in) wide, narrowly obovate, densely hairy.
Inflorescence: Dense, hairy cymes.
Flowers: *Pedicels* not longer than calyx. *Calyx* up to 5mm in fruit; lobes narrowly linear, twice as long as tube, hairy. *Corolla* deep blue with a white eye, limb about 6mm across, opening from violet-purple buds.
Nutlets: 1.5 x 1mm, ovoid, blunt, blackish-brown, rimmed.

GEOGRAPHY AND ECOLOGY
Distribution: Azores, Canary Islands, Algeria; naturalized in California. A rare plant species included in *The IUNC Red List of Threatened Plants*.
Habitat: Wet rocks.
Flowering: Summer.

CULTIVATION
Hardiness: Zone 9, but may survive in colder areas in a warm, sheltered position.
Sun/shade aspect: Sun.
Soil requirements: Gritty, well-drained soil that is moisture-retentive at the same time.
Garden uses: Alpine houses, or a very sheltered spot in a rock garden.
Spacing: 30cm (12in).
Propagation: From seed in late spring.
Availability: Rare but may be obtainable from suppliers.

Myosotis capitata Hook. f.

An unusual and beautiful tufted annual or short-lived perennial, 10–20cm (4–8in) tall, with rosettes of spoon-shaped to narrowly oblong, silky-hairy leaves, and clusters of striking, deep blue flowers, about 5mm across. Difficult to grow well, and rare in cultivation, this is a connoisseur plant, for special treatment in the alpine house.

'… One of the rarest and arguably the most beautiful of the New Zealand forget-me-nots … gorgeous iridescent purple-blue myosotis.' (Cliff Booker)

NAMING

Etymology: *capitata* – referring to the capitate (head-like) stigma of this plant.

DESCRIPTION

General: A tufted annual, biennial or short-lived perennial, usually with one rosette.
Stems: 10–20cm (4–8in), upright, branched.
Leaves: Basal narrowly oblong to spoon-shaped, rounded with a short narrow point at the tip, 3–12cm (1¼–4½in) long and 1–2.5cm wide (⁷⁄₁₆–1in), with a very broad stalk. Stem-leaves oblong, to 4cm (1½in) long, stalkless. All leaves silky-hairy.
Inflorescence: Short, usually branched cymes, normally without bracts, many-flowered.
Flowers: *Pedicels* very short. *Calyx* 3–5mm long, lobes more than half the length, broad, blunt, with a prominent nerve, silky-hairy. *Corolla* deep blue, tube 3–5mm long, limb flat, 4–8mm across. *Stamens* with very short filaments and anthers up to 1mm long. *Style* with a headlike stigma.
Nutlets: 1.2–2.5 × 1.2–1.5mm, black, pointed, keeled on inner surface.

GEOGRAPHY AND ECOLOGY

Distribution: New Zealand (Auckland Islands, Campbell Islands).
Habitat: Rocky places, gravelly banks, sea-level to 600m (1970ft).
Flowering: November to February in the wild, usually late spring to early summer in cultivation.

CULTIVATION

Hardiness: Zone 8.
Sun/shade aspect: Light shade, or sun with some shade provided during midday.
Soil requirements: Requires soil or compost that is continually moist throughout the growing season, but not stagnant.
Garden uses: Alpine house, or could try a sheltered cool position in a rock garden.
Spacing: 20cm (8in).
Propagation: From seed, as for other New Zealand forget-me-nots.

Availability: Not commonly available, though seed may appear on seed-exchange lists.
Problems: Leaves very susceptible to rotting in all but perfect conditions.

Myosotis colensoi (Kirk) Macbride

A creeping perennial forget-me-not to about 15cm (6in) high, with prostrate stems, softly hairy, grey-green, lanceolate leaves and relatively large, to 8mm (⅜in) across, white flowers borne from spring to summer. For alpine house or a sheltered spot outdoors. A rare and threatened plant in the wild.

NAMING

Etymology: colensoi – after William Colenso (1811–99), missionary and New Zealand botanist.
Synonyms: *Exarrhena colensoi* Kirk; *Myosotis decora* Kirk in Cheeseman.
Common name: *English*: Castle Hill Forget-me-not.

DESCRIPTION

General: A creeping, mat-forming perennial, spreading 15–20cm (6–8in) across.
Stems: Branched, lateral branches decumbent, up to 6cm (2⅓in) long.
Leaves: Basal leaves 2–3cm (¾–1¼in) long and 5–10mm (¼–⁷⁄₁₆in) wide, lanceolate, somewhat pointed, brownish, with a short, broad stalk. Stem leaves oblong, about 1cm (⁷⁄₁₆in) long, upper ones stalkless. All leaves hairy, more densely so above.
Inflorescence: Cymes usually simple, short, few- to many-flowered, with oblong bracts.
Flowers: *Pedicels* short. *Calyx* about 5mm long in flower, 7mm in fruit, lobes about one-third of the length of calyx, broad, somewhat pointed, hairy. *Corolla* white, with cylindrical tube to 5mm long, and flat limb about 8mm across, with rounded lobes. *Stamens* with short filaments, anthers about 2mm long, tips protruding above conspicuous scales. *Style* about twice the length of calyx, with a headlike stigma.
Nutlets: To 1.5 × 1mm, pointed.

GEOGRAPHY AND ECOLOGY

Distribution: New Zealand (Upper Waimakariri Valley, South Island). Included on the IUNC Red List as Vulnerable.
Habitat: Among limestone rocks, rare.
Flowering: November to December in the wild, in cultivation spring to summer.
Associated plants: In Castle Hill Basin, grows on limestone detritus with such plants as *Ranunculus paucifolius*.

CULTIVATION
Hardiness: Zone 8.
Sun/shade aspect: Light shade or sun with some shade at midday.
Soil requirements: Well-drained but moisture-retentive soil, ideally neutral to alkaline.
Garden uses: Rock garden, scree, alpine house.
Spacing: 20cm (8in).
Propagation: From seed or cuttings.
Availability: Sold by some alpine nurseries; seed may appear on exchange lists.

Myosotis explanata Cheeseman
A clump-forming, perennial forget-me-not to 15–20cm (6–8in) high, with rosettes of spoon-shaped leaves and terminal clusters of freesia-scented white flowers in late spring and early summer. A good subject for the alpine house.

'A stout hairy little plant, with solid white flowers.' (Anna N. Griffith)

NAMING
Etymology: *explanata* – flattened.

DESCRIPTION
General: A clump- or cushion-forming perennial.
Stems: To around 20cm (8in) high, ascending, with few upright lateral branches.
Leaves: Basal 3–7cm ($1^1/_4$–$2^3/_4$in) long and 1.5cm ($5/_8$in) wide, obovate to narrowly spoon-shaped, rounded at the tip and with a short, narrow point, hairy; the poorly defined stalk nearly as long as the blade. Upper stem leaves narrowly elliptical, pointed to 3cm ($1^1/_4$in) long, silky-hairy. All leaves grey-green.
Inflorescence: Cymes to 6cm ($2^1/_2$in) long, little branched, without bracts.
Flowers: *Pedicels* very short. *Calyx* to 10mm in fruit, lobes about half length, blunt, rather broad and keeled, covered with hairs, including a few weakly hooked ones. *Corolla* sweetly scented, white or rarely blue, with tube 6–10mm long, funnel-shaped, limb 7–10mm across, with flat, rounded lobes. *Stamens* with very short filaments, anthers nearly 2mm long, tips very slightly protruding beyond large throat-scales. *Style* longer than calyx in fruit, stigma headlike.
Nutlets: To 2.5mm long, oblong.

GEOGRAPHY AND ECOLOGY
Distribution: New Zealand (only above Arthur's Pass, South Island).
Habitat: Rocks by streams in mountains.
Flowering: December to February in the wild.

CULTIVATION
Hardiness: Zone 8.
Sun/shade aspect: Sun with some midday shade.
Soil requirements: Gritty, perfectly drained compost or soil, with sufficient humus.
Garden uses: Alpine house, rock garden, scree.
Spacing: 20cm (8in).
Propagation: As for other New Zealand forget-me-nots.
Availability: Sold by a few alpine nurseries, and can be found in seed-exchange lists and some catalogues.

Myosotis rakiura L.B. Moore
A clump-forming, perennial forget-me-not to 30cm (12in) high, occasionally forming mats as wide, with matt-green, oblong to spoon-shaped leaves, and branched clusters of white flowers in summer. Another nice plant for a rock garden, or perhaps the front of a well-drained border.

NAMING
Etymology: *rakiura* – the Maori name for Stewart Island, New Zealand.
Synonyms: *Myosotis albida* (Kirk) Cheeseman; *M. capitata* subsp. *albida* Kirk; *M. c.* var. *albiflora* Armstrong.

DESCRIPTION
General: A clump-forming, rather robust perennial, sometimes mat-forming.
Stems: To 30cm (12in) high, with few to many lateral branches, ascending to erect.
Leaves: Basal leaves up to 12cm ($4^1/_2$in) long and 2cm ($3/_4$in) wide, oblong-spoon-shaped, tips rounded, margins curved downwards and inwards; blade gradually narrowed into broad, flat stalk. Stem leaves many, about 2–3cm ($3/_4$–$1^1/_4$in) long and 1cm ($7/_{16}$in) wide, more or less oblong, blunt. All leaves covered in short, silky hairs.
Inflorescence: Cymes without bracts, usually several times branched, many-flowered, compact, reaching 4cm ($1^1/_2$in) long in fruit.
Flowers: *Pedicels* very short. *Calyx* 4–5mm long, lobed to more than half the length, lobes rather broad, blunt, with numerous hairs, some of which are irregularly hooked. *Corolla* white, about 6mm across, tube 3–4mm long, limb spreading, about 6mm across, with broadly oblong lobes. *Stamens* with short filaments, and anthers up to 1mm, protruding above the scales. *Style* with a headlike stigma.
Nutlets: Up to 2 x 1.3mm, somewhat pointed, slightly winged and keeled.

GEOGRAPHY AND ECOLOGY
Distribution: New Zealand (South Island, Snares Island, Stewart Island).
Habitat: Maritime rocks.
Flowering: November to January in the wild.

CULTIVATION
Hardiness: Zone 8.
Sun/shade aspect: Sun, with some midday shade.
Soil requirements: Gritty, well-drained but moist soil.
Garden uses: Rock garden, or could be tried at the front of a border, or in pots under cover in cold areas.
Propagation: As for the genus. Easy to propagate by division, and may even self-seed in suitable conditions.
Availability: Sold by some alpine nurseries, and seed regularly appears on seed-exchange lists.

Myosotis rehsteineri Wartmann
A miniature water forget-me-not, biennial or short-lived perennial, to 10cm (4in) high, with small oblong leaves, and bright blue, or occasionally pink, flowers to 10mm ($^7/_{16}$in) across, in late spring to summer. A nice little plant for the water's edge, or moist spots in a rock garden.

NAMING
Etymology: *rehsteineri* – after Hugo Rehsteiner (1864–1947), Swiss botanist.
Synonyms: *Myosotis scorpioides* subsp. *caespititia* (De Candolle) E. Baumann; *M. caespititia* (De Candolle) Kerner).
Common names: *English*: Lake Constance forget-me-not; *French*: Myosotis de Rehsteiner; *German*: Bodensee Vergißmeinnicht; *Italian*: Nontiscordardime di Rehsteiner.

DESCRIPTION
General: A tufted biennial or perennial, rather similar to *M. scorpioides*, but a shorter plant, forming mats 20–30cm (8–12in) across.
Stems: 2–10cm ($^3/_4$–4in) in flower, covered with upward-pointing hairs.
Leaves: 1.5–3cm ($^5/_8$–1$^1/_4$in) long and 5–10mm ($^1/_4$–$^7/_{16}$in) wide, oblong to oblong-lanceolate or oval, hairy on both surfaces.
Inflorescence: Cymes without bracts.
Flowers: *Pedicels* to about 5mm in fruit. *Calyx* 3–5mm in fruit, divided to half or occasionally to two-thirds, hairy or bristly. *Corolla* bright blue or sometimes pink, limb 6–12mm across.
Nutlets: About 1.8mm, ovoid-elliptical.
Chromosomes: 2n=22.

GEOGRAPHY AND ECOLOGY
Distribution: Alps – Austria, Germany, Switzerland, Italy, Liechtenstein. Included on *The IUCN Red List of Threatened Plants* as Endangered.
Habitat: Lake margins.
Flowering: Spring to early summer.

CULTIVATION
Hardiness: Zone 7.
Sun/shade aspect: Full sun is preferred.
Soil requirements: Any moist soil, preferably rich in humus.
Garden uses: Pondside, bog garden, a moist spot in a rock garden.
Spacing: 20–30cm (8–12in).
Propagation: Sow seeds in compost, which must be kept continuously moist. Divide mats into sections and pot up the divisions, standing the pots in a shallow tray of water.
Availability: Occasionally available from nurseries specializing in more unusual plants.

Myosotis scorpioides Linnaeus
A semi-evergreen, marginal aquatic perennial to about 45cm (18in) or taller, with oblong, hairy leaves and clear sky-blue, yellow-centred forget-me-not flowers opening from pink buds, from spring into summer. One of the indispensable garden pond plants.

> *'An ideal plant for helping to disguise where pool meets land and absolutely reliable in almost every situation.' (Philip Swindells and David Mason)*

NAMING
Etymology: *scorpioides* – like the scorpion (tail), referring to the shape of inflorescence.
Synonyms: *Myosotis palustris* (L.) Hill; *M. scorpioides* subsp. *palustris* (L.) F.Hermann; *M. praecox* Hülph.; *M. palustris* var. *praecox* (Hülph.) Hyl.; *M. scorpioides* subsp. *praecox* Almq. & Aspl.
Common names: *English*: Water forget-me-not, True forget-me-not; *Danish*; Eng-Forglemmigej; *Finnish*: Luhtalemmikki; *French*: Myosotis des marais; *German*: Sumpf-Vergißmeinnicht; *Icelandic*: Engjamunablóm; *Italian*: Nontiscordardime delle paludi; *Norwegian*: Engminneblom; *Russian*: nezabudka skorpionovidnaya; *Slovak*: nezábudka moËiarna; *Swedish*: (Äkta) förgätmigej, praktförgätmigej.

DESCRIPTION
General: An evergreen perennial, with a shortly creeping rhizome, often producing stolons (runners).

Myosotis scorpioides. (George McKay)

Stems: Usually 15–45cm (6–18in), occasionally to 100cm (40in), angular, ascending to upright, hairless or hairy.

Leaves: Basal leaves to 7–10cm (2³/₄–4in) long and 2cm (³/₄in) wide, oblong to oblong-lanceolate, usually shortly hairy or nearly hairless, almost stalkless. Stem leaves narrower and often with a small broad point at the apex.

Inflorescence: Cymes without bracts, sometimes containing up to 100 flowers.

Flowers: *Pedicels* to 10mm in fruit, spreading or reflexed. *Calyx* up to 6mm in fruit, bell-shaped, hairy, with triangular lobes, a quarter to a third the length of the calyx. *Corolla* sky-blue with a white, yellow or pink eye, rarely all white or pink, limb 3–10mm across, with flat, slightly notched lobes. *Style* often protruding.

Nutlets: About.1.5 x 1mm, narrowly ovoid, blunt, slightly bordered, black and shining.

Chromosomes: 2n=64.

GEOGRAPHY AND ECOLOGY

Distribution: Central and Northern Europe, Asia, North Africa; naturalized in North America and New Zealand.

Habitat: Wet places, pond and stream margins, ditches, often in water, to 2000m (6550ft).

Flowering: May to September in the wild in Europe.

Associated plants: In England grows in the same habitats as marsh bedstraw (*Galium palustre*), water mint (*Mentha aquatica*) and lady's smock (*Cardamine pratensis*).

CULTIVATION

Hardiness: Zone 5.

Sun/shade aspect: Sun or light shade.

Soil requirements: Moist soil to shallow water, to about 5cm (2in) deep, though it will tolerate up to 10cm (4in) of water depth; moderately acid to slightly alkaline, prefers soil of medium fertility. Does not stand drying out. Can grow on heavy clay.

Garden uses: Grow on a pond margin, in a bog, by a stream, in a wildflower garden.

Spacing: 15–30cm (6–12in).

Propagation: *Seed* – Sow in moist compost or straight into the flowering position, about 2mm (¹/₁₆in) deep. Germination may be slow; dormancy is said to be broken by dry storage at 20C (68°F). *Vegetative* – Propagate by division in late spring, replanting the portions back into the mud, or into containers of wet compost.

Availability: Widely available from aquatic nurseries and some wildflower nurseries.

RELATED FORMS AND CULTIVARS

'Blaqua' (syn. Maytime) – leaves broadly margined in cream.

'Mermaid' – more compact, with sturdy stems 15–23cm (6–9in) high, and dark green leaves, smallish bright blue flowers.

'Pinkie' – pink-flowered form.

'Semperflorens' – compact, to 20cm (8in) high.

'Snowflakes' – white-flowered form.

'Thuringen' – with sky-blue flowers.

RELATED SPECIES

Creeping forget-me-not, ***Myosotis secunda*** Murray (syn. *M. repens* Hook.) is an annual or biennial, with flowers 4–8mm across, found in wet habitats, on acid, often peaty soils, to 800m (2600ft) in Britain, Ireland, the Faroes, France, Spain and Portugal. Tufted forget-me-not, ***M. laxa*** subsp. ***caespitosa*** (K.F.Schultz) Hyl. (syn. *M. caespitosa*) is an annual or biennial, with flowers 2–5mm across, growing throughout Europe in wet habitats, marshes, stream and pond margins, ditches and dykes, stony and grassy ground to 1600m (5250ft). Both species can be grown in conditions similar to those for *M. scorpioides*, but are less garden-worthy due to their smaller flowers and shorter lifespan.

Myosotis sylvatica 'Rosylva'. (Howard Rice)

Myosotis sylvatica Hoffmann

A pretty, very variable, late spring- to early summer-flowering perennial often grown as a biennial, 15–50cm (6–20in) tall, with branched stems, hairy oval leaves and loose cymes of abundant, fragrant, bright sky-blue flowers with a yellow eye, usually opening from deep-pink buds. Numerous strains are grown for bedding, the species sometimes in a wildflower garden, in sun or moderate shade.

NAMING

Etymology: *sylvatica* – of woods, referring to the plant's habitat.
Synonyms: *Myosotis arvensis* var. *sylvatica* Persoon; *M. oblongata* Link.
Common names: *English*: Wood forget-me-not; Woodland forget-me-not (USA); *Danish*: Skov-Forglemmigej; *Dutch*: Bosvergeet mij nietje; *Finnish*: Puistolemmiki; *French*: Myosotis Des Forets; *German*: Wald-Vergißmeinnicht; *Italian*: Nontiscordardime dei boschi; *Norwegian*: Skogminneblom; *Swedish*: Skogsförgätmigej, trädgårdsförgätmigej; *Russian*: nezabudka lesnaya; *Slovak*: nezábudka lesná; *Welsh*: Llys Coffa'R Coed.

DESCRIPTION

General: A very variable, hairy biennial or short-lived perennial.
Stems: 15–50cm (6–20in), often much-branched, very leafy, hairy.
Leaves: Basal leaves to 11cm (4¹⁄₄in) long and 3cm (1¹⁄₄in) wide, elliptical-oblong, spoon-shaped, oblong-lanceolate or oval, rounded, with the base narrowing

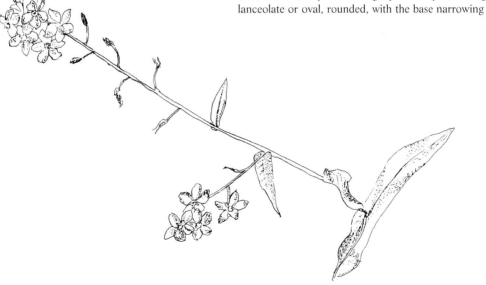

Myosotis sylvatica (Doreen Townley).

into a small, indistinct stalk; stem leaves are
lanceolate, pointed. All leaves grey-green, hairy,
especially densely above.
Inflorescence: Loose cymes, without bracts,
elongating to 15cm (6in) in fruit.
Flowers: Scented. *Pedicels* up to 10mm. *Calyx* to
5mm in fruit, with short hooked hairs, with linear to
narrowly triangular lobes and rounded base, open,
deciduous. *Corolla* bright sky blue, occasionally pink
or white, tube 2–3mm, limb 6–11mm across, flat,
with yellow scales in throat. Style 1.5–2.5mm long.
Nutlets: 1.7–2 x 1.2–2mm, ovoid, sharply pointed,
black brown, with an indistinct rim.
Chromosomes: 2n=18.

GEOGRAPHY AND ECOLOGY
Distribution: Most of Europe except south-west and
much of the north, North Africa, West Asia. Local in
the UK; naturalized in N America.
Habitat: Damp woodland and mountain grassland,
locally abundant, to 2000m (6550ft).
Flowering: April to July in the wild in Europe.

CULTIVATION
Hardiness: Zone 5.
Sun/shade aspect: Sun or light shade are best; will
tolerate moderate shade but the flowering may be
reduced.
Soil requirements: Tolerates any well-drained soil.
Garden uses: Grow in a wildflower garden, as an
easy filler in a border, in dappled shade of deciduous
trees and shrubs. The seed strains for spring bedding,
traditionally combined with tulips and wallflowers.
Good in a wildlife garden, as insects are attracted to
the flowers, and finches eat the seeds.
Spacing: 15cm (6in) apart, or more for taller cultivars.
Propagation: Sow any time from mid-spring to early
autumn. Bedding strains usually sown in summer.
Germination normally takes two to three weeks. If the
seed strains used for bedding are left to self-sow, after a
few generations they will revert back to wild type.
Germination improved in the dark.
Availability: Plants and seeds of the wild form are
available from many wildflower suppliers. Seed strains for
bedding or containers are offered by most seed merchants.
Problems: Powdery mildew. May self-seed more freely
than some would like, but in the right place it is a bonus.

RELATED FORMS AND CULTIVARS
There is a naturally occurring white-flowered form,
Myosotis sylvatica f. *lactea* (Boenn.) Duvign. (syn.
M. sylvatica 'Alba'), which is uncommon in
cultivation, but pretty and worth seeking out.
'Gold 'n' Sapphires' is a new, perennial cultivar with

golden-yellow foliage, from Terra Nova Nurseries
(USA), which has been erroneously listed as *M.
alpestris* form. The following seed strains are grown:
Ball Series – 15–20cm (6–8in), compact, ball-shaped:
    'Blue Ball' ♀ – azure-blue flowers; RHS Award
    of Garden Merit plant.
    'Pink Ball' – pink flowers.
    'Snowball' – pure white flowers.
'Blue Basket' – 25–30cm (10–12in), upright, with
deep azure-blue flowers.
'Bouquet' ♀ – RHS Award of Garden Merit plant;
does not appear to be in cultivation at present.
'Carmine King' – very deep pink flowers.
'Compindi' – 15–20cm (6–8in), compact habit, very
dark blue flowers.
'Music' – to 25cm (10in), vigorous and upright, with
large, bright blue flowers.
'Pompadour' – compact and ball-shaped, with large,
deep rose-pink flowers, 15–20cm (6–8in).
'Rosie' – 15cm (6in), rose-pink flowers.
'Rosylva' (Fleuroselect medal) – 15–20cm (6–8in),
large, clear rose-pink flowers.
'Royal Blue' – 30–40cm (12–16in), rich blue flowers;
tall stature makes it ideal for underplanting with tulips.
'Ultramarine' – dwarf and compact, with deep
indigo-blue flowers, to 15cm (6in).
Victoria Series – dwarf and compact, with white,
blue or pink flowers:
    'Victoria Rose' – bright rose-pink flowers, 10cm
    (4in).

Myosotis traversii J.D. Hooker
A cushion-forming perennial forget-me-not, to
about 10cm (4in) high, with pale green spoon-
shaped, hairy leaves, and dense heads of white or
pale yellow flowers, each to 4mm ($^3/_{16}$in) across,
borne in summer. An alpine house plant, not easy to
grow.

NAMING
Etymology: *traversii* – after W.T.L. Travers, the Irish
ornithologist and plant collector, who came to New
Zealand in 1849 and explored plants on South Island.

DESCRIPTION
General: A cushion-forming perennial, sometimes
forming a small mat.
Stems: To about 10cm (4in) high, pinkish-green,
branches upright or ascending.
Leaves: Basal leaves 2–7cm ($^3/_4$–$2^3/_4$in) long and
5–10mm ($^1/_4$–$^7/_{16}$in) wide, spoon-shaped, white-hairy,
with a broad stalk.
Inflorescence: Dense, headlike cymes, without bracts.

Flowers: *Calyx* to 5mm in flower, lobed to about one-third of the length, covered in short hairs and some long hooked hairs. *Corolla* lemon-yellow to white, sometimes with a yellow eye, tube to 5mm long, limb to 4mm across, with rounded lobes. *Stamens* with very short filaments.
Nutlets: 2–2.5 x 1–1.2mm, oblong, slightly winged near tip.

GEOGRAPHY AND ECOLOGY
Distribution: New Zealand (mountains of Nelson, Marlsborough, and North Canterbury, South Island).
Habitat: Scree and shingly places.
Flowering: December to March in the wild in New Zealand.

CULTIVATION
Hardiness: Zone 8.
Sun/shade aspect: Sun with some shade during midday.
Soil requirements: This plant is quite particular about soil conditions, as it must have very good drainage and should not be allowed to dry out; so a gritty but humus-rich mixture is necessary.
Garden uses: Alpine house, rock garden or scree.
Spacing: 20cm (8in).
Propagation: The species is quite difficult to grow, though it may be propagated from detached rosettes and seed.
Availability: Rarely obtainable, but may appear on seed-exchange lists.

New Zealand *Myosotis* species of interest

These species are not commonly in cultivation, but could crop up on seed-exchange lists. They should be hardy to Zone 8, but are best grown under the protection of an alpine house, in conditions similar to those for other New Zealand forget-me-nots.

Myosotis antarctica Hooker f.
Description: A tufted to small mat-forming perennial, about 5cm (2in) tall or more, with rosettes of silky-hairy obovate leaves 1–2cm ($^7/_{16}$–$^3/_4$in) long, and cymes of tiny blue or sometimes white flowers, to 3mm across.
Distribution: New Zealand (Campbell Island).
Habitat: Screes, stony slopes.

Myosotis arnoldii L.B. Moore
Description: A small perennial, 10–15cm (4–6in) high, forming a mat to about 10cm (4in) across, with silvery-hairy, narrowly spoon-shaped leaves 3–7cm ($1^1/_4$–$2^3/_4$in) long, and cymes of unusual, yellow-brown to nearly black flowers, each 8–10mm ($^5/_8$–$^7/_{16}$in) across, in summer.
Distribution: New Zealand (Nelson and Marlborough), rare.
Habitat: On limestone and marble in mountains.

Myosotis elderi L.B. Moore
Description: A perennial with decumbent lateral stems to 6cm ($2^1/_2$in) long, with obovate, blunt leaves 2–3cm ($^3/_4$–$1^1/_4$in) long on a winged stalk, and cymes of up to 12 white, blue or pinkish flowers, about 6mm ($^1/_4$in) across.
Distribution: New Zealand (N Tararua Range).
Habitat: Mountains.
Flowering: November to March in the wild.

Myosotis eximia Petrie
Description: A clump-forming perennial forget-me-not to 25cm (10in) high, with elliptical, hairy leaves to 10cm (4in) long, and widely funnel-shaped white flowers about 15mm ($^5/_8$in) across, in summer.
Distribution: New Zealand (Ruahine and Kaimanawa Ranges, North Island).
Habitat: Cliffs, usually of limestone, and screes, 480–1260m (1575–4130ft).

Myosotis lyallii Hooker f.
Description: A tufted perennial with slightly branched stems, decumbent branches to 8cm ($3^1/_4$in) long, rosettes of oval, spoon-shaped or elliptical leaves, to 3.5cm ($1^3/_8$in) long and 1cm ($^7/_{16}$in) wide, usually hairy, with stalk to 3.5cm ($1^3/_8$in), and few-flowered cymes of white flowers, to 8mm ($^3/_8$in) across.
Distribution: New Zealand.
Habitat: Rocks.

Myosotis macrantha (Hooker f.) Bentham & Hooker

'… Bears hanging funnel-shaped blooms of pale yellow which in the evening diffuse a soft sweet perfume.' (Roy Genders)

'Myosotis macrantha is extremely fickle in its colouring, varying from clear yellow, through

rather unpleasant mustard shades, to a pleasing milk-chocolate tint and a rich mahogany. I have even found plants with clear blue-green corollas.' (W.R. Philipson and D. Hearn)

Synonym: *Exarrhena macrantha* Hooker f.
Description: A tufted or clump-forming perennial 20–30cm (8–12in) high, with obovate to spoon-shaped, hairy leaves, to 12cm (5in) long, and flowers about 8mm ($^3/_8$in) across, varying from yellowish to brown-orange, commonly sulphur-yellow, in summer.
Distribution: New Zealand (South Island).
Habitat: Mountains, stony and rocky habitats, often in moist and shady places.
Flowering: Summer

Myosotis pulvinaris Hooker f., 1864

Synonym: *Myosotis hectori* Hooker.
Description: A small, tufted perennial forget-me-not, about 3cm (1$^1/_4$in) high, forming cushions, to 10cm (4in) across, of tiny soft grey hairy leaves, in summer covered in a profusion of stemless, solitary white flowers to 6mm ($^1/_4$in) across.
Distribution: New Zealand (South Island).
Habitat: Southern mountains of Central Otago, around 1800m (5900ft).
Flowering: Summer

Myosotis pygmaea Colenso

Description: A mat-forming perennial, spreading to 30cm (12in) across, with very slender, prostrate, leafy stems, many small rosettes of copper-tinged, obovate to spoon-shaped leaves, 1–3cm ($^7/_{16}$–1$^1/_4$in) long, somewhat hairy above, and tiny white flowers, only 1.5–3mm across.
Distribution: New Zealand.
Habitat: Heaths, other open habitats, dry, sandy and rocky places.
Flowering: Summer to autumn.

Myosotis saxosa Hooker f.

Synonym: *Exarrhena saxosa* Hooker.
Description: A cushion-forming perennial forget-me-not, only about 8cm (3$^1/_4$in) high, with crowded, broadly oval leaves 2–3cm ($^3/_4$–1$^1/_4$in) long, covered with long white hairs, and clusters of white flowers about 10mm across.
Distribution: New Zealand (North Island).
Habitat: On crags, only in Hawke's Bay.
Flowering: November to December in the wild.

Myosotis suavis Petrie

Description: Similar to *M. explanata,* this perennial forget-me-not is more hairy and compact, growing to 10cm (4in) high, and has sweetly scented flowers to 7mm ($^3/_8$in) across.
Distribution: New Zealand (South Island).
Habitat: Rocky places in mountains, 1000–2000m (3280–6550ft).
Flowering: December to February in the wild.

Myosotis uniflora Hooker, 1864

Description: A branched, tufted perennial, 1–5cm ($^7/_{16}$–2in) high, forming cushions to 10cm (4in) across, with upright stems, very thin, triangular, stiffly hairy leaves, and many solitary, sweetly scented yellow flowers, fading to white, 4–5mm ($^3/_{16}$in) across.
Distribution: New Zealand (Canterbury and Central Otago, South Island).
Habitat: Mountains, stony and gravelly river beds and flats.
Flowering: September to November in the wild.
Associated plants: On Pisa Flats, *M. uniflora* grows with plants such as *Rytiolosperma maculatum, Leucopagon colensoi, Leptinella, Craspedia* and *Galium.*

Other *Myosotis* species of interest

Myosotis decumbens Host.

Description: A plant very similar to *M. sylvatica,* but which is always perennial, rhizomatous and mat-forming, with calyx about half the length of the *corolla* tube.
Distribution: Northern, central, south-western Europe.
Habitat: Mountains, grassy and rocky places, to 2000m (6550ft).
Flowering: May to August.

Myosotis olympica Boissier

Description: A perennial to 15cm (6in) with few to many rosettes arising from short rhizomes, hairy stems, elliptical to spoon-shaped leaves to 10cm (4in) long, and clusters of deep-blue flowers to 8mm ($^3/_8$in) across.
Distribution: Turkey (north-west and east Anatolia), north-west Iran.

Habitat: Rocky slopes, grassy alpine meadows, 700–4100m (2300–13,450ft).
Flowering: May to September in the wild.
Associated plants: May be encountered with such plants as *Veronica caespitosa*, *Viola oreades*, *Draba* species and *Aethionema oppositifolia*.

Myosotis welwitschii Boissier & Reuter
Common name: Monchique forget-me-not.
Description: An annual to biennial, to 60cm (2ft), with robust stems, branched and hairy at base, hairy elliptical or oval-lanceolate leaves to 7cm (2³/₄in) long, and branched clusters of bright blue flowers, to 10mm (⁷/₁₆in) across, with a yellow-white eye. Hardy to Zone 9.
Distribution: Spain, Portugal, Morocco.
Habitat: Wet places, usually near the coast.

The genus *Nonea*
Medikus, 1789
Monk's-wort

NAMING
Etymology: *Nonea* – after J.P. Nonne of Erfurt (1729–72), a German writer on botany.
Synonyms: Sometimes incorrectly spelt '*Nonnea*'. Some species were previously included in the genera *Lycopsis* L. and *Anchusa* L.

GENUS CHARACTERISTICS AND BIOLOGY
Number of species: About 35.
Description: Annual or perennial herbs, often with glandular hairs. Basal and stem-leaves hairy, the former stalked, the latter stalkless. Flowers can be yellow, white, pink, violet, blue, purple or brown, borne in terminal cymes with bracts. *Calyx* divided up to a half, enlarged and becoming ovoid-spherical in fruit. *Corolla* either with long, slender tube and abruptly expanded limb or funnel-shaped; throat with 5 small semicircular, hairy scales cut into lobes, and sometimes a ring of hairs. Stamens included in the corolla. Nutlets downy or hairless, beaked.

Noneas superficially resemble pulmonarias, but can be distinguished by having 5 hairy scales in the corolla throat, and also the absence of a creeping rhizome.
Distribution and habitats: The Mediterranean, south-west and central Asia, to eastern Siberia; in a variety of dry habitats.

Life forms: Therophytes and hemicryptophytes.
Pollination: Bees.
Fruit dispersal: No specialized mechanism.

USES
No medicinal or culinary uses are known.

GARDEN
Noneas add variety to a mixed or herbaceous border, and the shorter species may be used in a rock garden.

CULTIVATION
Very few species of *Nonea* are grown, and only one is relatively easy to find, but it is likely that most will require similar conditions. These plants will thrive in sun, and are drought-tolerant, but will require moisture during the growing season. The soil should not be too fertile, otherwise foliage will be produced at the expense of flowers.

PROPAGATION
Propagation is by seed only, ideally as soon as ripe. Sow in pots or trays of gritty compost, placing these in a cold frame. On well-drained, sandy soil it is possible to sow *in situ*, and the plants may self-seed.

PROBLEMS
Plants are short-lived, and may look untidy when dying down, but it is necessary to leave some flowered stems in order to renew the plants from seed. Bristles may cause slight irritation to the skin.

REFERENCES
BLAMEY & GREY-WILSON, 1989, 1993; DAVIS, 1978; PHILLIPS & RIX, 1991, 1999; SHISHKIN & BOBROV, 1967–1977; STACE, 1992; TUTIN *et al*, 1972; WALTERS, 2000.

Nonea lutea (Desrousseaux) De Candolle in Lam. & DC
An unusual bristly annual, superficially looking like a yellow-flowered pulmonaria. It grows 10–60cm (4–24in) high, has lanceolate bristly leaves, and pale yellow, funnel-shaped flowers with a purple-tinged calyx. Easy to grow in well-drained soil in a sunny spot.

NAMING
Etymology: *lutea* – yellow, referring to the colour of flowers.
Synonyms: *Lycopsis lutea* Desrousseaux in Lam.; *Anchusa lutea* M.B. The author saw it sold as '*Nolina lutea*', which is in a genus belonging to a completely different family, Agavaceae.

Nonea lutea. (George McKay)

Common names: *English*: Yellow nonea, Yellow monk's-wort (USA); *French*: Nonnée Jaune; *German*: Gelbes Mönchskraut; *Italian*: Nonnea gialla; *Russian*: noneya zholtaya; *Slovak*: ostren tant.

DESCRIPTION

General: A bristly and glandular-hairy annual.
Stems: 10–60cm (4–24in), upright or ascending, branched.
Leaves: 20–70mm (^{3}/$_4$–2^3/$_4$in) long and 5–20mm (^{1}/$_4$–^{3}/$_4$in) wide, oblong-lanceolate to linear-lanceolate, with entire or weakly toothed margin, blunt or pointed; stem-leaves half-clasping.
Inflorescence: Short terminal cyme, with bracts; elongating in fruit.
Flowers: *Calyx* 6–10mm, to 20mm in fruit, divided to a quarter or half their length, lobes narrowly triangular, often violet-tinged. *Corolla* pale yellow, 7–18mm, limb erect with the lobes slightly spreading outwards to a quarter or a third of the length of the

tube; tube with 5 hairy scales and a ring of hairs.
Stamens with anthers 1.5–2mm long.
Nutlets: 3.6–6 × 2mm, erect, oblong-ellipsoid, downy, wrinkled, brown with white flecking and longitudinally ribbed.
Chromosomes: 2n=14.

GEOGRAPHY AND ECOLOGY
Distribution: South-eastern Russia, Ukraine, the Caucasus, northern Iran, north-east Turkey; naturalized in the former Yugoslavia.
Habitat: Field margins, rocky slopes, about 500m (1650ft) in Turkey.
Flowering: April to June in Turkey.

CULTIVATION
Hardiness: Zone 7
Sun/shade aspect: Full sun.
Soil requirements: Any well-drained soil.
Garden uses: Mixed border, any spot where a filler is required.
Spacing: 20–30cm (8–12in).
Propagation: As for the genus.
Availability: The only species that can be obtained with relative ease; plants sold by some herbaceous specialists, and the seed may appear in the exchange lists of plant societies.
Problems: After flowering looks very untidy, but cannot be tidied up if the seed is to be collected.

RELATED SPECIES
Nonea macrosperma Boiss. & Heldr., is another yellow-flowered species, 10–25cm (4–10in) high, from fields, meadows and steppes of Turkey. Unlike *N. lutea*, this is perennial; it is not in cultivation at present.

Nonea pulla (Linnaeus) De Candolle
A greyish annual or short-leaved perennial, around 30–40cm (12–16in) tall, with bristly lanceolate leaves and unusual, dark blackish-purple or reddish-brown flowers, with prominent calyces inflating in fruit. Nonea pulla is rare in cultivation, but it should be easy to grow in a well-drained, sunny position in a mixed border.

NAMING
Etymology: *pulla* – nearly black, referring to the colour of the flowers.
Synonyms: *Lycopsis pulla* Linnaeus, *Anchusa pulla* M.B., *Nonea atra* Grisebach (invalid).
Common names: *English*: Brown nonea; *Danish*: Sortrød Kosakurt; *Finnish*: Rusonunna; *German*: Braunes Mönchskraut; *Russian*: noneya tyomnaya,

noneya temnoburaya; *Slovak*: ostreňpočerný; *Swedish*: Svartnonnea.

DESCRIPTION

General: A grey-bristly and glandular-hairy annual or perennial.

Stems: 25–50cm (10–20in), upright, usually branched above.

Leaves: 3–12cm ($1^1/_4$–$4^1/_2$in) long and 7–20mm ($^3/_8$–$^3/_4$in) wide, lanceolate to linear-lanceolate, pointed, with entire margin.

Inflorescence: A corymb-like panicle of cymes, with short, lanceolate bracts.

Flowers: *Calyx* 6–8mm ($^1/_4$–$^3/_8$in), to 12mm ($^1/_2$in) in fruit, lobes one-third as long as the tube. *Corolla* dark reddish- or blackish-purple (occasionally yellowish), 9–14mm ($^3/_8$–$^1/_2$in) long, funnel-shaped, with a limb 5–8mm ($^1/_4$–$^3/_8$in) in diameter, divided to about a quarter into broadly oval lobes.

Nutlets: 2.5–3 × 3.5–5mm, obliquely ovoid, ribbed, with a ribbed collar-like basal ring.

Chromosomes: 2n=18, 20.

GEOGRAPHY AND ECOLOGY

Distribution: Eastern and east Central Europe (naturalized in other parts of Europe, including France, Finland, the Mediterranean); Turkey, Caucasus, Iraq, Iran.

Habitat: Dry, grassy habitats and stony places, at low altitudes in northern Europe, but to 2550m (8360ft) in Turkey, where it may be found in alpine meadows and damp places.

Flowering: April to August in the wild.

Associated plants: In the Altai, can be found in silver birch (*Betula pendula*) woodland, with plants such as *Rosa majalis*, *Iris ruthenica*, *Pulmonaria dacica*, *Eryngium planum*, angled Solomon's-seal (*Polygonatum odoratum*) and *Vicia unijuga*.

CULTIVATION

Hardiness: Zone 7.

Sun/shade aspect: Full sun.

Soil requirements: Any well-drained soil.

Garden uses: As a filler in a mixed or herbaceous border.

Spacing: 30cm (12in).

Propagation: As for the genus.

Availability: Not commonly available, but the seed may appear on the exchange lists of specialist societies.

RELATED SPECIES

Rose monk's-wort, **Nonea rosea** (Bieb.) Link, is rather similar to *N. pulla*, but has purplish or brownish flowers, 15–18mm ($^5/_8$–$^3/_4$in) long. It is native to the Caucasus, but naturalized in waste and grassy places in France, Holland, Belgium, Germany and parts of Scandinavia.

Red monk's-wort, **N. versicolor** (Steven) Sweet, which has been confused with the above species in the past, originates from the Caucasus and eastern Turkey, but naturalized in Denmark, Norway and Sweden. It differs from *N. pulla* in its slightly shorter stature, and purple to violet flowers, 12–17mm ($^1/_2$–$^5/_8$in) long. It is usually an annual but can be a short-lived perennial.

Neither of these plants is in general cultivation, but the seed may crop up on the exchange lists.

Other *Nonea* species of interest

These species do not appear to be in cultivation, but are worth trying if they can be obtained.

Nonea alpestris (Stev.) G. Don

Description: A glandular, hairy perennial or biennial, 15–30cm (6–12in), pale yellow flowers, 8–10mm ($^3/_8$–$^7/_{16}$in) across.

Distribution: Caucasus.

Habitat: Stony slopes.

Nonea intermedia Ledeb.

Description: A softly hairy perennial, 30–45cm (12–18in), flowers pale violet, tube about 5–6mm ($^1/_4$in) long, limb about 10–15mm ($^7/_{16}$–$^5/_8$in) across; flowering June to July in Turkey.

Distribution: Caucasus, north-east Turkey.

Habitat: Alpine meadows, at about 2000m (6550ft) in Turkey.

Nonea macrantha (H.Riedl) A. Baytop

Synonym: *N. pulmonarioides* Bornm. not Boiss. & Bal.

Description: A bristly, glandular perennial, 25–50cm (10–20in), with dark purple flowers, 12–18mm ($^1/_2$–$^3/_4$in) long, 10mm ($^7/_{16}$in) in diameter, flowering June to August in Turkey.

Distribution: Iraq, Iran, Turkey.

Habitat: Rocky slopes, in Turkey at 1800–3100m (5900–10,170ft).

Nonea pulmonarioides Boiss. & Bal. in Boiss.
Description: A softly hairy perennial, 30–40cm
(12–16in) high, with reddish-violet flowers, about
15mm (⁵/₈in) long, limb 10mm (⁷/₁₆in) across;
flowering July-August in the wild. A very beautiful
plant, resembling a pulmonaria. Rare, included in *The
IUCN Red List of Threatened Plants*.
Distribution: Turkey.
Habitat: Grassy screes, rocky slopes, snowmelt zones
at 2700–3400m (8850–11,150ft).

The genus *Omphalodes* Miller, 1754

Navelwort

NAMING
Etymology: *Omphalodes* – from Greek *omphalos*,
navel, and *-oides*, like, referring to the hollow on one
side of the nutlet, resembling a human navel, formed
by incurved wing.
Synonyms: Some species were originally described
under the genus *Cynoglossum* L. One species
described here has recently been separated into
another genus, *Sinojohnstonia* Hu.

GENUS CHARACTERISTICS AND BIOLOGY
Number of species: According to different
estimates, from 14 to around 30 species.
Description: A genus of annual, biennial or perennial
herbs, hairless or very finely hairy. Leaves simple,
alternate, or lower opposite, sometimes glaucous.
Flowers white or blue, rarely pink, usually in terminal
cymes, sometimes solitary in leaf-axils, sometimes
with bracts. *Calyx* 5-parted, divided to at least the
middle, usually enlarged in fruit with lobes spreading
or reflexed. *Corolla* 5-lobed, more or less wheel-
shaped/rotate, with a short tube and flat or concave
limb with spreading lobes. Throat with 5 scales that
are more or less notched at apex. Stamens 5, included
in the corolla tube, inserted near middle of tube. Style
included, with a headlike stigma. Nutlets 4,
depressed-globose, usually smooth, sometimes hairy,
with a wing that is a thickened ring, or can be broad
and incurving, with an entire or toothed margin.
Distribution and habitats: Europe, Mediterranean
regions, East Asia, North Africa, Mexico; in a wide
variety of habitats, from damp woodlands to high
mountain rocks and coastal sands.
Life forms: Therophytes (annuals, like *O. linifolia*)
and hemicryptophytes (most, such as *O.
cappadocica, O. verna*).

Pollination: By various bees.
Fruit dispersal: No specialized mechanism.

USES
No medicinal or culinary uses are known.

GARDEN
Most woodland species are prime candidates for
woodland gardens, groundcover under trees and
shrubs, or a partially shaded border. Some perennials,
like *O. cappadocica*, make a nice informal edging to a
lightly shaded path. Annual *O. linifolia* is an excellent
filler for a mixed border, or any sunny spots. Majority
of perennials and some annuals can be admitted to a
rock garden, and some, like *O. luciliae*, are alpine-
connoisseur's plants, and should ideally be offered
special treatment in the alpine house.

OTHER
Omphalodes linifolia can be used as a cut flower in
fresh arrangements, and lasts well in water.

CULTIVATION
Most species are hardy to at least Zone 7. Woodland
inhabitants, like *O. cappadocica*, *O. nitida* and
O. verna, thrive in moist but well-drained, humus-
rich soil, in dappled shade.
  Annuals, like *O. linifolia*, require sun and well-
drained soil. *Omphalodes luciliae* needs sun and
perfectly drained, gritty, but moist growing medium,
ideally with added limestone chippings, and
protection from winter wet.
  Water all species freely during hot weather. Most
omphalodes resent root disturbance, and should not
be moved unless necessary.

PROPAGATION
Seed Most omphalodes come readily from seed, sown
either fresh, or in spring. Hardy annuals can be in
their flowering positions, covering lightly. The rest
should be sown in pots or trays of good-quality
compost (with plenty of added grit for *O. luciliae*),
with a light cover of grit, and placed in a frame or
cold greenhouse. Many species will self-sow when
happy.
Vegetative Division is reasonably easy for woodland
perennials, such as *O. verna* and *O. nitida* during
early to mid-spring or late summer and autumn.
Replant the portions immediately in their flowering
positions or, to obtain large plants faster, pot them up
first. Well-established clumps of *O. luciliae* may be
divided with great care in early to midsummer.
Cuttings in a frame work well with *O. cappadocica* in
mid- to late summer.

Omphalodes cappadocica. (Howard Rice)

Omphalodes cappadocica 'Starry Eyes'. (Howard Rice)

REFERENCES
BECKETT, 1993–94; BRICKELL, 1996; CLAPHAM *et al.*, 1952; DAVIS, 1978; FISH, 1961; GRIFFITH, 1964; HEATH, 1964; OHWI, 1965; PHILIPS & RIX, 1993; POLUNIN, 1980; THOMAS, 1990, 1993; TUTIN *et al.*, 1972, WALTERS, 2001; WU & RAVEN, 1995.

Omphalodes cappadocica (Willdenow) De Candolle
A more or less evergreen, rhizomatous clump-forming perennial to 25cm (10in) with oval, pointed leaves, prominently veined, and bright azure-blue, white-eyed, large forget-me-not-like flowers in spring. Good groundcover for cool shade.

> '... makes a good clump of very elegant smooth pointed leaves, above which we have those dainty sprays of such very blue flowers. Why is it, I wonder, that blue flowers give such a feeling of innocence and simplicity?' (Margery Fish)

NAMING
Etymology: *cappadocica* – of Cappadocia, Turkey.
Synonyms: *Cynoglossum cappadocicum* Willd.; *Omphalodes cornifolia* Lehm.; *Omphalodes caucasica* Brand in Engler & Prantl.
Nomenclature note: Apparently, many plants sold under this name in fact belong to a different species, *O. lojkae* (Walters, 2000).
Common name: No universally accepted common name, but called 'Blue-eyed Betty' by Margery Fish.

DESCRIPTION
General: A clump-forming perennial, with a more or less horizontal, creeping rhizome.
Stems: 10–15cm (4–6in), sometimes to 30cm (12in), ascending or upright, slender, hairy; apart from flowering stems, shortly creeping vegetative stems are produced.
Leaves: Basal leaves 4.5–10cm (1³/₄–4in) long and 2.5–4.5cm (1–1³/₄in) wide, oval with a heart-shaped base, pointed, with a stalk to 15cm (6in) long. Stem-leaves only 3–7 per plant, shortly stalked or nearly stalkless. All leaves somewhat leathery, finely downy, dark green, with prominent lateral veins.
Inflorescence: Loose, few-flowered terminal cymes with no bracts, about 8 per plant.
Flowers: *Pedicel* about 3mm in flower and 10–30mm (⁷/₁₆–1¹/₄in) in fruit. *Calyx* lobes 4–5mm (³/₁₆in) in flower, to 20mm (³/₄in) in fruit, oval-oblong, pointed,

densely silvery-hairy inside and outside. *Corolla* bright blue, tube about 2mm (¹/₁₆in), limb more or less flat, about 10mm (⁷/₁₆in) across; throat scales white.
Nutlets: To 3.2 × 2mm, held horizontally, downy or almost hairless, with a thickened rim or a narrow incurved wing with toothed margin.

GEOGRAPHY AND ECOLOGY
Distribution: Turkey (north-east Anatolia), Caucasus (Georgia).
Habitat: Forests, especially of sweet chestnut (*Castanea sativa*), filbert (*Corylus maximus)* scrub, shady banks, damp rock ledges, by streams in Turkey, 20–1000m (65–3280ft).
Flowering: March to May in Turkey; April to June in cultivation in the UK.
Associated plants: In the Caucasus (Georgia), it grows with such plants as *Helleborus orientalis* subsp. *guttatus, Campanula alliarifolia, Epimedium colchicum, Ruscus colchicus* and *Cyclamen coum* subsp. *caucasicum.*

CULTIVATION
Hardiness: Zone 6.
Sun/shade aspect: Dappled shade is best, in a cool, sheltered position.
Soil requirements: Well-drained but moist soil, ideally enriched with leaf mould. It dislikes alkaline soils and is best on acid to neutral soils.
Garden uses: An excellent groundcover plant in shade under trees and shrubs; also good in mixed borders, or in a large rock garden.
Spacing: 45cm (18in), though can spread to 60cm (2ft).
Propagation: From seed or by division, as for the genus, and also from cuttings in mid- to late summer.
Availability: Widely available from many nurseries and garden centres, though the cultivars may need to be searched for. Seed can be found in some commercial catalogues and most seed-exchange lists.
Problems: Hard frosts may kill off the flowers and, rarely, the leaves, but the plants should recover in spring.

RELATED FORMS AND CULTIVARS
'Alba' – white-flowered.
'Anthea Bloom' – slightly grey-green leaves and lighter blue flowers; hardier than the species.
'Cherry Ingram' – a vigorous cultivar, to 25cm (10in), with deep-blue flowers rather larger than normal, and narrower dark green leaves.
'Lilac Mist' – with lilac-lavender flowers (a reversion of 'Starry Eyes'), from Terra Nova Nursery, USA.

'Parisian Skies' – large, deep blue flowers, from Terra Nova Nurseries, USA.
'Starry Eyes' – white flowers, tinged very pale blue, with a deep-blue 'star' in the middle of the flower. The seedlings of this plant come blue.

RELATED SPECIES
Omphalodes lojkae Sommier & Levier, from woodlands in the Caucasus is often sold under the name of *O. cappadocica.* It can be distinguished by the shape of basal leaves, which are never heart-shaped at the base in this species, but are gradually tapering. There are usually fewer cymes per plant (about 5) and the flowers are larger, to 20mm (³/₄in) across.
Omphalodes rupestris Rupr. in Boiss., is a similar, but smaller plant, densely tufted, with silver-white felted leaves, native to cliffs in the Caucasus. It is not in general cultivation, but worth looking out for. Probably needs alpine house conditions.

Omphalodes linifolia (Linnaeus) Moench ♟
A pretty, free-flowering, greyish-glaucous, upright annual to 30–40cm (12–16in) high, with spoon-shaped to linear or lanceolate leaves and long racemes of 5–12mm (¹/₄–¹/₂in) fragrant, white, forget-me-not flowers in summer. Easy to raise from seed, for a sunny, well-drained position. The RHS Award of Garden Merit plant.

NAMING
Etymology: *linifolia* – with leaves similar to those of *Linum* (flax).
Synonyms: *Omphalodes linifolia alba; Omphalodes linophyllum* St. Lag.; *Cynoglossum linifolium* L.
Common names: *English:* Venus's navelwort, Whiteflower navelwort; *Dutch:* Witte onschuld; *Finnish:* Kesäkaihonkukka; *French:* Petite bourrache; *Swedish:* Lammtunga.

DESCRIPTION
General: A nearly hairless, glaucous annual.
Stems: 5–40cm (2–16in), upright, simple or sometimes branched from the base.
Leaves: Basal leaves 1–10cm (⁷/₁₆–4in) long and 1–20mm wide, narrowly lanceolate to spoon- or wedge-shaped, stalked. Stem leaves smaller, linear to lanceolate, stalkless, half-clasping the stem. All leaves with a sparsely bristly-hairy margin.
Inflorescence: Branching, terminal, loose racemes of 5–15 flowers; bracts absent.
Flowers: *Pedicels* to 5mm in flower, to 20mm in fruit. *Calyx* lobes to 3mm in flower, to 7mm in fruit, oval to

lanceolate. *Corolla* white or very pale blue, 5–12mm across, with throat-scales the same colour as the corolla.

Nutlets: To 4mm, smooth or hairy, cupped, with an incurved, toothed wing, forming an navel-like depression.

Chromosomes: 2n=28.

GEOGRAPHY AND ECOLOGY

Distribution: South-west Europe (Iberian Peninsula, southern France) and north-west Africa. Naturalized in Ukraine and Crimea.

Habitat: Dry, open habitats, fields, cultivated and fallow land, stony places, roadsides; often on limestone.

Flowering: April to June in the wild; usually summer in cultivation.

CULTIVATION

Date of introduction: 1748.

Hardiness: Zone 7.

Sun/shade aspect: Full sun is required for good flowering.

Soil requirements: Any well-drained soil, tolerates high alkalinity. Soil that is too fertile may encourage foliage production at the expense of flowers.

Garden uses: Grow as an easy filler in borders, beds and containers; suitable for a large rock garden. Excellent for grey/white colour schemes.

Spacing: Thin the seedlings to about 15cm (6in) apart.

Propagation: From seed only. Sow directly in open ground between early spring and early summer. Germination within 2–3 weeks. Self-seeds easily in favourable conditions. Several successive sowings can provide flowering plants from early summer until mid-autumn. In the wild it germinates in autumn and overwinters, and from an autumn sowing will flower very early in the garden, but the seedlings may perish in a hard winter.

Availability: Mostly available as seed from a number of suppliers and many societies' exchange schemes, but also as a plant from a few nurseries. Seed named 'Omphalodes luciliae' and 'Omphalodes luciliae alba' on seed lists of plant societies usually turns out to be this species.

Problems: Some gardeners may find the habit of self-seeding annoying, others may find it delightful, but unwanted seedlings can be pulled out very easily. Slugs and snails are fond of the foliage, and young plants may suffer badly.

RELATED FORMS AND CULTIVARS

Plants named *O. linifolia alba* are identical to the species. The rarer ***O. linifolia caerulescens*** has pale blue flowers.

RELATED SPECIES

It is likely that some plants/seeds sold as *O. linifolia*, are in fact **O. commutata** López (*O. brassicifolia* misapplied not (Lagasca) Sweet), from the mountains of Spain and north-west Africa, which is very similar to *O. linifolia*, but is a more robust plant, with broader, clasping leaves, and hairless nutlets with a very narrow wing that is not incurved.

Other annual *Omphalodes* species that are not in general cultivation but would be of interest include: seaside omphalodes **O. littoralis** Lehm. from the Atlantic coast of France, also white-flowered; and blue-flowered **O. kuzinskyanae** Willk., a little plant only 5–15cm (2–6in) tall, from coastal rocks and sands of the Iberian Peninsula.

Omphalodes luciliae Boiss.

A difficult small perennial for a trough or an alpine house, tufted, to 10–15cm (4–6in), with oval or oblong, bluish-grey basal leaves and smaller stalkless leaves on stems. Flowers about 12mm ($^{1}/_{2}$in) across, in loose racemes, open pink and change to a beautiful sky-blue, between late spring and early autumn.

> 'This beautiful plant has narrowly oval leaves, and sprays of opalescent flowers, about $^{1}/_{2}$in across, of a lovely soft blue, developing from pink buds. The whole plant, leaves and stems are of an indescribable almost blue-grey, which makes a perfect foil for the flowers.'
> (Anna N. Griffith)

NAMING

Etymology: *luciliae* – after Lucile Boissier.

DESCRIPTION

General: A tufted, hairless perennial with almost vertical rootstock.

Stems: 5–25cm (2–10in), ascending to prostrate or trailing.

Leaves: Basal and lower stem leaves to 4.5cm ($1^{3}/_{4}$in) long and 3.5cm ($1^{3}/_{8}$in) wide, elliptic to oblong, obtuse or rounded, with stalk about 5cm (2in) long. Upper stem leaves smaller, stalkless.

Inflorescence: Terminal cymes of 3–15 flowers, a bract subtending each flower.

Flowers: *Calyx* lobes narrowly elliptical, somewhat pointed, hairless apart from tiny marginal bristles. *Corolla* at first rose-pink becoming sky blue on opening, with a yellow throat; tube to 4mm, limb to 8–15mm diameter, flat to concave.

Nutlets: To 2.5 × 1.5mm, with a narrow, untoothed wing forming a navel-like depression.

GEOGRAPHY AND ECOLOGY

Distribution: Southern and central Greece, Turkey, northern Iraq, western Iran.

Habitat: Shady rock crevices in high mountains, vertical cliffs, 1200–3350m (3900–10,990ft).

Flowering: May to September in cultivation, with an occasional flower at other times.

Associated plants: In Parnassos National Park, Greece, grows in the same habitat as other endemics, such as *Campanula aizoon*, *Verbascum delphicum*, *Thymus parnassicus* and *Lilium chalcedonicum*.

CULTIVATION

Hardiness: Zone 7.

Sun/shade aspect: Sun, but may need shade during the hottest weather at midday in an alpine house.

Soil requirements: Gritty, perfectly drained soil or compost, ideally with added limestone chippings, and kept reasonably moist during the growing season.

Garden uses: Best in alpine house or sunny rock crevices, or in tufa, but can be tried in a trough or scree in the open garden.

Spacing: 15cm (6in).

Propagation: From seed or by careful division, see the details for the genus.

Availability: Sold by a few alpine nurseries. Seed appearing under this name in seed-exchange lists often turns out to be the annual *Omphalodes linifolia*.

Problems: Not easy to grow unless all its needs are satisfied. Needs protection from slugs.

RELATED FORMS AND CULTIVARS

'Alba' is an uncommon, white-flowered form.
Omphalodes × *florairensis*, a hybrid between *O. luciliae* and *O. nitida*, was raised by Correvon near Geneva, but this is now thought extinct. There are several naturally occurring variants, or subspecies, including **O. luciliae subsp. cicilica** A. Brand **subsp. kurdica** Rech., and **subsp. scopulorum** J.R. Edmondson (this form has the most strongly glaucous leaves), which are very similar to the species.

Omphalodes nitida Hoffsmannsegg & Link

A clump-forming, upright perennial, usually growing to about 25cm (10in) though sometimes twice as high, with dark green lanceolate leaves with brownish stalks, and loose cymes of bright blue flowers with yellowish centres, borne in late spring and early summer. A good groundcover plant for partial shade or sun.

NAMING

Etymology: *nitida* – shining, referring to leaves, which is a slight exaggeration.

Synonym: *Omphalodes lusitanica* Pourreau in Lange, not Schrank.

DESCRIPTION

General: A clump-forming, sparingly hairy, perennial with rhizomes.

Stems: 20–30cm (8–12in), sometimes as tall as 65cm (26in), upright, branched, sparsely bristly.

Leaves: Basal leaves 7–20cm ($2^3/_4$–8in) long and 1–3.5cm ($7/_{16}$–$1^3/_8$in) wide, lanceolate or oblong-lanceolate, pointed, hairless and deep green above, slightly hairy below, with prominent lateral veins, and shortly hairy margins, long-stalked. Stem leaves smaller, short-stalked to stalkless.

Inflorescence: Terminal, loose, with bracts only near base.

Flowers: *Pedicels* 5–20mm ($1/_4$–$3/_4$in) in fruit, occasionally to more. *Calyx* lobes 2–4mm ($1/_{16}$–$3/_{16}$in) in flower, to 7mm ($3/_8$in) in fruit, lanceolate to oblong or oval. *Corolla* deep blue, 5–10mm ($1/_4$–$7/_{16}$in) across, blue, with yellowish scales in throat.

Nutlets: 2–3mm, hairy, wing toothed, forming a navel-like depression.

GEOGRAPHY AND ECOLOGY

Distribution: Northern Portugal and north-west Spain (Galicia).

Habitat: Damp, shaded places.

Flowering: Late spring and early summer.

Associated plants: In the wild can be found growing under cork oak (*Quercus suber*) and strawberry tree (*Arbutus unedo*), with tree heath (*Erica arborea*), greenweed (*Genista spp.*) and lauristinus (*Viburnum tinus*).

CULTIVATION

Hardiness: Zone 7.

Sun/shade aspect: Dappled shade is best, though will tolerate sun in cooler climates.

Soil requirements: Moist but well-drained, humus-rich soil.

Garden uses: Woodland garden, as groundcover under trees and shrubs, or in a mixed border.

Spacing: 30cm (12in).

Propagation: From seed or by division, as for the genus.

Availability: Plants sold by a few herbaceous nurseries, and seed can be found on seed-exchange lists and, occasionally, in some catalogues.

Omphalodes verna Moench

A fast-spreading, creeping perennial around 20cm (8in) tall, with oval, bright grass-green leaves, and abundant

Omphalodes verna. (Gordon Smith)

sprays of bright-blue, white-centred flowers, each
around 12mm (¹/₂in) across, in spring. An excellent
ground-covering plant for lightly shaded places.

'*Omphalodes verna is really an April flowerer but it
usually cannot wait till then. I often see a little blue
eye looking up at me in really wintry February weather
when all good children should be asleep. There is more
than a suggestion of pink about these early flowers and
when they come to their official opening they are the
bluest of the blue. … This is said to have been Marie
Antoinette's favourite flower.*' (*Margery Fish*)

NAMING
Etymology: *verna* – of spring.
Synonym: *Omphalodes repens* Schrank.
Common names: *English*: Blue-eyed Mary, Creeping
forget-me-not (USA); *Danish*: Kærminde; *Finnish*:
Kevätkaihonkukka; *French*: Omphalodes du
printemps, Petite Bourrache; *German:* Frühlings-
Gedenkemein, Nabelnuss; *Italian*: Borrana;
Norwegian: Vårkjærminne; *Russian*: pupochnik
vesennyi; *Slovak*: pupkovec jarny; *Swedish*: Ormöga.

DESCRIPTION
General: A shortly hairy, creeping perennial with
rhizomes and stolons/runners.
Stems: 5–30cm (2–12in) high, ascending or upright,
with sparse bristles and few leaves.
Leaves: Basal leaves 5–20cm (2–8in) long and 2–6cm
(³/₄–2¹/₂in) wide, oval or oval-lanceolate to heart-

shaped, pointed, or rounded with a short narrow
point, sparingly hairy, long-stalked. Stem leaves
smaller, oval to elliptical, short-stalked or stalkless.
Inflorescence: Loose, terminal, with bracts only at base.
Flowers: *Pedicels* 8–12mm (³/₈–¹/₂in), to 20mm (³/₄in)
in fruit. *Calyx* lobes about 4mm (³/₁₆in) long, elliptical,
hairy. *Corolla* bright blue, 8–12mm (³/₈–¹/₂in)across,
with blue scales in throat.
Nutlets: About 2mm, hairy, wing margins hairy,
forming a navel-like depression.
Chromosomes: 2n=42.

GEOGRAPHY AND ECOLOGY
Distribution: South-eastern Alps to northern
Apennines, Dolomites, Romania; naturalized in the UK,
France and elsewhere in Europe.
Habitat: Woodland, in the mountains.
Flowering: February to May in the wild.
Associated plants: In the wild, it grows in the company
of such plants as Christmas rose (*Helleborus niger*),
Helleborus dumetorum, wood anemone (*Anemone
nemorosa*), *A. ranunculoides*, alpine barrenwort
(*Epimedium alpinum*), *Euphorbia polychroma*, lesser
periwinkle (*Vinca minor*), mezereon (*Daphne
mezereum*), *Haquetia epipactis*, primrose (*Primula
vulgaris*) and snowdrop (*Galanthus nivalis*).

CULTIVATION
Hardiness: Zone 5.
Sun/shade aspect: Light to moderately deep shade.
Soil requirements: Well-drained but moist soil, rich in
organic matter. Lime-tolerant.
Garden uses: Makes excellent groundcover, in a
woodland garden, wild garden, under trees and shrubs,
with spring bulbs, such as dwarf narcissi.
Spacing: 30cm (12in).
Propagation: From seed or, the easiest, by division, as
for the genus.
Availability: Available from many herbaceous plant
nurseries, especially those specializing in woodland and
groundcover plants.
Problems: May become almost invasive, spreading
rapidly by stolons, so needs to be allocated enough space
to achieve its potential without choking other plants.

RELATED FORMS AND CULTIVARS
'**Alba**' – pure white flowers with yellowish scales in
throat.
'**Elfenauge**' – a selection of 'Grandiflora' with pale
blue flowers from pale pink buds. Very pretty.
'**Grandiflora**' – a larger-flowered form, generally
unavailable in the UK, but better known in
continental Europe.

Other *Omphalodes* species of interest

These are not generally in cultivation, but are worth seeking out. They are mostly hardy to Zone 7.

Omphalodes japonica (Thunb.) Maxim.
Description: A clump-forming, hairy perennial with a short, stout rhizome, few stems 12–20cm (6–8in), oblanceolate basal leaves 7–15cm ($2^3/_4$–6in) long, and racemes of blue flowers, to 10mm ($^7/_{16}$in) across, on *pedicels* to 15mm ($^5/_8$in). Flowers April to May in the wild. There is a white-flowered form, **O. japonica f. albiflora** S.Okamoto.
Distribution: Japan (Hinshu, Shikoku, Kyushu).
Habitat: Mountain woods.

Omphalodes krameri Franch. & Savat.
Description: A clump-forming, hairy perennial, with a short, stout rhizome, few upright stems 25–40cm (10–16in), long-stalked, broadly oblanceolate basal leaves 7–15cm ($2^3/_4$–6in) long, and forked, loose racemes of blue flowers 10–15mm ($^7/_{16}$–$^5/_8$in) across, on 10–15mm ($^7/_{16}$–$^5/_8$in) *pedicels*. Flowers April to June in the wild. There is a white-flowered form, **O. krameri f. alba** (T. Ito) Hara.
Distribution: Japan (Hokkaido, Honshu).
Habitat: Mountain woods.

Omphalodes moupinensis Franchet
Synonym: *Sinojohnstonia moupinensis* (Franchet) Z.Y. Zhang
Description: A rhizomatous perennial with trailing stems to 20cm (8in), with a few broadly heart-shaped basal leaves and pale blue flowers about 8mm ($^3/_8$in) across, in late spring. Currently considered to belong to a separate genus, *Sinojohnstonia* Hu.
Distribution: West China (Sichuan, Shanxi, Yunnan).
Habitat: Scrub, damp shady places at about 2000m (6550ft).

Omphalodes prolifera Ohwi
Description: A hairy, clump- or mat-forming perennial, with slender rhizomes, oblanceolate basal leaves 12–25cm (6–10in) long, and terminal racemes of 4 to 8 rose-pink flowers, about 8mm ($^3/_8$in) across. The lowest bract is proliferous in the axil when the plant is in fruit. Flowers in May in the wild. Listed as Vulnerable in the IUCN Red List.

Distribution: Japan (Honshu, rare).
Habitat: Woods.

Omphalodes ripleyana P.H. Davis
Description: A tufted perennial, resembling *O. luciliae*, but with stems to 30cm (12in) long, bluish-green leaves to 7cm ($2^3/_4$in) long, and white flowers.
Distribution: Turkey. Rare, included on the IUCN Red List.
Habitat: Deeply shaded limestone formations.

The genus *Onosma* Linnaeus, 1762
Golden Drop

NAMING
Etymology: From the Greek *onos*, an ass, and *osme*, smell; the roots of some species are said to smell like an ass. An alternative explanation is that the plant is liked by asses.
Synonyms: *Podonosma* Boissier; *Colsmannia* Lehm. Some species have been allocated to a separate genus, *Maharanga* A. DC.
Nomenclature note: Previously, the generic name *Onosma* has been treated as of neuter gender; but it was suggested by William Stearn (1993) that it should be considered feminine, like other generic names ending in -*osma*. Hence, the specific names previously ending in -*um* (neuter), have been changed to end in -*a*, to indicate feminine gender.

GENUS CHARACTERISTICS AND BIOLOGY
Number of species: About 150.
Description: A genus of biennial or perennial, bristly herbs. Leaves usually long, narrow and very bristly. Flowers usually drooping, borne in terminal, usually branched cymes with bracts. *Calyx* lobed almost to base. *Corolla* most often yellow, sometimes whitish- or purplish-pink, tubular to tubular-bell-shaped, with 5 short lobes, with a basal ring of hairs. Stamens included or protruding from corolla tube. Nutlets often 4, ovoid or triangular, smooth or warty, flat-based.

Identification of *Onosma* species is difficult, and the keys rely on features such as the structure and distribution of bristles on leaves and calyces.
Distribution and habitats: From the Mediterranean to south-west and central Asia; north to Switzerland; mostly in dry habitats, often in mountains, in rock and wall crevices, usually on sandy or limestone soils.
Life form: Chamaephytes.
Pollination: By long-tongued bees.
Fruit dispersal: No specialized mechanism.

Onosma polyphylla. (Howard Rice)

USES

GARDEN

Grow in a rock garden, in a vertical wall crevice or top of a wall, in a raised bed, scree, sand bed, or in pots in an alpine house. Some of the taller species are suited to the front of a sunny border. Flowers in many species possess strong almond- or honey-like fragrance, so onosmas can be usefully included in special collections and gardens of scented plants.

MEDICINAL

Onosma echiodes and *O. bracteata* are known to be used in herbal medicine. The powdered leaves of *O. echioides* have a laxative effect, the flowers are used to treat rheumatism and palpitations, while the root is applied to skin eruptions. A decoction of *O. bracteata* is used by local people to treat syphilis, rheumatism and leprosy.

CULINARY

The root of *O. hispida* is used as a flavouring.

ECONOMIC

A red dye is obtained from the roots of *O. echioides* and *O. frutescens*, and used as an alkanna substitute.

Onosma alborosea (Doreen Townley).

CULTIVATION

Many onosmas are hardy to at least -15—-10°C (5–14°F), but some Himalayan species may need protection from hard frosts. All onosmas need full sun in the warmest position possible. A very well-drained, gritty soil is a must, but they are not fussy about the pH and can withstand a high degree of alkalinity, as well as moderate acidity. Best planting time is late spring, so that plants could establish before the onset of cold weather. All species hate getting water on their hairy leaves and may rot in wet weather, so in climates with high rainfall in winter it may be necessary to protect the plants grown outdoors with a pane of glass. Plants grown under glass in the alpine house need a freely draining mixture – one part loam, one part grit and one part leaf mould – and must be watered very carefully, trying not to splash water on leaves. Most onosmas are drought-tolerant once established.

PROPAGATION

Seed Onosma seed germinates quite easily, without any special treatment, although germination may be quite slow. The seed is difficult to collect because of irritating hairs covering the plants. Panayoti Kelaidis said: *'They are possibly the least pleasant group of plants I know of to clean seed of, or to cut back, since their tiny hairs are extremely irritating – almost like glochids on an* Opuntia'. Sow fresh, or in spring, in pots or trays of gritty, perfectly drained compost, covering lightly, and germinate in a garden frame or a cold greenhouse. Damping off can be a problem, so maintain good hygiene and careful watering. Transplant seedlings as soon as they are large enough to handle. Where happy, onosmas may self-seed prolifically to the point of becoming a nuisance.
Vegetative Although seed is the easiest propagation method for this genus, it is also possible to multiply onosmas by taking soft or semi-ripe cuttings of small non-flowering shoots from mid- to late summer, after flowering has finished, which is the common means of propagation used by nurseries. Insert the cuttings into pots of gritty compost and root in a cold frame or propagator at about 15°C (59°F), providing some midday shade during the first couple of weeks. Pot the young plants up in autumn, and overwinter in a cold frame or greenhouse, keeping them fairly dry.

PROBLEMS

The bristles covering onosmas can cause unpleasant skin irritation, so people with sensitive skin may need to wear gloves when handling the plants. Where happy, their habit of self-seeding may become a little more than annoying, as Robert Nold (Colorado) notes:

Onosma alborosea subsp. *sanguinolenta*. (Erich Pasche)

'I have about 400,000 seedlings of O. tauricum, O. nanum, and others more or less everywhere in the garden. The seedlings are no fun to pull out … You can approximate the effect by running a belt-sander over the tips of your fingers.'

Plants grown under glass may suffer from aphid or whitefly infestation. Some species are short-lived.

REFERENCES

BLAMEY & GREY-WILSON, 1989, 1993; DAVIS, 1978; FEINBRUN-DOTHAN, 1978; GENDERS, 1994; GRIFFITH, 1964; HEATH, 1964; PFAF, 1992–02; PHILLIPS & RIX, 1991b; POLUNIN, 1980; POLUNIN & STAINTON, 1997; RECHINGER, 1964; STEARN, 1993; TUTIN *et al.*, 1972; WALTERS, 2001.

Onosma alborosea Fischer & Meyer

A semi-evergreen, tufted perennial to 30cm (12in) high, with oblong, silver-hairy leaves and nodding heads of tubular or trumpet-shaped flowers, around 2.5cm (1in) long, opening at first white and then ageing to rose-pink.

Onosma alborosea. (Erich Pasche)

Naming

Etymology: *alborosea* – white and pink, referring to the flower colour.
Synonyms: *Onosma albopilosa* hort.; *O. cinerea*.
Nomenclature note: The plant grown in gardens under the name of '*Onosma albopilosa*' is identical to this species.

Description

General: A tufted, bristly, hairy perennial.
Stems: 10–20cm (4–8in), upright, usually unbranched.
Leaves: 2.5–6cm (1–2½in) long and 2–12mm (¹⁄₁₆–½in) wide, obovate to oblong or lanceolate to spoon-shaped, densely covered in hairs and bristles.
Inflorescence: Of one or more cymes.
Flowers: *Calyx* 14–16mm (½–⅝in), densely hairy. *Corolla* whitish, becoming pink and rose-purple, 18–30mm (¾–1¼in) long, hairless or shortly hairy.

Geography and ecology

Distribution: Turkey, northern Syria and northern Iraq.
Habitat: Rocky slopes and cliffs on limestone, at up to 2550m (8360ft).
Flowering: April to July in the wild.
Associated plants: May be found in company with *Helleborus vesicarius, Scilla sibiria, Cyclamen pseudoibericum* and *Crocus* spp. in Turkey.

Cultivation

Hardiness: Zone 7 (to -15°C/5°F).
Spacing: 20cm (8in).
Availability: One of the more commonly grown species, this is sold by specialist alpine nurseries, and the seed is frequently found on seed-exchange lists. For other details of garden uses, cultivation and propagation see under the genus description.

Related forms and cultivars

***Onosma alborosea* subsp. *sanguinolenta* Vatke**
From slopes, steppes and oak forests, in Turkey, up to 2800m (9180ft); differs in its flowers ageing from white to intense purple.

Onosma arenaria Waldstein & Kitaibel

A rather robust, tufted, bristly perennial, or occasionally biennial, of variable height, with bristly spoon-shaped leaves, and branched clusters of pale yellow, tubular flowers. An uncommon, short-lived species suitable for a sunny mixed border.

Naming

Etymology: *arenaria* – of sands, referring to the plant's preference for sandy soils.
Common names: *English:* Golden drop; *Hungarian:* homoki vérto; *Russian:* onosma peschanaya; *Slovak:* rumenica piesoāná.

Description

General: A rather robust perennial or sometimes biennial with non-flowering rosettes and usually a single flowering stem.
Stems: 15–70cm (6–28in), usually single, with numerous branches, hairy and covered with bristles 2–3mm (⅛in) long.
Leaves: Basal leaves 4.5–18cm (1¾–7in) long and 4–13mm (³⁄₁₆–½in)wide, oblong to spoon-shaped, sparsely to densely hairy, and with both simple and stellate bristles, 1–2mm (¹⁄₁₆in) long.
Inflorescence: Branched, with bracts not longer than calyx.
Flowers: *Pedicels* 0–4mm (0–³⁄₁₆in). *Calyx* 6–12mm (¼–½in) in flower, up to 18mm (¾in) in fruit. *Corolla* pale yellow, 12–19mm (½–¾in), hairless or hairy.
Nutlets: 2.5–3mm, smooth, shining.
Chromosomes: 2n=12.

Geography and ecology

Distribution: South-eastern and central Europe, north to southern Germany.
Habitat: Rocky and stony places, sometimes grassy places; to 1700m (5770ft).

Flowering: March to July in the wild.
Associated plants: Can be found growing with plants such as maiden pink (*Dianthus deltoides*), childing pink (*Petrorhagia prolifera*), Spanish catchfly (*Silene otites*) and *Helichrysum arenarium*.

CULTIVATION
Hardiness: Zone 6.
Garden uses: Can grow too large for a rock garden, but useful for a sunny mixed border.
Spacing: 30cm (12in).
Availability: Not generally available, but seeds may appear on exchange lists.

For other details of garden uses, cultivation and propagation, see under the genus description.

Onosma echioides Linnaeus
A tufted, hairy, upright perennial, around 20cm (8in) high, with grey-green, narrowly oblong leaves covered in stellate bristles, and pendent clusters of pale yellow flowers, in summer. A good rock garden plant.

NAMING
Etymology: *echioides* – possibly referring to another plant of the borage family, *Arnebia pulchra* (syn. *Echioides longiflorum*).
Synonym: *Onosma aucheriana* subsp. *javorkae* (Simonkai) Hayek.
Common names: *English*: Golden drop; *Spanish*: Orcaneta amarilla.

DESCRIPTION
General: More or less tufted perennial with several erect flowering stems.
Stems: 10–30cm (4–12in), simple or with a few branches, hairy and stellate-bristly.
Leaves: Basal leaves 2–6cm ($^3/_4$–$2^1/_2$in) long and 0.1–0.7mm, linear or linear-oblong, sparsely to densely covered with stellate bristles, each bristle 1–1.5mm, white, grey or yellow, with rays 0.3–0.4mm.
Inflorescence: Cymes with bracts no longer than calyx.
Flowers: *Pedicels* 0–2mm ($^1/_{16}$in), *calyx* about 10mm ($^7/_{16}$in) in flower, to 15mm ($^5/_8$in) in fruit, with simple and some stellate bristles. *Corolla* pale yellow, 18–25mm ($^3/_4$–1in), finely hairy, about twice as long as calyx, cylindrical at the apex, then tapering towards the base.
Nutlets: About 2.5mm, smooth, shining.
Chromosomes: 2n=14.

GEOGRAPHY AND ECOLOGY
Distribution: Italy, Sicily, west of Balkan Peninsula (Albania, the former Yugoslavia).
Habitat: Dry rocky and stony habitats, limestone slopes, crevices in rocks and cliffs, to 1600m (5250ft).
Flowering: April to July in the wild.
Associated plants: In Italy, it may be found growing with *Jurinea mollis*, *Hippocrepis glauca*, *Polygala nicaeensis*, *Dianthus ciliatus*, *Iris illyrica*, *Gentiana tergestina* and *Centaurea rupestris*.

CULTIVATION
Hardiness: Zone 7.
Spacing: 30cm (12in).
Availability: Grown quite a lot in the past, but now less common; may still be found in some specialist catalogues, and as seed on many societies' lists.

For other details of garden uses, cultivation and propagation see under the genus description.

Onosma frutescens Lamarck
A bristly, more or less upright perennial, to 25cm (10in) high, with stems somewhat woody at the base, oblong-lanceolate leaves, and clusters of tubular-bell-shaped yellow flowers, about 20mm long ($^3/_4$in), ageing to orange- or reddish-brown, in summer. A nice plant for a hot, sunny crevice.

NAMING
Etymology: *frutescens* – becoming shrubby, referring to the woody base of the plant.

DESCRIPTION
General: A bristly perennial with somewhat woody base.
Stems: 10–25cm (4–10in), upright or ascending, unbranched, covered with hairs and bristles 1–3mm ($^1/_8$in) long.
Leaves: Basal leaves 2–7cm ($^3/_4$–$2^3/_4$in) long and 4–10mm ($^3/_{16}$–$^7/_{16}$in) wide, linear- or oblong-lanceolate, grey-hairy and bristly.
Inflorescence: Unbranched or with few short branches; bracts lanceolate, to about 15mm ($^5/_8$in) long.
Flowers: *Pedicels* 5–8mm ($^1/_4$–$^3/_8$in) long. *Calyx* 10–15mm ($^7/_{16}$–$^5/_8$in) in flower, to 16mm ($^5/_8$in) in fruit, with yellow bristles. *Corolla* pale yellow, turning reddish-purple or orange brown, 16–21mm ($^5/_8$–$^7/_8$in), widened above, hairless. *Stamens* with anthers longer than filaments, shortly protruding from the corolla.
Nutlets: 3–5mm, ovoid, glossy, often marbled.
Chromosomes: 2n=14.

GEOGRAPHY AND ECOLOGY

Distribution: Greece, Turkey, Syria.
Habitat: Limestone slopes and rocks at up to 1600m (5250ft), also on old walls. In Syria it grows in maquis and *Pinus halepensis* forest.
Flowering: April to June in the wild.
Associated plants: In Mistras (Greek Peloponnese), it grows on sunny walls together with *Campanula andrewsii* and *Aurinia saxatilis*.

CULTIVATION

Hardiness: Zone 8.
Spacing: 45cm (18in).
Availability: Rarely available, but seed may appear on exchange lists of specialist societies.

For other details of garden uses, cultivation and propagation, see under the genus description.

Onosma helvetica (A. De Candolle) Boissier

A tufted perennial around 30cm (12in) high, with basal rosettes of oblong, spoon-shaped leaves, covered with straight and star-shaped hairs, and nodding clusters of finely hairy, pale yellow flowers, in late spring and early summer.

NAMING

Etymology: *helvetica* – of Switzerland, one of the countries where this plant is native.
Common names: *English*: Swiss golden drop; *French*: Onosma de Suisse; *German*: Schweizer Lotwurz; *Italian*: Viperina Elvetica.

DESCRIPTION

General: A tufted perennial, usually with non-flowering rosettes and several erect flowering stems.
Stems: 20–50cm (8–20in), simple or with up to 6 branches above, hairy and bristly, the bristles 2–3mm, usually stellate, with rays about 0.1mm.
Leaves: Basal leaves 3–7cm (1¼–2¾in) long and 2–6mm wide, oblong to spoon-shaped, covered in simple hairs, and stellate bristles 1–1.5mm long, with rays 0.1–0.3mm.
Inflorescence: Cymes, with bract shorter than calyx.
Flowers: *Pedicels* 0–2mm (¹⁄₁₆in) . *Calyx* 9–12mm (³⁄₈–¹⁄₂in) in flower, up to 17mm (⁵⁄₈in) in fruit, with stellate bristles. *Corolla* pale yellow, 16–24mm (⁵⁄₈–¹⁵⁄₁₆in), downy. *Stamens* with anthers 6–8mm long.
Nutlets: 2–4mm, smooth, shining.
Chromosomes: 2n=26, 28.

GEOGRAPHY AND ECOLOGY

Distribution: South-western Alps (France, Switzerland, Italy).

Habitat: Stony and sandy places, to 2500m (8200ft).
Flowering: Summer.

CULTIVATION

Hardiness: Zone 5.
Spacing: 30cm (12in).
Availability: Occasionally may be available from alpine nursery specialists, or as seed on exchange lists.

For other details of garden uses, cultivation and propagation, see under the genus description.

Onosma nana De Candolle

A bristly, tufted perennial, 10–20cm (4–8in) high, with densely hairy, narrow, spoon-shaped leaves, and clusters of white or pale yellow tubular flowers, becoming blue or pink as they age, in late spring and early summer. A nice little plant for a rock garden, scree, crevice, or pot in an alpine house.

NAMING

Etymology: *nana* – dwarf, small.
Synonym: *Onosma microsperma* Stev.

DESCRIPTION

General: A more or less cushion-forming, bristly perennial.
Stems: 8–18cm (3¼–7in), upright, unbranched, bristly.
Leaves: Basal leaves 2.5–5.5cm (1–2¼in) long, linear to spoon-shaped, densely hairy.
Inflorescence: A few-flowered cyme.
Flowers: *Calyx* 10–20mm (⁷⁄₁₆–³⁄₄in) long, a half to four fifths as long as the corolla. *Corolla* white to pale or bright yellow at first, becoming blue or pink, 18–25mm (³⁄₄–1in) long.
Nutlets: About 3mm.

GEOGRAPHY AND ECOLOGY

Distribution: Turkey, Eastern Mediterranean area.
Habitat: Limestone ledges, screes and slopes, to 3200m (10,500ft).
Flowering: Late spring to summer.

CULTIVATION

Hardiness: Zone 8.
Spacing: 15–20cm (6–8in) apart.
Availability: Occasionally sold by some alpine nurseries, and seed may be found on the specialist societies' exchange lists.

For other details of garden uses, cultivation and propagation, see under the genus description.

Onosma sericea Willdenow

A semi-evergreen, silvery-hairy perennial, to 30cm (12in) high, with narrow leaves and racemes of dropping creamy-yellow flowers in summer. A lovely onosma for a rock garden, alpine house or crevice.

'... Produces stunning rosettes of a silky, satiny texture and silvery color that are truly delightful – particularly since it is virtually evergreen.' (Panayoti Kelaidis)

Naming
Etymology: *sericea* – silvery-hairy.
Synonym: *Onosma flava* (Lehm.) Vatke.

Description
General: Hairy perennial.
Stems: 30–50cm (12–20in), yellowish, with procumbent branches, the sterile densely leafy, the fertile sparingly leafy.
Leaves: Lower stalked, obovate to spoon-shaped, upper oblong-lanceolate to lanceolate, stalkless.
Inflorescence: Terminal raceme or few-flowered panicle.
Flowers: *Calyx* lobes 5, or by adhesion 4 to 3, to 15–20mm ($^5/_8$–1in) long in fruit, with linear or narrowly oblong lobes. *Corolla* cream, velvety at apex, a third to a quarter longer than calyx, club-shaped. Stamens included in the corolla, filaments shorter than anthers.
Nutlets: 5mm long, warty or smooth.

Geography and ecology
Distribution: Asia Minor, Iran, Iraq.
Habitat: Dry places.
Flowering: Summer.
Associated plants: In Lebanon, may be found growing with *Notobasis syriaca, Euphorbia macroclada, Stachys nivea, Convolvulus derychium.*

Cultivation
Hardiness: Zone 7.
Spacing: 30cm (12in).
Availability: Not commonly available, but seed may appear on exchange lists.

For other details of garden uses, cultivation and propagation, see under the genus description.

Onosma stellulata Waldstein & Kitaibel

A semi-evergreen, woody-based perennial, around 20cm (8in) high, covered in stellate/star-shaped bristles, with oblong leaves, and clusters of yellow flowers, each around 15mm ($^5/_8$in) long.

Naming
Etymology: *stellulata* – starry, referring to the shape of hairs covering the plant.

Description
General: A bristly perennial with branched, woody base.
Stems: 10–25cm (4–10in), unbranched, covered with hairs and stellate bristles.
Leaves: Basal leaves 4–14cm ($1^1/_2$–$5^1/_2$in) long and 7–15mm ($^3/_8$–$^5/_8$in) wide, oblong to spoon-shaped, sparsely covered with stellate bristles 1–1.5mm long, with rays about 0.4mm.
Inflorescence: Cymes with bracts that are shorter than calyces.
Flowers: *Pedicels* 6–14mm ($^1/_4$–$^1/_2$in) in flower; bracts shorter than calyx. *Calyx* 7–9mm ($^3/_8$in), with simple hairs and sometimes with short-rayed stellate bristles. *Corolla* pale yellow, 5–18mm ($^1/_4$–$^3/_4$in), about twice as long as calyx, hairless.
Nutlets: 2.5–3mm, smooth, shining.
Chromosomes: 2n=22.

Geography and ecology
Distribution: Western parts of the former Yugoslavia.
Habitat: Dry places.
Flowering: Summer.

Cultivation
Hardiness: Zone 6.
Spacing: 20cm (8in).
Availability: Occasionally sold by alpine and rock plant nurseries, and seed regularly appears on societies' exchange lists.

For other details of garden uses, cultivation and propagation, see under the genus description.

Onosma taurica Pallas in Willdenow ♔

A grey-hairy, tufted perennial, about 20cm (8in) tall, woody at the base, with bristly oblong to narrowly lanceolate leaves, bearing nodding clusters of pale yellow or cream tubular flowers, 2–3cm long, over a long period from late spring to summer. The RHS Award of Garden Merit plant.

'One of the most beautiful of all alpine plants. The flowers appear in continuous succession for about five months and all the time diffuse their heavenly honey perfume.' (Roy Genders)

Naming
Etymology: *taurica* – of Taurus Mountains (Crimea).

Synonym: *Onosma erecta* Sibth.
Common names: *English*: Golden drop; *Russian*: onosma krymskaya.

DESCRIPTION

General: A cushion-forming, grey, woody-based perennial.
Stems: 10–40cm (4–16in), several, simple, rather thick, with downy grey hairs and stellate/star-shaped bristles.
Leaves: Basal leaves 4–12cm (1^1/$_2$–4^1/$_2$in) long and 3–8mm wide, linear-oblong, with stellate-rayed bristles, each bristle 1–2mm long, with rays up to 0.5mm.
Inflorescence: Cymes, with lower bracts longer than calyces.
Flowers: *Pedicels* 0–2mm. *Calyx* 10–13mm (^{7}/$_{16}$–^{1}/$_2$in) in flower, up to 18mm (^{3}/$_4$in) in fruit, with usually simple bristles. *Corolla* pale yellow, 2–3cm (^{3}/$_4$–1^1/$_4$in) long, 2–3 times as long as calyx, hairless. *Stamens* with filaments slightly shorter than anthers.
Nutlets: 3.5–4mm, smooth, shining.
Chromosomes: 2n=14.

GEOGRAPHY AND ECOLOGY

Distribution: Caucasus, Crimea, Balkans, Greece, Turkey.
Habitat: Rocks and cliffs, stony places.
Flowering: April to August.

CULTIVATION

Date of introduction: 1800.
Hardiness: Zone 6.
Spacing: 30cm (12in).
Availability: One of the more commonly available species, sold by alpine nurseries. Seed can be found in some seed catalogues and in most seed-exchange lists.

For other details of garden uses, cultivation and propagation, see under the genus description.

Other *Onosma* species of interest

These are not in general cultivation, but may appear on seed-exchange lists, and most are worth trying. The majority on this list are hardy to Zone 7.

Onosma aucheriana De Candolle

Synonym: *O. pallida* Boissier.
Description: A bristly perennial 15–30cm (6–12in), with oblong or spoon-shaped leaves, and

Onosma aucheriana in north-western Turkey.
(Erich Pasche)

broadly cylindrical flowers, to 18mm (^{3}/$_4$in) long, varying in colour from white and cream to primrose or sulphur yellow.
Distribution: Turkey, southern Balkans.
Habitat: A wide variety of habitats, including steppes, deciduous and coniferous forests and cornfields, mostly on alkaline soils.

Onosma bourgaei Boiss.

Description: A hairy perennial, 30–50cm (12–20in) high, with narrow to spoon-shaped leaves, and drooping clusters of white, cream or pale yellow, fragrant flowers 15mm (^{5}/$_8$in) long.
Distribution: Turkey, Armenia.
Habitat: Limestone slopes, steppes, grassland, fir (*Abies*) forests, to 2000m (6550ft).

Onosma bracteata Wallich

Description: A very hairy, stout perennial, to 40cm (16in), with narrowly lanceolate leaves to 15cm (6in) long, and dense clusters of dark red to mauve flowers, to 15mm (^{5}/$_8$in) long, surrounded by long, narrow, woolly bracts that almost conceal the flowers.
Distribution: East Asia.
Habitat: Rocky slopes, in Himalayas at 3300–5000m (10,800–16,400ft).

Onosma erecta Sibthorp & Smith
Synonym: *O. laconica* Boiss.
Description: Tufted perennial, 15–25cm (6–10in), densely covered with starry hairs, oblong to narrowly lanceolate leaves to 6cm (2½in) long, and pale yellow tubular flowers to 24mm (¹⁵⁄₁₆in) long in summer.
Distribution: Southern Greece, Crete.
Habitat: Cliffs, mountain rocks.

Onosma heterophylla Grisebach
Synonym: *O. paradoxa* Janka.
Description: A bristly perennial, 20–40cm (8–16in), with oblong or linear leaves, 2.5–15cm (1–6in) long, and pale yellow flowers, 20–30mm (¾–1¼in) long. flowers April to June in the wild.
Distribution: Balkan Peninsula, Romania, Turkey.
Habitat: Sandy coastal habitats and rocks.

Onosma hispida Wallich in G. Don
Description: A very bristly perennial, with several stems 20–50cm (8–20in) high, with narrow basal leaves to 13cm (5in) long, and pale yellow tubular flowers to 3cm (1¼in) long.
Distribution: East Asia.
Habitat: Scrub, open slopes, rocks, savannah, 1000–4000m (3280–13,100ft) in the Himalayas.

Onosma syriaca. (Masha Bennett)

Onosma syriaca Labill.
Synonyms: *Onosma orientalis,* (L.) L.; *Podonosma syriaca* (Labill.) Boiss.
Description: A glandular-hairy, woody-based perennial, 20–40cm (8–16in) high, with numerous brittle stems, which may be ascending or prostrate (often hanging off vertical rock surfaces). Covered in many stalkless, lanceolate leaves, and tubular, 10–13mm (⁷⁄₁₆–½in) long, pale blue flowers, often with a pale yellow edge to the corolla, thus giving a bi-coloured effect.
Distribution: Middle East.
Habitat: Crevices in walls and rocks.
Associated plants: On the walls of old Jerusalem, this grows in abundance with henbane *Hyosciamus aurea* and pellitory-of-the-wall *Parietaria judaica*.

Onosma tornensis Jav.
Description: A bristly perennial 15–30cm (6–12in) high, with narrowly lanceolate basal leaves, and pale yellow tubular flowers in loose clusters. Flowers July to August in the wild.
Distribution: Slovakia and Hungary; rare, included on the IUCN Red List as Indeterminate.
Habitat: Dry and stony places, often on limestone.
Associated plants: In the ruins of Turòa Castle in Slovakia, it grows with nearby plants such as *Ajuga laxmannii* and *Thalictrum foetidum*.

Onosma tricerosperma Lag.
Description: A bristly, upright perennial, 20–40cm (8–16in) high, with reddish-purple stems, oblong to spoon-shaped leaves to 15cm (6in) long, and clusters of pale yellow flowers 16–23mm (⅝–⅞in) long. Nutlets usually with 3 horns, hence the specific name. *Onosma t.* subsp. *hispanica* P.W. Ball differs by greenish-yellow stems and greyish, rather than plain green, leaves.
Distribution: Central, southern and south-eastern Spain.
Habitat: Dry places.

Onosma visianii Clem.
Common name: *German:* Visianis Lotwurz.
Description: A bristly biennial, 30–60cm (1–2ft) tall, with narrowly lanceolate leaves, the basal to 25cm (10in) long, and creamy-yellow flowers about 2cm (¾in) long, in summer.
Distribution: Eastern, central and south-eastern Europe.
Habitat: Dry grassland.
Associated plants: In Austria, grows on dolomite slopes with *Campanula sibirica, Seseli osseum* and *Fumana procumbens*.

The genus *Pentaglottis*
Tausch, 1829 ♀
Green Alkanet

NAMING
Etymology: *Pentaglottis* – five-tongued, referring to scales in throat.
Synonym: *Caryolopha* Fischer & Trautv. Previously included within *Anchusa* L.

GENUS CHARACTERISTICS AND BIOLOGY
Number of species: 1 species.
Description: See the description of *P. sempervirens*.
Distribution and habitats: Western Europe; damp and shady places.
Life form: Chamaephyte.
Pollination: Bees.
Fruit dispersal: No specialized mechanism.

USES
GARDEN
Grow in a woodland garden, wildflower garden, on a hedge bank or in any wilder corners, including deep shade.

CULINARY
Flowers are edible, with a mild flavour, and can be eaten raw. They have a mucilaginous texture and are mainly used for decoration in fruit drinks and salads.

CULTIVATION
Evergreen alkanet is hardy and very easy to grow in almost any soil or position. It prefers a lightly shaded situation on moist, well-drained soil, but will adapt to most conditions.

PROPAGATION
Seed The seed can be sown as soon as ripe, or in spring, either in pots in a cold frame, or directly into the permanent position. Once you have a clump of *P. sempervirens*, it is likely to perpetuate itself by self-seeding cheerfully all over the place.
Vegetative Division or root cuttings in spring are easy, and can be planted straight into permanent positions, or potted up.

PROBLEMS
It can become invasive by self-seeding and, if dug out, any fragments of roots remaining in the soil are likely to regrow.

REFERENCES
BRICKELL, 1996; CLAPHAM *et al.*, 1952; FISH, 1961; HUXLEY, 1992; STACE, 1991; TUTIN *et al.*, 1972; WALTERS 2001.

Pentaglottis sempervirens (Linnaeus) Tausch in Bailey
A stout, bristly perennial, 30–100cm (12–40in) high, with deep green, oval leaves, and clusters of forget-me-not-like flowers of azure-blue, in late spring and summer. Very easy to grow and can be invasive, but a nice plant for wilder areas in the garden.

> '... a welcome guest with its hairy leaves and twinkling flowers.' (Margery Fish)

NAMING
Etymology: *sempervirens* – evergreen.
Synonyms: *Anchusa sempervirens* L.; *Caryolopha sempervirens* (L.) Fischer & Trautv.
Common names: *English*: Green alkanet; Bird's Eye; Evergreen bugloss; *Dutch*: Overblijvende Ossetong; *French*: Buglosse Toujours Verte.

DESCRIPTION
General: A hairy evergreen perennial.
Stems: 30–100cm (12–40in), ascending to upright, branched.
Leaves: Basal leaves 10–30cm (4–12in), oval, pointed, narrowing into a long stalk. Stem leaves are stalkless, sharply pointed. All very bristly.
Inflorescence: Long-stalked, dense cymes in axillary pairs, with 5–15 flowers in a cyme; each branch is subtended by a large, leaflike bract, 1.5–3cm ($^5/_8$–$1^1/_4$in) long.
Flowers: *Pedicels* very short. *Calyx* lobes linear-lanceolate, to 5mm, enlarging to 8mm in fruit. *Corolla* bright blue, 8–10mm across, with a funnel-shaped or cylindrical tube 4–6mm long, and a spreading limb; 5 white, hairy scales close the mouth of the tube. *Stamens* included in the corolla tube. *Style* also included, with a head-like stigma.
Nutlets: 1.5–2mm, rough-netted, blackish, ovoid, concave.
Chromosomes: 2n=22.

GEOGRAPHY AND ECOLOGY
Distribution: South-west Europe (from central Portugal to south-west France); naturalized in Britain, Belgium, Ireland, Italy.
Habitat: Moist and shaded habitats, in the UK naturalized in hedgerows and woodland margins, possibly native in a few places in SW England.

Pentaglottis sempervirens. (George McKay)

Flowering: May to June in the wild in the UK, sometimes into August.

CULTIVATION

Hardiness: Zone 7.

Sun/shade aspect: Tolerates full sun to moderately deep shade. The best effect is in dappled shade.

Soil requirements: Will grow on any soil, though prefers moisture-retentive, humus-rich soil.

Garden uses: Grow in a woodland garden, on a hedge bank, in a mixed border.

Spacing: 30cm (12in).

Propagation: See under the genus description.

Availability: Quite widely available from herb and wildflower nurseries.

RELATED FORMS AND CULTIVARS

No cultivars are known, but some interesting variations occur in the wild: white-flowered with a blue eye, gold-variegated on leaf margins, and forms with variable silver spotting on leaves, sometimes strong enough to resemble a *Pulmonaria* leaf. These forms do not appear to be in general cultivation at present.

The genus *Symphytum* Linnaeus, 1753
Comfrey

NAMING

Etymology: *Symphytum* – from the Greek *syn*, together, and *phyton*, plant, referring to the healing properties of the plants, i.e. 'knitting the bones together'.

Synonym: *Procopiania* Gusul.

GENUS CHARACTERISTICS AND BIOLOGY

Number of species: About 35.

Description: A genus of stiffly hairy perennials, usually with rhizomes, some far creeping. Leaves often rough, hairy, oval or elliptical with a heart- or wedge-shaped base, the basal long-stalked, and the stem leaves shortly stalked or stalkless. Flowers nodding, borne in dense, spiralled cymes without bracts. *Calyx* bell-shaped or tubular, lobed from one fifth to nine tenths of their length. *Corolla* blue, pink, white or yellowish, funnel- or bell-shaped, widest above the middle, with 5 triangular or rounded short lobes. Throat with 5 linear or lanceolate scales, forming a cone. Stamens 5, usually included in the corolla. Style protruding from the corolla, with a small, headlike stigma. Nutlets egg-shaped, smooth or finely warty, often wrinkled, base with a thickened collar-like ring.

Distribution and habitats: Europe and western Asia, north to Siberia; mostly damp habitats.

Life forms: Geophytes and hemicryptophytes.

Pollination: Long-tongued bees.

Fruit dispersal: No specialized mechanism.

Animal feeders: Moths *Diachrysia chrysitis*, *Ethmia lugubris* and *E. quadrillella quadrillella* are known to feed on *S. officinale*; and *Anthophila fabriciana* on *S. tuberosum*.

USES

CULINARY

Comfrey has been used as a salad green and pot herb, but pyrrolizidine alkaloids contained within the plants have been found to be carcinogenic, and have a cumulative toxic effect on the liver. Although it would probably be necessary to consume very large amounts of the plant to cause any harm, it is best not to use it as food, to be on the safe side.

WARNING: Do not eat any garden or wild plant without prior consultation with a doctor.

GARDEN
Comfreys can be grown in a herb garden, herbaceous or mixed border, as groundcover, in a woodland garden, or in any wild corner. They should not be used in small spaces or positioned near small and delicate plants.

MEDICINAL
Comfrey is perhaps one of the oldest medicinal herbs known to man. It has been cultivated since about 400BC as a healing herb. Greeks and Romans used comfrey to stop heavy bleeding, treat bronchial problems, and heal wounds and broken bones. Wild comfrey was brought to America by English immigrants for medicinal uses.

It is one of the very few non-animal sources of vitamin B12. The substance allantoin, contained in comfrey leaves and roots, affects the rate of cell multiplication, which explains the popularity of comfrey in the treatment of wounds, sores, burns, broken bones and swollen tissue. The mucilage of the plant acts to moisturize and soothe the skin.

OTHER
A number of comfrey species, especially *S. asperum*, has been cultivated for forage, and fed to pigs, sheep, horses, goats and poultry. In gardens, comfrey foliage can be used as mulch, and made into a liquid fertilizer.

CULTIVATION
Comfreys are most valuable for their tolerance of even deep shade though, for the optimum flower and foliage display, dappled or partial shade is probably the best. Many symphytums, such as *S. asperum*, *S. caucasicum*, *S. officinale* and *S. × uplandicum*, are also happy in full sun, as long as the soil is continuously moist. The ideal soil would be moist, fertile, humus-rich, with a deep root run, though these plants are exceptionally tough and will survive on most soils, including heavy clay. When comfrey is cultivated for use in compost, as fertilizer or mulch, it should be grown in an open, sunny site, on fertile, deep soil.

If growing variegated cultivars primarily for their leaf colour, it is worth removing developing flower stems to improve the show of foliage.

Some species and varieties, such as *S. asperum* and *S. × uplandicum* forms, may flower for a second time in the season if stems are cut down after flowering, and may even repeat up to four times a year.

It is beneficial to divide comfrey clumps every 4–5 years, discarding the old parts, and replanting the younger portions into their permanent positions.

PROPAGATION
Seed Seeds sown in late spring or in autumn in an outdoor seed bed. Will normally take two years to reach flowering size if grown from seed. Some forms will self-seed all too happily, while others rarely produce viable seed.
Vegetative It is very easy to divide comfreys, at almost any time of the year, but ideally in spring or autumn, replanting the sections immediately in their flowering positions, or potting them up to grow on. Root cuttings can be taken in early or midwinter, inserted in pots or trays of compost and rooted in a frame (but this is not a suitable method for variegated forms). Keep young plants moist until well established.

PROBLEMS
Comfreys can be highly invasive, spreading both by seed and vegetatively, and can overtake a small garden unless their growth is carefully monitored. Comfrey rust fungus (*Melampsorella symphyti*) is a problem in large-scale plantings of comfrey in Britain, and reduces the yields.

REFERENCES
BRICKELL, 1996; CLAPHAM *et al.*, 1952; DAVIES, 1978; HUXLEY, 1992; PFAF, 1992–02; PHILLIPS & RIX, 1991; POLUNIN, 1987; RANDUSAHKA *et al.*, 1990; STACE, 1991; THOMAS, 1990, 1993; TUTIN *et al.*, 1972; WALTERS, 2001.

Symphytum asperum Lepechin
A roughly hairy perennial, up to 1.5m (5ft) tall, though often only 90cm (3ft), its leaves very rough to touch, with branching clusters of nodding, rich blue flowers in spring and summer. Nice groundcover in dappled shade or sun.

NAMING
Etymology: *asperum* – rough.
Synonyms: *Symphytum asperrimum* Donn; *S. sepulcrale* Boiss. & Bal. in Boiss.
Common names: *English*: Rough comfrey, Prickly comfrey; *Danish*: Ru Kulsukker; *Dutch*: Ruwe Smeerwortel; *Finnish*: Tarharaunioyrtti; *French*: Consoude Rude; *German*: Rauher Beinwell, Rauhe Wallwurz; *Italian*: Consolida aspra; *Norwegian*: Fôrvalurt; *Russian*: okopnik sherohovaty; *Swedish*: Fodervallört, taggvallört.

DESCRIPTION
General: A roughly hairy perennial with thick, vertical, branched rootstock.
Stems: To 1.5–1.8m (5–6ft), much branched, covered with short, stout, hooked bristles.

Leaves: Oval or elliptical, at least on the midrib covered in stiff bristles; basal leaves 15–19cm (6–7¹/₂in), with a heart-shaped or rounded base; upper leaves very shortly stalked, cuneate at base.
Inflorescence: Cymes with 10–20 flowers.
Flowers: *Calyx* 3–5mm, lobed to two-thirds or three-quarters of its length, covered with short, stout bristles; lobes linear-oblong, rounded, enlarging in fruit. *Corolla* pink-red in bud, becoming clear blue after opening, 9–17mm, gradually widening towards the top. Scales lanceolate, about equalling stamens. *Stamens* with filaments narrower than anthers and about as long.
Nutlets: 4mm, roughly netted and finely warted, brown.
Chromosomes: 2n=32, 36, 40.

GEOGRAPHY AND ECOLOGY
Distribution: The Caucasus, north-eastern Turkey, northern and north-western Iran; naturalized in Europe, including Austria, Belgium, Britain, Denmark, Finland, France, Switzerland, Norway, Russia, Sweden. Classified as a noxious weed in California.
Habitat: In Turkey grows in spruce (*Picea*) forests, meadows, on streambanks, at 700–2200m (2300–7200ft); scattered throughout Britain in waste places and on margins of cultivated land (formerly widely grown as forage.)
Flowering: May to August in the wild.

CULTIVATION
Date of introduction: 1799.
Hardiness: Zone 5.
Sun/shade aspect: Sun to shade.
Soil requirements: Any soil, though prefers moist, fertile, deep soil.
Garden uses: As for the genus.
Spacing: 60cm (2ft).
Propagation: As for the genus.
Availability: Sold by many nurseries.

Symphytum bulbosum C. Schimper
A creeping, hairy perennial, with 30cm (12in) stems arising from tuberous rootstock, with oval or oblong leaves and nodding clusters of tubular, pale yellow flowers. For any wilder corners of the garden.

NAMING
Etymology: *bulbosum* – referring to bulblike tubers produced on rhizomes.
Synonyms: incl. *Symphytum zeyheri* C. Schimper
Common names: *English*: Bulbous comfrey; *French*: Consoude bulbeuse; *German*: Kleinblütige Wallwurtz; *Italian*: Consolida Minore.

Symphytum asperum in north-eastern Turkey.
(Erich Pasche)

DESCRIPTION
General: A hairy perennial, with a slender, creeping rhizome, producing rounded tubers.
Stems: 15–50cm (6–20in), simple or little-branched; with dense, very small hooked hairs and scattered bristles up to 1.5–2mm.
Leaves: Covered with hairs and bristles like the stem. Lower leaves oval to elliptical-lanceolate or spoon-shaped, gradually or abruptly narrowed into a long stalk. Middle leaves larger, oblong-oval, shortly decurrent. Uppermost leaves stalkless, slightly decurrent.
Inflorescence: Paired cymes.
Flowers: *Calyx* 4–5mm, lobed from about one third to six sevenths of their length. *Corolla* 7–12mm, pale yellow, with erect lobes that are ¹/₆ to ¹/₃ as long as the tube. Scales 5–10mm, protruding for 1–5mm from the corolla, pointed, narrowly lanceolate. *Stamens* with anthers 2.5–4mm, twice as long as filaments.
Nutlets: 3–3.5mm, curved, wrinkled, warty.

GEOGRAPHY AND ECOLOGY
Distribution: Southern Europe from Corsica to Greece and western Balkans, including north-west Turkey.
Habitat: Damp habitats, woods, forests, scrub, stream banks. Naturalized in central and southern Britain, in woods and by streams.
Flowering: April in the wild in Turkey.

CULTIVATION
Hardiness: Zone 5.
Spacing: 50cm (20in).

Symphytum caucasium. (Erich Pasche)

Availability: Occasionally sold by some herbaceous and herb nurseries, but uncommon.

Other details of cultivation and propagation as for the genus.

RELATED SPECIES
A similar species is the Turkish comfrey, **Symphytum ottomanum** Friv. (syn. *S. euboicum* (Runemark) Runemark), native to woods in the Balkans and Turkey, differing mainly in its smaller flower size, only 5–7mm long, and a thick spindle-shaped rootstock. It grows 30–80cm (12–32in) tall and is rarely found in cultivation.

Symphytum caucasicum Bieberstein
A clump-forming perennial comfrey, to about 60cm (2ft) tall, with hairy leaves and drooping clusters of azure-blue tubular flowers in spring. A very attractive, but invasive plant for wilder parts of the garden.

> *'This plant will smother anything but the largest perennials, but is very pretty for a long season, flowering again in late summer if the old stems are cut down.'* (Roger Phillips and Martin Rix).

NAMING
Etymology: *caucasicum* – of the Caucasus.
Common names: *English*: Caucasian comfrey, Blue comfrey; *French*: consoude de Caucase; *Russian*: okopnik kavkazsky.

DESCRIPTION
General: A finely downy perennial with spreading, branched rhizomes.
Stems: 40–60cm (16–24in), branched, upright, finely downy.
Leaves: Basal leaves to 20cm (8in) long and 4–6cm (1½–2½in) wide, oblong-ovate, narrowing to winged stalks, withered at flowering. Stem leaves oval-lanceolate to oblong-lanceolate, tapered at base, shortly decurrent.
Inflorescence: Paired cymes.
Flowers: *Calyx* 4–6mm, to 9mm in fruit, lobes triangular or broadly linear, blunt. *Corolla* red-purple at first, becoming blue, 13–17mm long, funnel-shaped, with narrow, triangular lobes; scales linear, blunt, equalling stamens. *Stamens* with anthers as long as filaments.
Nutlets: Pale, minutely warty.

GEOGRAPHY AND ECOLOGY
Distribution: Caucasus, Iran.
Habitat: Wet meadows, stream-sides; in hedgerows and shady places. Occasionally naturalized in south-east England.
Flowering: April to June.

CULTIVATION
Date of introduction: Before 1816.
Hardiness: Zone 4.
Sun/shade aspect: Sun or partial shade.
Soil requirements: Any good soil.
Garden uses: As for the genus.
Spacing: 45–60cm (18–24in).
Propagation: As for the genus.
Availability: Sold by numerous nurseries, specializing in herbaceous plants and herbs.
Problems: Occasionally tall stems may require staking. Very invasive.

RELATED FORMS AND CULTIVARS
'Eminence' – 30–45cm (12–18in) high, low spreading habit, greyish-green, softly hairy, narrower leaves, flowers rich blue, somewhat smaller than in the species, in late spring and early summer; often will bloom again if stems are cut down after flowering. Collected by Bill Baker in the Caucasus. Bob Brown of Cotswold Garden Flowers commented, *'I spent a long time wondering why I grow it until suddenly squiffy flowers turn to burgeoning blue.'*
'Norwich Sky' – pretty blue flowers, but extremely invasive.

Symphytum ibericum Steven

A creeping perennial comfrey, to 30cm (12in) high, with oval leaves and pale yellow, tubular flowers opening from red buds in spring. A nice, fast ground-covering plant.

NAMING

Etymology: *ibericum* – from Iberia.

Synonym: *Symphytum grandiflorum* invalid, not De Candolle.

Nomenclature note: Though commonly referred to as *S. grandiflorum*, it is currently considered that the plant in cultivation should be called *S. ibericum* Steven; the true **S. grandiflorum** De Candolle is a different species, and can be distinguished mainly by its larger flowers, with corolla 20–24mm ($^3/_4$–$^7/_{16}$in) long. To make matters more complicated, it is possible that this is not the true *S. ibericum* either, and that some cultivars attributed to this species may be unrelated to it.

Common names: *English*: Creeping comfrey; Dwarf comfrey.

DESCRIPTION

General: A perennial with decumbent sterile stems in first year and upright fertile stems in second year to 30cm (12in) from spreading, branched rhizomes with tuberous swellings.

Stems: Weak, bearing 8–10 leaves with prickles to 3mm long, and very short hairs.

Leaves: Broadly oval, sharply pointed, with heart-shaped base and long, winged stalks, shortly decurrent.

Inflorescence: Cymes dense, with about 20 flowers.

Flowers: *Calyx* 4–7 ($^3/_{16}$–$^3/_8$in) mm, lobed from $^2/_3$ to $^5/_6$ of the length, enlarged in fruit; lobes narrowly lanceolate, blunt. *Corolla* reddish in bud, pale yellow after opening, 14–20mm ($^1/_2$–$^3/_4$in) long, cylindrical, tube exceeding calyx. Throat-scales equalling stamens, tongue-shaped, blunt. *Stamens* with anthers twice as long as filaments.

Nutlets: 3mm, finely warted.

GEOGRAPHY AND ECOLOGY

Distribution: Caucasus and north-east Turkey, naturalized in Britain.

Habitat: In Turkey, grows on shady banks and in *Rhododendron* scrub, up to 1350m (4430ft); in Britain, naturalized in woods, hedges and grassland from Midlands southwards.

Flowering: March to July in the wild in Turkey; May to June in Britain; in cultivation, may start in late winter, with a peak in mid-spring, and continuing sporadically until midsummer.

Symphytum ibericum. (Erich Pasche)

CULTIVATION

Hardiness: Zone 5.

Sun/shade aspect: As for the genus.

Soil requirements: As for the genus.

Garden uses: A rapidly spreading species that makes excellent groundcover under trees and shrubs.

Spacing: 45cm (18in), though can spread wider.

Propagation: As for the genus.

Availability: Sold by numerous nurseries specializing in herbaceous plants.

RELATED FORMS AND CULTIVARS

'All Gold' – 75cm (2$^1/_2$ft), rich golden-yellow leaves in spring, and again later if cut back. Summer leaves green; flowers pinky-mauve. Although listed under *S. ibericum*, this plant is totally different in habit and form, and should probably belong elsewhere.

'Blaueglocken' – 38cm (15in), from Herr Pagels in Germany; coral-red buds open into narrow, tubular flowers of light blue, early spring to early summer. Probably a hybrid between *S. ibericum* and *S.* × *uplandicum*. Drought-tolerant.

'Gold in Spring' – 20–30cm (8–12in), new foliage is golden, especially when plants are growing strongly; pink and cream flowers, sometimes with a tinge of blue, from mid-spring to early summer. Slower spreading than the species. Found by Alan Leslie and named by Joe Sharman.

'Goldsmith' see *Symphytum* 'Goldsmith' under *Symphytum* hybrids, see page 186.

'Jubilee' see *Symphytum* 'Goldsmith'.

'Lilacinum' – form with lilac flowers; a hybrid between *S. ibericum* and *S.* × *uplandicum*.

'Pink Robins' – very narrow tubular flowers, pink with a slight tinge of blue. Named by Joe Sharman after the Robinsons who owned Hyde Hall, where this plant originated as a seeding.

'Variegatum' see *Symphytum* 'Goldsmith'.

'Wisley Blue' – form with blue flowers; it has been suggested that it is a hybrid between *S. ibericum* and *S.* × *uplandicum*.

Symphytum officinale Linnaeus

A perennial comfrey, of variable height, with hairy, oval-lanceolate leaves, and drooping clusters of many tubular flowers, which may vary in colour from purple-violet, to pinkish or white. This is one of the most ancient medicinal plants, a classic for a herb garden, or any wilder areas.

NAMING

Etymology: *officinale* – sold as a herb.

Common names: *English*: Common comfrey; Knitbone; Boneset; Slippery root; *Danish*: Læge-Kulsukker; *Dutch*: Gewone Smeerwortel; *Finnish*: Rohtoraunioyrtti; *French*: Consoude Officinale, Consoude Grande; *German*: Echter Beinwell; *Italian*: Consolida Maggiore; *Norwegian*: Valurt; *Polish*: Zywokost lekarski; *Russian*: okopnik lekarstvennyi; *Slovak*: kostihoj lekársky; *Spanish*: Consuelda; *Swedish*: (Äkta) vallört; *Turkish*: Buyuk Karakafesotu.

DESCRIPTION

General: An upright, roughly hairy perennial, with thick, fleshy, spindle-shaped roots.

Stems: 30–120cm (1–4ft), robust, branched, clothed with long hairs and bristles.

Leaves: Lower leaves 15–25cm (6–10in) long, oval-lanceolate, pointed, stalked. Upper leaves oblong-lanceolate, stalkless, broadly decurrent at base.

Inflorescence: Cymes many-flowered, drooping, later straightening.

Flowers: *Pedicels* 5–10mm ($^1/_4$–$^7/_{16}$in), shortly bristly. *Calyx* 7–13mm ($^3/_8$–$^1/_2$in), lobes lanceolate, pointed, 2–3 times as long as tube. *Corolla* 8–20mm ($^3/_8$–$^3/_4$in), usually 15–17mm ($^5/_8$in), white, creamy, red-purple or pink; silky hairy, with broad, triangular, deflexed lobes. Scales triangular, scarcely longer than the stamens. *Stamens* with anthers longer than filaments.

Nutlets: About 5–6mm, black, very smooth, shining.

Chromosomes: 2n=24, 36, 48.

Symphytum officinale (Doreen Townley).

GEOGRAPHY AND ECOLOGY

Distribution: Much of Europe, but rare in the extreme south, and not native but naturalized in much of the north, including Scandinavia and northern Russia.

Habitat: Damp places, especially beside rivers and streams, more rarely in drier rough, grassy places. Throughout the UK, though less common in the north and probably not native there.

Flowering: May to June in the wild in the UK.

Associated plants: In the UK, grows with other vigorous species such as *Angelica archangelica*, great hairy willowherb (*Epilobium hirsutum*), greater bindweed (*Calystegia sepium*) and reed-grass (*Phalaris arundinacea*).

CULTIVATION

Hardiness: Zone 5.

Spacing: 60cm (2ft), though can spread a lot wider, to over 1.8m (6ft).

Availability: Widely available, mostly from herb nurseries, but also some herbaceous suppliers.

For other details of cultivation and propagation, see the genus description.

RELATED FORMS AND CULTIVARS

Named colour variants are **S. officinale var. ochroleucum** De Candolle, with creamy-yellow flowers, and **var. purpureum** Persoon, with carmine-coloured flowers. These hybridize and may result in bi-coloured flowers.

Symphytum officinale subsp. *bohemicum* (Schmidt) Célakovsky is a smaller and more slender plant, to 1m (3ft) high, with finer hairs, and corolla to 14mm ($^1/_2$in), usually creamy-yellow but occasionally white. This subspecies is scattered throughout Europe, including eastern England. It has a requirement for moist soil and is rarely cultivated.

Subspecies uliginosum (Kerner) Nyman (syn. *S. uliginosum* Kerner) has rough and prickly stems up to 70cm (28in), and deep-purple corolla, 16–19mm. ($^5/_8$–$^3/_4$in) Mainly eastern Europe but probably throughout central Europe as far west as The Netherlands, but absent from Britain.

Symphytum orientale Linnaeus

A short-lived, softly hairy perennial comfrey, to about 60cm (2ft) high, with oval leaves and tubular, pure white flowers in late spring and early summer. A nice plant suitable for a mixed border, because it does not spread as widely as some.

NAMING

Etymology: *orientale* – oriental, eastern.

Synonyms: *Symphytum tauricum* Sims not Willdenow; *S. jacquinianum* Tausch.

Common name: *English*: White comfrey.

DESCRIPTION

General: A softly hairy perennial, with thick, spindle-shaped, branched rootstock.

Stems: To 60–70cm (24–28in) high, sparsely downy, branched.

Leaves: Softly pubescent, oval or oblong, somewhat pointed, with a heart-shaped, truncate or rounded base. Lower leaves to 14cm ($5^1/_2$in) long, often less, with stalk narrowly winged at top. Upper leaves short-stalked or stalkless, oblong.

Inflorescence: Cymes with about 20 flowers.

Flowers: *Calyx* 6–10mm ($^1/_4$–$^7/_{16}$in), in fruit to 19mm ($^3/_4$in), tubular, lobes half length of the tube, oval or oblong, obtuse. *Corolla* pure white, 14–20mm ($^1/_2$–$^3/_4$in), funnel-shaped. Throat scales triangular-lanceolate, slightly exceeding the stamens.

Nutlets: Erect or slightly curved, dark brown, warty.

GEOGRAPHY AND ECOLOGY

Distribution: West and north-west Turkey, south-western Ukraine; introduced in the UK, France, Italy. In Britain naturalized in hedgebanks in grassy places.

Habitat: Shady stream banks, woodlands such as black pine (*Pinus nigra*) forest in Turkey, up to 1500m (4900ft).

Flowering: April to June in the wild in Turkey.

CULTIVATION

Date of introduction: Before 1752.

Hardiness: Zone 5.

Spacing: 35–40cm (14–16in)

Availability: Only from a few specialist herbaceous nurseries, but excellent in dry shade under trees.
For other details on cultivation and propagation, see the genus description.

Symphytum tuberosum Linnaeus

A bristly, perennial comfrey, to 30–60cm (1–2ft) tall, arising from a curiously shaped rhizome with oval leaves and nodding clusters of pale yellow or cream flowers in spring and early summer. It is invasive and is more suitable for wilder parts of the garden.

NAMING

Etymology: *tuberosum* – tuberous, referring to the rootstock.

Common names: *English*: Tuberous comfrey; *French*:

Consoude Tubéreuse; *German*: Knollige Wallwurz; *Italian*: Consolida Femmina; *Russian*: okopnik klubnenosny; *Slovak*: kostihoj hl'uznaty.

DESCRIPTION

General: A stiffly hairy perennial, arising from a stout creeping rhizome with alternating thick tuberous and thin sections.

Stems: 20–60cm (8–24in), covered with reflexed bristles, simple or with one or two short branches near the top.

Leaves: All with bristly hairs swollen at the base. Basal leaves small, oval to spoon-shaped, with a winged stalk; they die down at flowering time. Middle stem leaves oval-lanceolate to elliptical, 10–14cm (4–5$\frac{1}{2}$in) long and 3–6cm (1$\frac{1}{4}$–2$\frac{1}{2}$in) wide, shortly stalked; upper stem leaves stalkless.

Inflorescence: Cymes with 4–16 flowers, occasionally more.

Flowers: *Pedicels* shortly downy, drooping. *Calyx* 5–8mm ($\frac{1}{4}$–$\frac{3}{8}$in), lobed to a third to nine tenths of its length; lobes linear-lanceolate, pointed. *Corolla* 12–19mm ($\frac{1}{2}$–$\frac{3}{4}$in), pale yellow or cream, with deflexed lobes. Throat scales triangular-lanceolate, included in the corolla. *Stamens* with anthers 3–5mm, about twice as long as filaments.

Nutlets: Slightly curved, wrinkled, very finely warty.

Chromosomes: 2n=18

GEOGRAPHY AND ECOLOGY

Distribution: Widely distributed in western, central and southern Europe and north-west Turkey. Scattered throughout Great Britain, commoner in the north, and occurs as an escape from cultivation in Ireland.

Habitat: Damp woods and meadows, hedgebanks, streamsides, other moist and shady habitats.

Flowering: June to July in the wild in Britain, April to June in southern Europe.

Associated plants: In the Carpathians, may be found with *Pulmonaria obscura*, *Symphytum cordatum*, *Cardamine trifolia*, *C. enneaphyllos*, *Dryopteris villarii*.

CULTIVATION

Hardiness: Zone 5.

Sun/shade aspect: Light to deep shade, though will tolerate sun if the soil is not allowed to dry out.

Soil requirements: Any moist soil.

Garden uses: As for the genus. It becomes dormant by mid-summer in the UK.

Spacing: 60cm (2ft), though may spread even wider.

Propagation: As for the genus.

Availability: Available from many herb nurseries and some herbaceous perennial specialists.

RELATED SPECIES

Symphytum cordatum Waldst. & Kit in Willd. is another comfrey with pale-yellow flowers, endemic to the Carpathians, where it inhabits moist woodland. It grows 20–40cm (8–16in) high, and has large, softly hairy, heart-shaped leaves arising from a creeping rhizome. An attractive plant, but even more invasive than *S. tuberosum*.

Yet another pale-yellow-flowered species is the Crimean comfrey, **S. tauricum** Wildenow, native to Crimea, Bulgaria, Romania and Turkey. It grows to 60cm (2ft), and has densely hairy, triangular leaves, and a thick, spindle-shaped rootstock. This plant has been naturalized on a hedgebank in Cambridgeshire, UK. Both of these species are rare in cultivation.

Symphytum × *uplandicum* Nyman

A robust, hairy perennial comfrey, with branching stems to 1.2m (4ft) high, large leaves and drooping clusters of tubular flowers, very variable in colour – from blue and pale mauve to rich purple, in spring and summer. This hybrid between *S. asperum* and *S. officinale* is very decorative, but invasive.

NAMING

Etymology: *uplandicum* – of Uppland, Sweden.

Synonym: *Symphytum peregrinum* Ledeb.

Common names: *English*: Russian comfrey, Quaker comfrey (Canada); *Dutch*: Bastaardsmeerwortel; *Finnish*: Ruotsinraunioyrtti; *German*: Wallwurz; *Swedish*: uppländsk vallört.

DESCRIPTION

General: A hairy perennial, intermediate between the parents, *S. asperum* and *S. officinale*, less rough and prickly than the former.

Stems: 40–140cm (16–55in), branched.

Leaves: Oval-lanceolate, the upper stalkless, shortly decurrent or more or less clasping.

Inflorescence: Cymes.

Flowers: *Calyx* 5–7mm, lobes usually pointed. *Corolla* pink turning blue, or persistently violet or purple, 12–18mm long. Scales longer than stamens, triangular-lanceolate with broad, rounded tip. *Stamens* 5–6mm long, with filaments equalling or longer than anthers.

Nutlets: Dull brown, many non-viable, though some fertile nutlets may form.

Chromosomes: 2n=36.

GEOGRAPHY AND ECOLOGY

Distribution: Naturalized widely in western and northern Europe.

Symphytum x *uplandicum* 'Variegatum'. (Howard Rice)

Habitat: Shady, moist habitats.
Flowering: Spring and summer.

CULTIVATION
Date of introduction: 1815.
Hardiness: Zone 4.
Spacing: 90cm (3ft).
Propagation: Mostly vegetative, as this hybrid does not often set viable seed.
Problems: Variegated forms may revert to green, and it is necessary to keep removing the reverted parts of the plant to stop them from taking over.
Availability: Widely available from many nurseries.
Other details on cultivation as for the genus.

RELATED FORMS AND CULTIVARS
'Axminster Gold' – 1.2 (4 ft) variegated foliage with broad bright yellow margins to leaves, blue flowers from pink buds in summer.
'Bocking 14' – is a sterile clone that is less invasive.
'Jenny Swales' – 75–90cm (2½–3ft) tall, large leaves variegated with yellow stripes and splashes; flowers dark purple. A sport from *S.* x *uplandicum* 'Bocking 14', sterile (from Monksilver Nursery). Flowers in spring, with a second flush if cut down.
'Mereworth' – 1.2m (4ft) high, large leaves, to 50cm (20in) long, irregularly variegated: either yellow-spotted or splashed, or with a pale green central splash. Flowers purple, in summer.

'Variegatum' ♀ – 60–90cm (2–3ft) tall, with evergreen, greyish-green leaves that are broadly margined with cream, flowers pale mauve or blue, opening from lilac buds, in late spring and early summer. Variegation may revert to dark green. Cutting the stems down after flowering will encourage more fresh foliage, and repeat flowering. Propagation only by division, as root cuttings would not produce variegated plants. Graham S. Thomas comments: *'… the large leaves are a foot or more in length, broadly margined with cream and grey. It is a rare plant, slowly increasing, of remarkable beauty.'* RHS Award of Garden Merit plant.

Symphytum hybrids

The parentage of the majority of cultivars below is uncertain, so they are listed separately here.

Symphytum 'Belsay Gold'
Probably a form of *S.* x *uplandicum*, with upper leaves golden-yellow. Available from one or two herbaceous nurseries.

Symphytum 'Denford Variegated'
45cm (18in) high; leaves streaked and blotched in cream and yellow, flowers pale pink and blue. Sold by a few herbaceous specialists.

Symphytum 'Goldsmith' (syn. 'Variegatum', 'Jubilee').
15–30cm (6–12in) high; low, evergreen, with heart-shaped, dark green, hairy leaves with broad creamy-yellow edges; tubular pink, blue and white flowers in spring to early summer. The best-known and most common comfrey, selected by Eric Smith, available from numerous nurseries and some garden centres. Easier to control than most other comfreys, as it is quite shallow-rooted. It is currently listed in *The RHS Plant Finder* as a hybrid but should really belong under *S. grandiflorum* hort. (here *S. ibericum*).

Symphytum 'Hidcote Blue' (*S. grandiflorum* x ?*S uplandicum* hybrid)
About 50cm (20in) high, with ascending to upright stems, spreading by stolons. Leaves oval to elliptical, rough, to 25cm (10in) long. Flowers to 15mm (⅝in) long, pale blue opening from pink-red buds, and fading gradually, in mid- to late spring. Naturalized in

hedges and woodland in the UK in Surrey and Shropshire, and in Guernsey. One of the most popular cultivars, sold by many nurseries.

Symphytum 'Hidcote Pink' (syn. 'Roseum')
To 45cm (18in) high, similar to 'Hidcote Blue', but flowers pink and white, fading with age.

Sold by many herbaceous plant nurseries.

Symphytum 'Langthorns Pink'
Probably a hybrid of *S. officinale* 'Rubrum' and *S.* x *uplandicum*. 1–1.2m (3–4ft) high, vigorous, with clusters of pink flowers in early summer. May bloom for the second time if cut down when flowers have finished. Sold by a number of nurseries specializing in herbaceous perennials.

Symphytum 'Rubrum'
Possibly a hybrid between *S. officinale coccineum* and *S. grandiflorum*. To 45cm (18in) tall, with hairy, dark green leaves and deep crimson-red, tubular flowers from late spring into summer. Not as invasive as some other comfreys. Widely available.

The genus *Trachystemon* D. Don, 1832

NAMING
Etymology: *Trachystemon* – from Greek *trachys*, rough, and *stemon*, a stamen, referring to the rough stamens of one species.
Synonym: *Nordmannia* Ledeb, *Psilostemon* DC.

GENUS CHARACTERISTICS AND BIOLOGY
Number of species: 1 or 2. The second species, *T. creticum* D. Don, from Greece and Bulgaria, is sometimes included within *T. orientalis*.
Description: See the description of *T. orientalis*.
Distribution and habitats: Caucasus, Turkey, Greece, Bulgaria; shady places.
Life form: Geophyte.
Pollination: By bumblebees.
Fruit dispersal: No specialized mechanism.

USES
No known medicinal or culinary uses.

GARDEN
Effective groundcover for wilder, more informal parts of the garden, under trees and shrubs, on stream banks.

CULTIVATION
This plant is very easy to grow in most soils and situations, including dry shade, though it prefers dappled shade, moist soil and a sheltered position. See other details on cultivation under *T. orientalis*.

PROPAGATION
See under *T. orientalis*.

REFERENCES
BLAMEY & GREY-WILSON, 1989, 1993; BRICKELL, 1996; CLAPHAM *et al.*, 1952; DAVIES, 1978; HUXLEY, 1992; STACE, 1991; THOMAS, 1993; WALTERS, 2001.

Trachystemon orientalis (Linnaeus) G. Don
An impressive, invasive, hairy perennial, to 45cm (18in) high, with heart-shaped leaves to 50cm (20in) long, and branched clusters of purplish-blue, borage-like flowers in spring. An excellent groundcover plant for wilder areas, tolerant of almost any conditions.

> *'Immense, hairy, heart-shaped leaves … grow up very quickly in late spring and create a handsome ground cover. Branching heads of blue starry-flowers appear before then, carried on hairy stems, with purplish calyces … For large areas, where it will dominate handsomely.'* (Graham S. Thomas)

NAMING
Etymology: *orientalis* – eastern.
Synonyms: *Borago orientalis* L.; *B. constantinopolitana* Hill; *Nordmannia cordifolia* Steven; *N. orientalis* (L.) Steven; *Psilostemon orientalis* (L.) De Candolle.
Common names: *English*: Eastern borage; Abraham, Isaac and Jacob.

DESCRIPTION
General: A hairy perennial, with a far-creeping rhizome to 5cm (2in) in diameter.
Stems: 20–60cm (8–24in), usually to 40cm (16in), upright, few-branched, sparsely covered in coarse hairs.
Leaves: Basal leaves oval with a heart-shaped base, 15–50cm (6–20in) long, with 10–25cm (4–10in) long stalk; stem leaves oval to lanceolate, stalkless, clasping; all sparsely covered in coarse hairs.

Inflorescence: Loose panicle of cymes, with elliptical-lanceolate bracts; each cyme with 5–15 flowers.
Flowers: *Calyx* 3–6mm in flower, to 9mm in fruit, lobes triangular, blunt, coarsely hairy. *Corolla* purplish-blue, tube 4–7mm long, limb 9–12mm, with linear lobes recurved; there are 2 series of 5 scales, at the middle of the tube and at the apex, upper scales white. *Stamens* hairy, with pink filaments, anthers long-protruding from the corolla. *Style* long-exserted.
Nutlets: About 2 × 4mm, wrinkled, each with a slightly ribbed ring at the base.
Chromosomes: 2n=56.

GEOGRAPHY AND ECOLOGY
Distribution: Caucasus, Turkey (N Anatolia), eastern Bulgaria.
Habitat: In Turkey, grows in beech (*Fagus*) forest, shady riverbanks, moist ravines, 50–1000m (160–3280ft). Naturalized in damp woods in the UK, in Devon, Yorkshire and elsewhere.
Flowering: March to May in the wild in Turkey; April to May in the UK.
Associated plants: In Georgia (Caucasus) may be found growing in the company of such plants as

Omphalodes cappadocica, *Helleborus orientalis* subsp. *guttatus*, *Epimedium colchicum* subsp. *caucasicum* and *Ruscus ponticus*.

CULTIVATION
Date of introduction: 1752.
Hardiness: Zone 5.
Sun/shade aspect: Sun or shade, though dappled shade is best.
Soil requirements: Any type of soil is tolerated, unless continuously waterlogged.
Garden uses: Excellent groundcover, though deciduous. For wild and woodland gardens, under trees and shrubs, streamsides, and anywhere where a fast ground-cover is required for a large area, especially in dry shade.
Spacing: 60cm (2ft).
Propagation: Easily divided in autumn or spring, replanting in permanent positions. Sow seed in spring in pots or trays of compost, and germinate in a cold frame.
Availability: Widely available from herbaceous plant suppliers.
Problems: Invasive, and will smother smaller plants.

Trachystemon orientalis. (Erich Pasche)

CHAPTER 3
GENERA RARE IN CULTIVATION

This chapter gives accounts of Boraginaceae genera and species that only rarely, if at all, appear in general cultivation. I hope that this information will arouse curiosity in gardeners, who might try to seek out some of these plants, many of which are fascinating and unusual. Some of the plants described in this chapter are rare because they are very difficult or near-impossible to cultivate (such as *Chionocharis hookeri*), others may be widely grown in warmer regions but are little known to gardeners in temperate climates, being too tender for outdoor cultivation (for example *Cordia* and most *Ehretia* species). Certain Boraginaceae have only limited ornamental qualities (such as some *Cryptantha* species). Some plants, however, are both attractive and perfectly growable, but still only rarely cultivated, for no apparent reason (for example, *Caccinia* species), and deserve more attention.

In addition to the plants listed in this chapter, other Boraginaceae genera that may be worth trying are: *Argusia* Boehm., *Coldenia* L., *Craniospermum* Lehm., *Dasynotus* I.M.Johnst., *Halacsya* Dorfl., *Lappula* Moench, *Lepechinella* Popov, *Mallotonia* (Griseb.) Britton, *Paracaryum* (A.DC) Boiss., *Rochelia* Riechb., *Tianschaniella* B.Fedtsch. and *Tournefortia* L. Although none of these plants are in general cultivation, it is possible that at least some may appear on specialist societies' seed lists.

Alkanna strigosa. (Masha Bennett)

The genus *Alkanna* Tausch, 1824

Alkanets

NAMING

Etymology: *Alkanna* – from Spanish *alcanna*, derived from the Arabic word for henna.

Synonyms: *Baphorhiza* Link; *Camptocarpus* C. Koch; *Campylocaryum* DC.

GENUS CHARACTERISTICS AND BIOLOGY

Number of species: From 30 to 40 according to different estimates.

Description: A genus of evergreen perennials or sometimes annuals, clump-forming, hairy or bristly, with basal tufts or rosettes of narrow hairy leaves and upright leafy flowering stems. Inflorescence consists of one or more terminal cymes with bracts, the *corolla* funnel- or salver-shaped, with a ring of hairs in the throat, often purplish-blue, but sometimes yellow, orange or white. Stamens and style are included in the corolla tube. The root is often very large in relation to the green parts of the plant. Nutlets are usually two, kidney-shaped to obliquely ovoid, beaked, granular to warty.

Distribution and habitats: From southern Europe, throughout the whole Mediterranean region, to Iran, especially widespread in Turkey. Mostly dry places, particularly rocks and screes. 32 *Alkanna* species and forms are included in *The IUCN Red List of Threatened Plants*.

Life forms: Chamaephytes and some therophytes (annual species).

Pollination: By bees, such as *Anthophora pauperata* for *Alkanna orientalis*.

Fruit dispersal: No specialized mechanisms.

USES

GARDEN

Rock garden, crevices, scree, raised beds, troughs, herb garden (*A. tinctoria*), groundcover for sunny dry banks, but some species need protection of an alpine house (for example, *A. aucheriana*).

MEDICINAL

Roots of *A. tinctoria* are astringent and anti-bacterial, encouraging healing and relieving itching when used externally for varicose and indolent ulcers, bedsores and itching rashes. Culpeper said of this plant:

'It is an herb under the dominion of Venus, and indeed one of her darlings, though somewhat hard to come by. It helps old ulcers, hot inflammations, burnings by common fire and St. Anthony's fire ... for these uses your best way is to make it into an ointment also if you make a vinegar of it, as you make a vinegar of roses, it helps the morphy and leprosy ... it helps the yellow jaundice, spleen, and gravel in the kidneys. Dioscorides saith, it helps such as are bitten by venomous beasts, whether it be taken inwardly or applied to the wound, nay, he saith further, if any that hath newly eaten it do but spit into the mouth of a serpent, the serpent instantly dies. ... Its decoction made in wine and drank, strengthens the back, and easeth the pains thereof. It helps bruises and falls, and is as gallant a remedy to drive out the smallpox and measles as any is.'

ECONOMIC

Roots of *A. tinctoria* are used as a red colorant for foodstuffs, pharmaceutical products and cosmetics. It has also been used to tint inferior port and to colour thermometer fluids and chemists' shop bottles. Wood can be stained with alkanna to give it an appearance of rosewood or mahogany. To obtain grey, lavender and purple dye, alkanna roots should be simmered in water for at least an hour (this produces an unpleasant odour so is best done outdoors).

CULTIVATION

Alkannas are hardy to Zone 8. They all require full sun (though *A. tinctoria* may put up with very light shade, especially in warmer climates) and gritty, perfectly drained soil or compost, and many will thrive in alkaline conditions. Some, like *A. aucheriana*, can grow in tufa. Under glass, give moderate amounts of water during the growing season, keeping only slightly moist in winter. Try to keep the water off the leaves. Outdoors, protect from winter wet by a pane of glass.

PROPAGATION

Seed In autumn, seed can be sown in trays or pots of gritty compost, and placed in a cold frame to overwinter. Perfect drainage is essential.

Vegetative Softwood cuttings can be taken in summer and inserted into gritty compost; root cuttings can be tried in autumn. Mark McDonough (of Pepperell, Massachusetts, USA – USDA Zone 5) had success with rooting the cuttings of *A. aucheriana* in moistened coarse sand mixed with perlite, with cuttings uncovered but frequently watered.

PROBLEMS

When grown under glass, alkannas are susceptible to attack by aphids and red spider mite. Unless they have the protection of an alpine house, they are inclined to rot in wet weather.

AVAILABILITY
Alkannas are rarely available in the trade, apart from perhaps *A. tinctoria,* which may be sold by a few herb nurseries. Seed of rock garden species may appear on the lists of specialist alpine societies. Sometimes grown in botanic gardens, including the Royal Botanic Garden Edinburgh, and may be available as seed to the garden's 'friends'.

REFERENCES
BLAMEY & GREY-WILSON, 1993; DAVIS, 1978; DEAN, 1999; FEINBRUN-DOTHAN, 1978; HALLIWELL, 1992; HARKNESS, 1993; INGWERSEN, 1991; POLUNIN, 1980; SLABY, 2001; TUTIN *et al.,* 1972.; WALTERS, 2000.

Alkanna aucheriana A. D.C.
Description: A cushion-forming perennial, felted with soft white or grey hairs, 5–15cm (2–6in) tall, with oblong or linear leaves 2–6cm (^{3}/$_4$–2^1/$_2$in) long, and cymes of blue or purple flowers to 10mm (^{7}/$_{16}$in) wide, borne from spring to summer.
Distribution: Turkey.
Habitat: Limestone crevices in oak and pine forest and garrigue, at 500–1000m (1650–3280ft).
Related species: Another attractive blue-flowered species native to Turkey is **Alkanna saxicola** Huber-Morath., a rare plant of limestone rocks at 1200–1290m (3900–4150ft). A great alpine gardener, Panayoti Kelaidis, who gardens in Denver, Colorado (Zone 5), describes it thus: *'This makes a tiny tuft barely 4 inches tall and not much more across of very silvery leaves with heavenly blue flowers. … I have grown it on a dry scree for nearly ten years now: every winter it looks completely dead and it recovers magnificently each spring.'* A closely related plant is **A. oreodoxa** Huber-Morath., on limestone rocks at 470–700m (1540–2300ft) in Turkey; its corolla lobes are white and the tube may be blue, pink or yellowish-brown. Both species are rare in cultivation but may occasionally be found in botanical gardens and in specialist collections.

Alkanna graeca Boissier & Sprun.

'Isn't in the same league as A. aucheriana *but a nice quiet "perennial" with yellow flowers in vaguely crozier shaped spikes. It sows around a bit, but has a long flowering season.'* (Geoffrey Charlesworth)

Synonym: *Alkanna baeotica* has sometimes been considered a separate species, but is now included within *A. graeca* as a subspecies.
Description: A bristly perennial with ascending stems 15–45cm (6–18in), lanceolate or narrowly oblong leaves 6–10cm (2^1/$_2$–4in) long, and cymes of yellow flowers 8–10mm across and 5–6mm long (10–12mm long in subspecies *baeotica*).
Distribution: Southern Greece, former Yugoslavia, Albania.
Habitat: Mountain slopes and rocks: subsp. *graeca* from rocky slopes below 1500m (4900ft), subsp. *baeotica* on mountain rocks above 1500m (4900ft).
Related species: Another attractive, yellow-flowered species is **A. primuliflora** Griseb., native to dry, rocky slopes of Bulgaria, Greece and Turkey. It is a glandular-hairy plant growing to 15–30cm (6–12in), with golden-yellow flowers to about 10mm (^{7}/$_{16}$in) across, and is suited for the alpine house. Yet another interesting plant is **A. stribnyi** Velen., with ascending to semi-upright stems 30–50cm (12–20in) long; comes from similar habitats in Bulgaria and former Yugoslavia. It has deep orange, or sometimes violet or greyish flowers – due to its larger size it is probably more suitable for a sunny, dry border. Both of these species are included in *The IUCN Red List of Threatened Plants,* so make sure that any propagation material or plants you may obtain are from legitimate sources.

Alkanna incana Boissier

'This is an exceptional species. From basal tufts of grey-green, roughly bristly leaves rise short, stiff stems carrying several deep blue flowers … Should be treated austerely in full sun, or kept in the alpine house.' (Will Ingwersen)

Description: A beautiful, evergreen, mound-forming, grey-hairy perennial, with stems 5–25cm (2–10in) high, grey-green lanceolate leaves 3–6cm (1^1/$_4$–2^1/$_2$in) long, and cymes of bright blue flowers about 8–10mm (^{7}/$_{16}$in) across.
Distribution: Turkey.
Habitat: Among limestone rocks and scrub.
Related species: Another garden-worthy, blue-flowered alkanna native to Turkey is **A. megacarpa** A. DC, growing to 15–25cm (6–10in), and distinguished by its very large nutlets, 4–5mm in diameter. It grows on limestone slopes and screes and among scrub above 720m (2360ft). Likely to need alpine house treatment.

Alkanna orientalis (Linnaeus) Boissier

Synonyms. *Anchusa orientalis* L.; *Lithospermum orientale* (L.) L.

Common names: *English*: Oriental alkanet; *Arabic*: Libbeid, lebbd, anzurn, uzz waran.

Description: A bristly, glandular-hairy perennial with upright or ascending stems 25–50cm (10–20in), oblong to lanceolate leaves to 20cm (8in) long and 4cm (1½in) wide, and dense racemes of sweetly scented, bright golden-yellow flowers, 9–12mm (⅜–½in)across and to 13mm (½in) long, borne from spring into summer.

Distribution: From southern Greece eastwards – Turkey, eastern Mediterranean to south-west Asia; not in Cyprus.

Habitat: Rocky places, stony slopes, to 2500m (8200ft).

Related forms and cultivars: *Alkanna orientalis* **var. *leucantha*** (Bornm.) A. Huber-Morath., (syn. *A. leucantha*) has white or ivory-coloured flowers. It is endemic to Turkey and probably not in cultivation.

Alkanna tinctoria Tausch.

Synonyms: *Alkanna lehmannii* (Tineo) A. DC; *A. tuberculata* (Forssk.) Meikle. *Lithospermum tinctorum* L.; *Anchusa tinctoria* invalid, not L.

Common names: *English*: Dyer's alkanet, Dyer's bugloss, Spanish bugloss; *French*: Racines d'alkanna; *German*: Ochsenzunge; *Polish*: Alkanna barwierska; *Russian*: Alkanna krasilnaya; *Slovak*: Alkana farbiarska; *Spanish*: Onoquiles; *Turkish*: Havaciva.

Description: A bristly, sprawling, woody-based perennial with stems 10–30cm (4–12in) long clothed in lanceolate leaves to 5cm (2in) long, in spring bearing bright blue flowers about 10mm (⁷⁄₁₆in) across, its stout purple-brown roots containing a red dye. In the past has been important as a dye plant and in cosmetics.

Distribution: Throughout the Mediterranean, not the Balearic Islands or Corsica.

Habitat: Rocky garrigue and other sandy and rocky habitats, waste ground, roadsides, often close to the sea. Native to limestone screes, pine forests, and coastal sands in eastern Mediterranean areas.

The genus *Amblynotus* (A. De Candolle) I.M. Johnston, 1924

NAMING

Etymology: *Amblynotus* – literally, 'round back', referring to the shape of the nutlets.

Synonyms: Previously included within *Eritrichium* Schrad. and *Myosotis* L.

GENUS CHARACTERISTICS AND BIOLOGY

Number of species: Only one species.

Description: See the description of *A. rupestris* below.

Distribution: Siberia to Mongolia.

Habitats: Stony slopes in mountains and steppes.

Life form: Chamaephyte.

Pollination: Probably bees.

Fruit dispersal: No specialized mechanism.

USES

GARDEN

If available, *Amblynotus* would probably make a suitable subject for an alpine house or, possibly, a rock garden.

No medicinal or culinary uses known.

CULTIVATION AND PROPAGATION

Almost nothing is known about the cultivation requirements of this plant, but it is likely to be extremely hardy and require treatment similar to that for *Eritrichium* species: protection from winter wet, ideally in alpine house, full light and a perfectly drained growing medium. Propagation should be possible from fresh seed – in the same way as for eritrichiums.

AVAILABILITY

Unavailable at present.

Amblynotus rupestris (Pallas in Georgi) M.Popov in Serg.

Synonyms: *Amblynotus obovatus* (Ledeb.) Johnst.; *Myosotis rupestris* Pallas in Georgi; *Eritrichium rupestris* (Pallas in Georgi) Bunge; *E. dahuricum* (Pallas) Brand.

Common name: *Russian*: Kruglospinnik skal'nyi.

Description: A cushion-forming perennial arising from thick vertical roots, with many thin, upright or ascending stems 5–20cm (2–8in) high, branched only in the inflorescence. Leaves densely hairy, basal

1–2cm ($^{7}/_{16}$–$^{3}/_{4}$in) long, oblong-spoon-shaped, stem leaves narrowly oblong. Flowers borne in loose cymes, on upright pedicels 2–10mm long, with *calyx* 1–3mm long, and blue *corolla* 3–7mm across. Nutlets 1.5–2mm, white, shiny, oval, with a rounded dorsal surface.

Distribution: Western Siberia, Manchuria, Mongolia, northern China.

Habitats: Open, stony and gravelly slopes in mountains and steppes.

Associated plants: In Siberia, grows in steppe communities dominated by grasses or sedges and *Gypsophila patrinii*, with such plants as *Dracocephalum peregrinum*, *Aquilegia buriatica*, *Saussurea schanginiana*, *Dianthus versicolor*, *Iris humilis*, *Artemisia commutata* and *Pedicularis achilleifolia*.

REFERENCES
MALYSHEV, 1997; SLABY, 2001; WU & RAVEN, 1995.

The genus *Amsinckia* Lehmann, 1831
Fiddleneck

NAMING
Etymology: *Amsinckia* – after Wilhelm Amsinck, who developed the botanic gardens at Hamburg in the nineteenth century.

Synonyms: Some members of the genus were previously included in the genus *Lithospermum* L.

GENUS CHARACTERISTICS AND BIOLOGY
Number of species: 15 species.

Description: A genus of bristly annual herbs. Leaves linear to oblong or lanceolate, stalkless. Flowers borne in terminal, spiralled cymes with or without bracts. *Calyx* divided nearly to base. *Corolla* yellow to orange, with tube longer than limb. Stamens equal, included in the corolla. Style simple, also included. Nutlets keeled, warty.

Distribution and habitats: Western USA and temperate South America, mostly in dry places. 3 *Amsinckia* species are included on *The IUCN Red List of Threatened Plants*.

Life form: Therophytes.

Pollination: Bees.

Fruit dispersal: No specialized mechanism.

Animal feeders: Caterpillars of several species of moths belonging to family Ethmiidae feed on *Amsinckia* species.

USES
GARDEN
Wildflower gardens in North America, otherwise they can be included in a sunny mixed border as a curiosity.

CULINARY
The fresh shoots of *A. lycopsoides* and *A. tesselata* were traditionally eaten by Native Americans, and the seeds were ground to make cakes.

CULTIVATION
Most amsinckias should be hardy to Zone 7. They can be grown in full sun in any well-drained soil, and are tolerant of drought. In general, their requirements are similar to those of anchusas.

PROPAGATION
From seed only: sow in late spring directly into flowering positions. In cold areas, seeds could be sown in pots or trays of compost under cover in early spring.

PROBLEMS
In warmer, dry climates they are likely to become excessively weedy.

AVAILABILITY
Generally unavailable. May appear on exchange lists.

REFERENCES
BLAMEY & GREY-WILSON, 1989; HITCHCOCK, 1955; HUXLEY, 1992; MOERMAN, 1998; STACE, 1991; TUTIN *et al.*, 1972.

Amsinckia lycopsoides Fisch. & C.A. Mey.
Synonym: *Amsinckia menziesii* (Lehm.) A. Nelson & J. F. Macbr.

Common names: *English*: Scarce fiddleneck, Tarweed fiddleneck, Bugloss fiddleneck; *Danish*: Hønse-Gulurt; *Norwegian*: Hønsegullurt; *Swedish*: Hönsgullört.

Description: An upright, bristly annual, to 70cm (28in) tall, often branched, with linear to elliptical leaves, and coiled cymes of deep orange-yellow, funnel-shaped flowers 5–8mm ($^{1}/_{4}$–$^{3}/_{8}$in)long, from spring to summer.

Distribution: Western USA, naturalized on the Farne Islands off the north-east coast of Britain, and now in various places in England, also in France and parts of Scandinavia; in Australia it is listed as a noxious weed.

Habitat: Dry open places, often on disturbed soils.

The genus *Bourreria*
P. Browne, 1756

NAMING
Etymology: Not known.
Synonyms: Some species have been included in genus *Ehretia* P.Browne.
Nomenclature note: This genus is sometimes separated into the family Ehretiaceae, together with the related genera such as *Ehretia* and *Cordia.*

GENUS CHARACTERISTICS AND BIOLOGY
Number of species: From 30 to 50 species, according to various estimates.
Description: A genus of trees and shrubs, hairless or shortly and softly hairy. Leaves alternate, elliptical, obovate or nearly rounded, stalked, with entire margins. Flowers numerous, borne in terminal corymb-like or dichotomously branched cymes. *Calyx* rounded or ovoid, 2–5-lobed, with a short tube that expands at apex. *Corolla* white, with broad spreading lobes. Stamens can be included in the corolla or protrude from it; bases of filaments sometimes hairy. Style with a divided tip. Fruit is a drupe containing 4-ridged, bony seeds.
Distribution and habitats: Central America and warm regions of North and South America, East Africa, Madagascar, Mascarene Islands, forests. 8 species included on *The IUCN Red List of Threatened Plants.*
Life form: Phanerophytes.
Pollination: By insects.
Fruit dispersal: By birds.

USES
GARDEN
Can be grown only in frost-free gardens, or, in colder climates, as a pot plant in a conservatory or greenhouse. Some species, such as *B. ovata*, are especially attractive to butterflies.

MEDICINAL
Leaves, twigs and bark of *B. ovata* are used medicinally for a variety of complaints, including kidney problems, diarrhoea, and to stop bleeding.

ECONOMIC
Fragrant flowers of *B. huanita* are added to tobacco and drinks.

CULTIVATION AND PROPAGATION
Bourrerias do not tolerate frost and cannot be grown outdoors in areas colder than Zone 10. Their cultivation requirements are similar to those of *Cordia* species.
Availability: Generally unavailable.

Bourreria ovata Miers
Common names: *English:* Strongbark, Bahama strongbark.
Description: A shrub to 3.5m (11$\frac{1}{2}$ft) high, with upright, crooked branches with grey-grown bark. Leaves 3.6–7.5cm (1$\frac{1}{2}$–3in) long, green to yellowish-green, glossy above, paler and shortly hairy underneath, with yellow midrib, the tip pointed or rounded, with base cuneate or narrowing into a stalk to 3.7cm (1$\frac{1}{2}$in) long. Flowers borne in corymbs of 15 to 20, white, to 12mm ($\frac{1}{2}$in) across, with protruding stamens, in summer, followed by rounded, orange-red, shiny fruit to 8.5mm ($\frac{3}{8}$in) in diameter.
Distribution: West Indies.

Bourreria succulenta Jacq.
Synonym: *Ehretia bourreria* L.
Common names: *English*: Bodywood, Currant tree; *Spanish*: Ateje de costa, Roble guayo, Fruta de catey.
Description: A shrub or small tree to 10m (33ft) high, with hairless or softly hairy branches, elliptical, obovate or rounded leaves 5–12cm (2–4$\frac{1}{2}$in) long, rough above, on stalks to 2cm ($\frac{3}{4}$in) long. Flowers borne in corymbs of 10 to 60, white, 7–10mm ($\frac{3}{8}$–$\frac{7}{16}$in) in diameter, followed by almost rounded, orange or red fruit 8–11mm ($\frac{3}{8}$–$\frac{7}{16}$in) in diameter.
Distribution: West Indies.
Associated plants: May grow with *Zanthoxylum thomasianum, Guettarda elliptica, Acacia muricata* and *Coccoloba macrostachya.*

REFERENCES
HUXLEY, 1992; MABBERLEY, 1997.

The Genus *Caccinia*
Savi, 1832

NAMING
Etymology: *Caccinia* – after Mateo Caccini of Florence, Italy, a seventeenth-century grower of rare plants.
Synonyms: *Anisanthera* Raf. Some species were previously included in the genus *Borago* L.

GENUS CHARACTERISTICS AND BIOLOGY
Number of species: 5 or 6 species.

Description: A genus of bushy, sprawling, biennial or perennial herbs. Stems hairless. Leaves glaucous, hairless but with numerous white tubercles. Flowers in paniculate inflorescences, the ultimate branches appearing raceme-like; each flower subtended by a leaflike bract. *Calyx* almost spherical, with 10 warty keels in flower, with the lobes clasping the apex of the corolla-tube; fruiting calyx-lobes separating to near their middle to form a 5-angled, 10-keeled structure. *Corolla* with a narrow tube that equals the calyx; lobes spreading, linear-oblong; throat-scales pyramidal, protruding from the corolla. Stamens and style also protruding, the former with unequal anthers: 1 large, 4 small, each with 2 forked appendages. Nutlets 1 or 2, rarely more, depressed-ovoid or circular, with narrow, toothed margins.

Distribution and habitats: South-western and central Asia, in dry places.

Life form: Hemicryptophytes.

Pollination: Probably by bees.

Fruit dispersal: No specialized mechanism.

USES

MEDICINAL
Caccinia crassifolia is used to treat rheumatism and syphilis in Iraq.

GARDEN
Can be grown in a herbaceous border, or a large rock garden. Useful in a seaside garden, as it is tolerant of salt-spray.

CULTIVATION
Those caccinias that are likely to appear in cultivation should probably be hardy to Zone 5. They need full sun and well-drained soil that is not too rich, and are tolerant of drought once established. Do not move plants unless really necessary, as they deeply resent root disturbance. *Caccinia kotschyi* may need protection from winter wet. Cacinneas can become untidy, so cut off the stems after flowering unless the seed is required in propagation.

PROPAGATION
Easy from seed. Sow each large seed in an individual pot; this will avoid later root disturbance.

AVAILABILITY
Not generally available, although seed may occasionally be offered for sale or appear on the seed-exchange lists. Grown in some botanic gardens.

REFERENCES
HUXLEY, 1992; PHILLIPS & RIX, 1991; RECHINGER, 1964; SLABY, 2001; WALTERS, 2000.

Caccinia macranthera (Russel) A. Brand
Synonyms: *Borago macranthera* Banks & Soland., *Caccinia glauca* Savi, *Caccinia russelii* invalid.

Description: An upright or ascending biennial, with branching, leafy, stems 20–50cm (8–20in), sometimes to 90cm (3ft), obovate or elliptical basal leaves to 10cm (4in) long and 4cm (1$^{1}/_{2}$in) wide, and narrowly-lanceolate stem leaves, with tiny prickles on margins. In spring and summer bears panicles of purple-blue flowers with lanceolate calyx lobes 7–12mm ($^{3}/_{8}$–$^{1}/_{2}$in) long and corolla consisting of a tube 9–15mm ($^{3}/_{8}$–$^{5}/_{8}$in) long and lobes 5–9mm ($^{1}/_{4}$–$^{3}/_{8}$in) long.

Distribution: Syria, northern Iraq, Iran, eastern Turkey, Afghanistan, Pakistan and Armenia.

Habitat: Banks, dry hillsides, abandoned fields, to 1900m (6230ft).

Related forms: The form that normally appears in cultivation is **C. macranthera var. crassifolia** (Vent.) A. Brand, that differs by being perennial, having thicker lanceolate leaves, inflorescence branches arising only in the upper part of the plant, larger calyx (to 20mm/$^{3}/_{4}$ in fruit) with linear lobes, corolla tube 15–18mm ($^{5}/_{8}$–$^{3}/_{4}$in) long with lobes 6–12mm ($^{1}/_{4}$–$^{1}/_{2}$in) long, and usually producing only one nutlet per flower as opposed to 4.

Caccinia strigosa Boissier

'Has marvellous, warty, blue-grey leaves and typical borage blue/pink flowers. After some time in flower or in seed stems are so long and untidy I cut them off.' (Helen Dillon)

Description: A glaucous perennial to 40cm (16in) tall, with ascending or decumbent, unbranched stems, obovate or oblong, blunt, almost unstalked leaves to 17cm (6$^{1}/_{2}$in) long, and terminal, bractless panicles of blue flowers, each to 12mm ($^{1}/_{2}$in) long.

Distribution: South-west Asia to Iran.

Habitats: Dry places.

Related species: Another plant of interest is **Caccinia kotschyi** Boissier, from rocky subalpine slopes in Iran. It is a tufted perennial 15–20cm (6–8in) high, with decumbent stems, oval-oblong green leaves and small clusters of blue flowers. It is very rarely, if ever, grown, but would make a good rock garden plant.

The genus *Carmona* Cavanilles, 1799

NAMING
Etymology: *Carmona* – meaning uncertain, named by Bruno Salvatori.
Synonyms: Sometimes included in genus *Ehretia* P. Browne.

GENUS CHARACTERISTICS AND BIOLOGY
Number of species: 1 only.
Description: See the description of *C. retusa* below.
Distribution and habitats: East Asia, in forests and scrub.
Life form: Phanerophyte.
Pollination: By insects.
Fruit dispersal: Birds ingesting the berries may carry the seeds long distances away – this is how *C. retusa* became naturalized in Hawaii.

USES
GARDEN
In temperate climates only grown as a bonsai, and especially popular as such in China. In Malaya, used for hedging.

MEDICINAL
Carmona retusa is used as a remedy for coughs and stomach upsets in the Philippines, to treat cachexia and syphilis in India and for fever in Malaya.

CULTIVATION
Carmona retusa is not frost hardy, and does not tolerate prolonged exposure to temperatures below 10°C (50°F), so in temperate climates it must be cultivated under cover, where it is normally grown as a bonsai. In winter, the temperatures should ideally be maintained between 15 and 20°C (59 and 68°F). Grow in a draught-free position, in bright light, with up to one hour of direct sunlight a day, but protection from intense heat is recommended. Avoid overwatering or allowing the compost to become bone dry. The plants should be fed with a special bonsai fertilizer, or a half-strength general purpose fertilizer fortnightly throughout the growing season, and repotted once every 2–3 years in spring.
Refer to specialist publications on bonsai for further details of cultivation and training of the plants.

PROPAGATION
Sow seed in spring. Take softwood cuttings in spring or summer, rooting them in a propagator with bottom heat.

AVAILABILITY
Carmona retusa is infrequent in cultivation in Europe, but is more commonly grown in the USA. Plants are only likely to be sold by specialist bonsai suppliers, and seed may occasionally appear in some catalogues.

PROBLEMS
Red spider mites and mealy bugs can be a problem on bonsai trees grown under cover, and so can aphids and scale insects. The plants are sensitive to insecticides so, ideally, cultural or biological control methods should be used. *Carmona retusa* is not the easiest bonsai to grow, and is sensitive to drastic changes in temperature, light regime, irregular watering and over-feeding. The plants may drop their leaves when stressed, but these usually grow again within a few weeks.

REFERENCES
bonsaiweb.com/care/faq/ehretia.html
HUXLEY, 1992; WU & RAVEN, 1995.

Carmona retusa (Vahl) Masamune
Synonyms: *Carmona microphylla* (Lam.) G. Don; *C. heterophylla* Cavanilles; *Ehretia retusa* Vahl; *E. microphylla* Lamarck; *E. buxifolia* Roxburgh; *E. dentata* Courchet.
Common names: *English*: Fukien tea, Philippine tea; *Chinese*: ji ji shu.
Description: A slow-growing shrub or small tree, from 1–10m (3–33ft) high in the wild, most commonly grown as a bonsai. Branches hairy and bristly, becoming hairless with age. Leaves resemble those of box (*Buxus sempervirens*), obovate, blunt or pointed, very variable in size, 1–10cm ($^7/_{16}$–4in) long and 5–40mm ($^1/_4$–1$^1/_2$in) wide (smaller in bonsai specimens), glossy green above, duller below. White flowers to 9mm ($^3/_8$in) across are borne in loose cymes, and followed by red or yellow fruits, 3–6mm ($^1/_8$–$^1/_4$in) in diameter.
Distribution: Southern China, Taiwan, Japan (Ryukyu Islands), Indonesia; naturalized in Hawaii.
Habitat: Forests and scrub.

Chinocharis hookeri. (Harry Jans)

The genus *Chionocharis* I.M. Johnston, 1924

NAMING
Etymology: *Chionocharis* – literally, 'grace of the snow'.
Synonyms: Previously included within *Myosotis* L. and *Eritrichium* Schrad.

GENUS CHARACTERISTICS AND BIOLOGY
Number of species: 1 species only.
Description: A genus of one cushion-forming perennial herb. Leaves alternate, overlapping. Flowers solitary, terminal. *Calyx* 5–parted to base, lobes narrowly spoon-shaped. *Corolla* wheel-shaped, tube as long as calyx, with 5 appendages in throat; lobes of limb spreading. Stamens included in the corolla, with very short filaments and oval anthers. Ovary 4-parted. Style short, included in the corolla, stigma headlike. Nutlets oval, shortly hairy.
Distribution and habitats: China, India, Bhutan, Nepal; rocky slopes at high altitudes in mountains.
Life form: Chamaephyte.
Pollination: Probably by bumblebees.
Fruit dispersal: No specialized mechanism, but nutlets likely to be dispersed by melting snow.

CULTIVATION
There is hardly any mention of cultivation of

C. hookeri in the literature, but, from observations of the plants in their habitat (see the account by Bodil Larsen below), it should be possible to work out what environment we must provide to have any chance of success: gritty acid soil, plenty of light and, ideally, snow cover! This plant must be extremely hardy, and will need to be protected from winter wet. The alpine house is probably the best bet, in conditions similar to those for eritrichiums.

PROPAGATION
The seed apparently germinates without any special treatment, but growing seedings on is likely to present a challenge.

AVAILABILITY
Seed may appear on specialist societies' exchange lists.

REFERENCES
BECKETT, 1993–1994; POLUNIN & STAINTON, 1997; WU & RAVEN, 1995.

Chionocharis hookeri (C.B. Clarke) I.M. Johnston
Synonyms: *Myosotis hookeri* C.B. Clarke in J.D. Hooker; *Eritrichium hookeri* (C.B. Clarke) Brand.
Description: A dwarf perennial, forming cushions 15–40cm (6–16in) across, with crowded, much-branched stems about 3 cm (1¼in) high, clothed in pointed, white-hairy leaves 7–12mm (³⁄₈–¹⁄₂in) long and 4–6mm (³⁄₁₆–¹⁄₄in) wide, and bearing profuse light blue, forget-me-not-like flowers about 7–8mm (³⁄₈in) across.
Distribution: China (Sichuan, Xizang, Yunnan), Bhutan, north-east India, Nepal.
Habitats: Rocky slopes, precipices; at 3500–5000m (11,500–16,400ft).
Associated plants: Bodil Larsen describes the habitat of *C. hookeri* and associated plants, observed on the Beima Shan (near Dechen) during the Alpine Garden Society trip to Yunnan in 1998:
'It was growing at 4670m [15,300ft], about 600m [1970ft] up the side of a valley and the terrain had started to level out, so although drainage was probably good it was not extreme. It was above tree line and there were patches of snow both higher up and lower down (June 16). We started our hike at about 4000m [13,100ft] where there were lots of waist-high rhodos. Where *C. hookeri* was growing, few things were more than a foot high; protected between boulders the rhodos could perhaps reach a little more. Around *C. hookeri* were growing *Oxygraphis glacialis, Androsace delavayi, Diapensia purpurea, Cassiope pectinata, Primula dryadifolia* and a very

beautiful large-flowered *P. chionantha* subsp. *sinoplantaginea*. I believe the rhodo was *R. saluensis* v. *chameunum*. The soil looked like your typical ericaceous soil, probably not very deep though. The *C. hookeri* cushion was over a foot across. It would be snow-covered in the winter but certainly tolerated summer rain (we got drenched!) On sunny days it would get sun all day and it was fully exposed to wind.'

The Genus *Cordia* Linnaeus, 1753

NAMING
Etymology: *Cordia* – for Euricius Cordius (1486–1535) and his son Valerius (1515–44), German botanists and pharmacists.
Synonyms: *Cordiada* Vell.; *Cerdana* Ruiz & Pav.; *Cordiopsis* Desv.; *Gerascanthus* P. Browne; *Lithocardium* Kuntze; *Rhabdocalyx* Lindl.

GENUS CHARACTERISTICS AND BIOLOGY
Number of species: About 320 species.
Description: A genus of deciduous or evergreen shrubs or trees, or occasionally climbers. Leaves usually alternate or rarely opposite, rough, downy or hairy, sometimes hairless, with stalks channelled above. Flowers borne in terminal cymes, panicles or spikes. *Calyx* usually 3–5–lobed, rarely 10–lobed, usually persistent, smooth or ribbed. *Corolla* white, yellow, red or orange, mostly 5–lobed, but can be 4–18 lobed, funnel- to bell-shaped, sometimes tubular. Stamens same in number as the corolla lobes, protruding from the corolla or included within it, with filaments fused to the tube in lower half. Style 1, cleft, usually with 4 stigmas. Fruit is a fleshy drupe or 1-seeded dry nut.
Distribution and habitats: Central and Southern America, tropical Africa, Middle East, Asia; forests, dry hills, coasts. 37 *Cordia* species are included in *The IUCN Red List of Threatened Plants*.
Life form: Phanerophytes.
Pollination: Mostly by moths.
Fruit dispersal: Fleshy fruits are eaten and the ingested seeds are dispersed by birds; dry fruits are carried by wind, with the dry corolla acting as a parachute. Unusually, *C. subcordata* fruit is carried by sea currents.
Symbiosis: *Cordia* include some of the very few trees to have a symbiotic relationship with *Azteca* ants, who inhabit special swellings – domatia – within the branches of *C. alliodora* in Costa Rica, and protect their host plant from insect pests and other animals.

USES
GARDEN
In temperate climates, *Cordia* species are suitable only for heated glasshouses and conservatories. They are especially valued for their flowers, which are often intensely fragrant.

MEDICINAL
A decoction, made from the leaves and fruits of *C. boisseri*, is used to treat colds. Fruits of *C. myxa* and *C. sebestena* are also used medicinally, the latter for breathing difficulties, dysentery, venereal disease, malaria and spider bites.

EDIBLE
The fruit of some species, including *C. dodecandra*, *C. gharaf*, *C. rothii*, *C. sebestena* and *C. collocca*, is edible. The leaves of *C. globosa* are used for seasoning armadillo dishes.

ECONOMIC
Several species, including *Cordia geraschanthus*, *C. dodecandra*, *C. goeldiana* and *C. alliodora*, are valuable timber sources, used for furniture, doors and beams in houses in Central and South America. The leaves of *C. dodecandra* are so rough that they are used as sandpaper. *Cordia myxa* is used for furniture and the fruit is made into bird-lime.

CULTIVATION
Cordias cannot be grown outdoors in areas colder than Zone 10. In temperate climates, they could be accommodated in a large glasshouse or conservatory where, although these trees will not be able to achieve their full potential, they will nevertheless make attractive container plants. Cordias require full light and most species need high humidity and minimum temperatures of 15–18°C (59–64°F); however, *C. boisseri*, *C. geraschanthus*, *C. greggii* and *C. sebestena* prefer lower winter temperatures, 10–13°C (50–55°F), and may be able to tolerate very brief spells just below 0°C (32°F). They need a very well-drained but moisture-retentive compost mix, of sterilized loam, leaf mould and sharp sand. Water the plants moderately during summer and less at other times.

PROPAGATION
Can be propagated from seed or cuttings.

REFERENCES
HEYWOOD, 1993; HUXLEY, 1992; MABBERLEY, 1997; PHILLIPS & RIX, 1997; SCOFIELD, 1996–98; WU & RAVEN, 1995.

Cordia africana Lam.
Synonyms: *Cordia abyssinica* R. Brown, *C. sebestena* Poir. not L.
Common names: Mukumari, Muringa.
Description: A semi-evergreen tree to 24m (78ft) high, with fissured grey-brown bark, nearly round or oval, alternate leaves to 30cm (12in) long and 22cm (8$\frac{1}{2}$in) wide, rusty-hairy underneath, and panicles of fragrant, pure white flowers, with a tube to 24mm ($\frac{15}{16}$in) long and lobes 30 × 25mm (1$\frac{1}{4}$ × 1in), that are followed by yellow fruits to 12 × 9mm ($\frac{1}{2}$ × $\frac{3}{8}$in).
Distribution: Tropical Africa, Saudi Arabia, Yemen.

Cordia alliodora (Ruiz & Pavon) Cham.
Synonyms: *Cerdana alliodora* Ruiz & Pavon.
Common names: Cyp, Cype, Ecuador laurel, Salmwood.
Description: A semi-evergreen tree to 30m (100ft) high, with grey-green or black-green (at maturity) smooth bark, pointed oval leaves to 18cm (7in) long and 5cm (2in) wide, and abundant creamy-white flowers about 12mm ($\frac{1}{2}$in) long, followed by dry, papery fruits. Young branches of this species bear star-shaped hairs, and develop gall-like swellings that are inhabited by ants of genus *Azteca*.
Distribution: Tropical America.

Cordia boissieri A. De Candolle
Common names: *English:* Mexican Olive; *Spanish:* Ancahuita.
Description: An evergreen shrub or tree to about 8m (26ft) high, with elliptical to oval, blunt leaves, rough-hairy above and felty-hairy underneath, and terminal clusters of 5-lobed, white, yellow- or orange-centred flowers, to about 25mm (1in) across, followed by red-brown fleshy fruits about 13mm ($\frac{1}{2}$in) in diameter.
Distribution: South-western USA, Mexico.
Associated plants: It can be found growing with Texas ebony (*Pithecellobium ebano*), mesquite (*Prosopis glandulosa*), Texas persimmon (*Diospiros texana*) and bold cypress (*Taxodium mucronatum*).

Cordia collococca Linnaeus
Common name: Manjack.
Description: A deciduous tree to 22m (72ft) high, with smooth, grey-brown bark, gradually becoming rough, pointed oval leaves, to 24cm (9$\frac{1}{2}$in) long and 10cm (4in) across, bristly underneath, and dense clusters of white flowers appearing before the leaves, male and female borne on separate trees, the female ones followed by edible, cherry-like red fruit.
Distribution: West Indies.

Cordia dodecandra De Candolle
Common name: Ziricote.
Description: An evergreen tree to 16m (52ft) high, with oblong to almost rounded, blunt leaves, to 15cm (6in) long and 8cm (3$\frac{1}{4}$in) wide, which are bristly and very rough to touch underneath (even used as sandpaper locally), and compact panicles of bright orange, 12–16-lobed, funnel-shaped flowers, followed by pale yellow, edible fruits, about 5cm (2in).
Distribution: Mexico, Guatemala.
Associated plants: Near Veracruz in Mexico, can be found growing in a dry forest with plants such as *Tabebuia chrysantha*, *T. rosea*, *Ehretia tenuifolia*, Jamaica dogwood (*Piscidia piscipula*), *Crescentia alata* and *Enterolobium cyclocarpum*.

Cordia geraschanthus L.
Synonym: *Cerdana geraschanthus* (Linnaeus) Mold.
Common names: Prince wood, Spanish elm, Baria.
Description: A semi-evergreen tree to 30m (100ft) high, similar to *C. alliodora* but without ant swellings on branches and lacking star-shaped hairs, with oblong-lanceolate, pointed leaves, and terminal panicles of creamy-white flowers, followed by dry, papery fruit.
Distribution: Mexico, West Indies.

Cordia greggii Poir.
Description: A compact shrub to 2.5m (8ft) high, with oval, softly hairy, toothed leaves about 1.8cm ($\frac{3}{4}$in) long, and terminal panicles of fragrant, pure white, funnel-shaped flowers, to 3cm (1$\frac{1}{4}$in) across, followed by fleshy fruits.
Distribution: Southern USA (New Mexico, California).

Cordia monoica Roxburgh
Common names: Snot berry, Marer Deylab, Mareer Docol.
Description: An evergreen tree or shrub 4–8m (13–26ft), with pale grey, peeling bark, and oval or almost round, often uneven leaves to 9.5cm (3$\frac{3}{4}$in) long and 6.5cm (2$\frac{5}{8}$in) wide, rough above and softly hairy underneath. Dense panicles of pale yellow or white flowers, each to 15mm ($\frac{5}{8}$in) long (which are usually hermaphrodite or, sometimes, unisexual and borne on separate trees), the flowers followed by elongated fruits to 18mm ($\frac{3}{4}$in) long.
Distribution: Tropical Africa, Asia, Arabia.

Cordia myxa Linnaeus

Synonyms: *Sebestena officinalis* Gaertn.; *Cordia sebestena* Forssk. not L.

Common names: Sudan teak, Sebesten plum, Assyrian plum, Selu.

Description: An evergreen shrub or tree to 12m (40ft), with oval to heart-shaped, toothed leaves, sometimes hairy underneath, and loose panicles of white or cream flowers, male and female flowers borne on different plants: male flowers 5–lobed, with tube to 4.5mm and lobes 5 × 2mm, and female flowers 4–6-lobed, with tube to 6.5mm and lobes to 7mm; the latter followed by yellow-orange to black fruit to 3.5cm (1³/₈in) in diameter.

Distribution: Tropical Asia, Africa, Arabia.

Cordia sebestena Linnaeus

Common name: *English:* Geiger tree.

Description: An evergreen shrub or tree to 8m (26ft) high, with oval, pointed leaves to 20cm (8in) long and 12cm (5¹/₂in) wide, rough above, and cymes up to 12cm (5¹/₂in) in diameter, of bright orange-red, 5–7–lobed, funnel-shaped flowers to 5.8cm (2³/₈in) long, followed by dry fruit enclosed in white fleshy calyx.

Distribution: West Indies to Venezuela.

Cordia subcordata Lam.

Common names: Marer, Mareer.

Description: A tree to 15cm (6in), with grey-brown fissured bark, softly-hairy leaves of variable shape, to 20cm (8in) long and 15cm (6in) wide, and loose terminal clusters of white, orange or red flowers with a tube to 25mm (1in) long and lobes to 15 × 13mm (⁵/₈ × ¹/₂in), followed by a dry fruit (nut) enclosed in calyx which, at least in certain populations, is dispersed by sea.

Distribution: Tropical Asia, Indian Ocean, South Pacific, East Africa.

The Genus *Cryptantha* Lehmann, 1837

White forget-me-nots, Cat's-eyes

NAMING

Etymology: *Cryptantha* – from Greek *krypto*, hidden, and *anthos*, flower.

Synonyms: *Wheelerella* G.B. Grant (SUS); some species may be separated into genus *Oreocarya* Greene, and some have been included in *Eritrichium* Schrad.

Cryptantha virgata in Rocky Mountain National Park, Colorado. (Masha Bennett)

GENUS CHARACTERISTICS AND BIOLOGY

Number of species: About 100 species.

Description: A genus of hairy annual and perennial herbs. Leaves opposite at base and alternate above, entire, linear or spoon-shaped to oblanceolate. Flowers borne in a spike, raceme or thyrse, often with bracts. *Calyx* enlarging in fruit, deciduous, 5-lobed, with oval-lanceolate or linear lobes. *Corolla* yellow or white, funnel-shaped, salverform or wheel-shaped, often very small, 5-lobed, hairless, with 5 small appendages in throat. Stamens 5, included in the corolla, inserted below middle of tube. Style short, simple. Nutlets 1–4, rounded to lanceolate, winged or margined, smooth or finely warty.

Distribution and habitats: Western North America, mostly dry, open habitats. 51 species and forms are included on *The IUCN Red List of Threatened Plants*.

Life forms: Therophytes and hemicryptophytes.

Fruit dispersal: Nutlets in some species have a wing that facilitates dispersal by wind.

USES

No known medicinal or culinary uses.

GARDEN

Though rarely grown, smaller cryptanthas are well worth trying in a rock garden or alpine house, while taller species will make interesting additions to a mixed border, or a native plant collection in western North America.

CULTIVATION

There is little information in the literature on cultivation of this genus. Generalizing from the environment in their natural habitats, they will require full sun and very well-drained soil. High alpine species and those from semi-deserts are probably safest in the alpine house, in pots of very gritty compost.

PROPAGATION

Little is known about propagation of cryptanthas. The seed should probably be sown as soon as ripe or in spring. If it does not germinate within a few weeks, six weeks' cold stratification may help. Some species, such as *C. paradoxa*, may need light for germination.

The AGS *Encyclopedia of Alpines* (BECKETT, 1993–94) also suggests basal cuttings in late spring or late summer as a possible means of propagation.

AVAILABILITY

Occasionally seed may appear on exchange lists or in some catalogues. A few alpine nurseries have offered plants in the past.

REFERENCES

BECKETT, 1993–94; HITCHCOCK, 1955; HUXLEY, 1992; RYDBERG, 1954; SLABY, 2001; WEBBER, 1988; WEBBER & WITTMAN, 1996.

Cryptantha barbigera (Gray) Greene

Common name: *English:* Bearded forget-me-not.
Description: An upright perennial to 40cm (16in), with one or several branched, hairy stems, narrowly lanceolate or oblong hairy leaves to 7cm (2³/₄in) long and 1.5cm (⁵/₈in) wide, and a spike to 15cm (6in) long, of tiny, white, almost unstalked, forget-me-not-like flowers.
Distribution: South-western USA.

Cryptantha capituliflora Reiche.

Description: A tufted perennial to 15cm (6in) high, oblanceolate or obovate leaves to 10cm (4in) long, covered in short hairs with swellings at their bases. White flowers, 3–4mm(¹/₈in) across, are borne in dense white-woolly cymes.
Distribution: Argentina (Central Cordilleras).
Habitat: Rocky slopes in high mountains.

Cryptantha celosioides Payson

Synonyms: *Cryptantha bradburyana* Payson, *C. sheldonii* (Brand) Payson, *Oreocarya celosioides* Eastwood.
Common name: *English:* Northern miner's-candle.
Description: A perennial or sometimes biennial, 15–35cm/6–14in (occasionally to 50cm/20in) high, with densely hairy, oblanceolate or spoon-shaped basal leaves to 8cm (3¹/₄in) long, and, in summer, dense clusters of white flowers, each 8–12mm (³/₈–¹/₂in) across, subtended by bracts.
Distribution: North-western USA.
Habitat: Dry places.

Cryptantha flava Payson

Synonyms: *Oreocarya flava* A. Nelson.
Common name: *English:* Golden cryptantha, yellow cryptantha.
Description: A tufted, upright perennial, 15–40cm (6–16in) high, densely hairy, with crowded, oblanceolate basal leaves to 9cm (3¹/₂in) long and smaller, stalkless stem leaves, and, in spring, cylindrical clusters of bright yellow flowers 7–11mm (³/₈–¹/₂in) across.
Distribution: Western USA (from Wyoming and Utah through western Colorado to New Mexico and Arizona).
Habitat: Dry fields and slopes, rocks, up to 2300m (7540ft).

Cryptantha intermedia (Gray) Greene

Common name: *English:* Common cryptantha.
Synonym: *Eritrichium intermedia* Gray.
Description: An upright annual 15–30cm (6–12in) high, or more, with one or several stems, hairy lanceolate or linear leaves to 5cm (2in) long, and white flowers 5–8mm (¹/₄–³/₈in)across, carried in slender branching clusters to 15cm (6in) long.
Distribution: USA (California to Oregon and Idaho).
Habitat: Dry, open slopes, up to 2000m (6550ft).

Cryptantha interrupta Payson

Synonyms: *C. macounii* Payson; *C. spiculifera* Payson.
Description: A hairy perennial or sometimes biennial, 15–30cm (6–12in) high, with oblanceolate basal leaves to 6cm (2¹/₂in) long, and white flowers 5–8mm (¹/₄–³/₈in) across, carried in a dense cluster, with bracts.
Distribution: North-western USA.
Habitats: Semi-deserts at 1500–2500m (4900–8200ft).

Cryptantha johnstonii L.C.Higgins
Common name: *English:* Johnston Cat's-eye.
Description: A tufted, upright, hairy perennial, 10–35cm (4–14in) high, with oblanceolate leaves to 6.5cm (2⅝in) long and, in late spring, bearing open, cylindrical clusters of white flowers 12–17mm (½–⅝in) across.
Distribution: Western USA (Great Basin).
Habitat: Semi-deserts, at 1500–1900m (4900–6230ft).

Cryptantha nubigena Payson
Synonym: *Cryptantha hypsophila* I.M. Johnston.
Common name: *English:* Crater Lake Cat's-eye.
Description: A compact perennial or sometimes biennial to 10cm (4in) or somewhat more in height, with narrowly oblanceolate or spoon-shaped basal leaves to 4cm (1½in) long, and dense clusters of white, yellow-throated flowers about 5mm (¼in) across, subtended with bracts.
Distribution: USA (California to Washington).
Habitats: Subalpine and alpine fellfields, at 2700–4100m (8850–13,450ft).

Cryptantha paradoxa (A. Nelson) Payson

'Tufts of green, spathulate leaves produce masses of intensely fragrant, white, yellow-throated, waxy blossoms.' (Alan Bradshaw)

Synonym: *Oreocarya paradoxa* Nelson.
Common names: *English:* Paradox Valley cryptantha, Handsome Cat's-eye.
Description: A small, hairy perennial only about 6cm (2½in) high, with spoon-shaped basal leaves and clusters of scented white flowers.
Distribution: USA (endemic to western Colorado and Utah).
Habitat: Only on gypsum-rich soils of the Paradox Valley, often on outcrops of pure gypsum, and sometimes on clay.

Cryptantha virgata Porter
Synonyms: *Oreocarya virgata* (Porter) Greene
Common name: *English:* Miner's-candle.
Description: An upright, stiffly-hairy perennial, usually 30–50cm (12–20in) high, sometimes higher, with linear leaves and a tall, narrow spike of white flowers, each about 10mm (⁷⁄₁₆in) across, interspersed with many narrow, long bracts.
Distribution: North-western USA.
Habitat: Hillsides, gravelly slopes, meadows, dry fields, especially on granite.

Cryptantha virgata. (Masha Bennett)

Associated plants: In Estes Park, Colorado, can be found growing with lanceleaf bluebells (*Mertensia lanceolata*), pasqueflower (*Pulsatilla patens* subsp. *multifida*), bitterbrush (*Purschia tridentata*), *Erysimum capitatum*, *Penstemon virgatum* subsp. *asagrayi* and *Geranium richardsonii*.

The genus *Ehretia* P. Browne, 1756

NAMING
Etymology: *Ehretia* – after Georg Dionysius Ehret, 1708–70, a German botanical artist.
Synonyms: *Traxilum* Raf.; *Carmona retusa* was previously included in this genus.

GENUS CHARACTERISTICS AND BIOLOGY
Number of species: About 50.
Nomenclature Note: The genus *Ehretia* is often separated from Boraginaceae into a different family, Ehretiaceae, which also includes such genera as *Carmona* and *Cordia*.
Description: A genus of deciduous or evergreen trees or shrubs to 25m (80ft). Leaves alternate or clustered on older branches, hairy or hairless, obovate, elliptical, oval to lanceolate or oblong, stalked entire or sharply toothed. Flowers borne in a terminal or axillary cymes or panicles. *Calyx* 5-lobed, tube short. *Corolla* white to yellow, or rarely blue, sometimes

fragrant, tubular bell-shaped, with spreading or reflexed lobes. Stamens 5, usually protruding from the corolla tube, with oblong anthers. Style divided, with 2 stigmas. Fruit a rounded, hairless drupe to 20mm ($^3/_4$in), often fleshy, yellow or orange to red; with one or two stones each containing 1–2 seeds.

Distribution and habitats: Africa, Asia, North and South America; mostly in forests. 3 species are included on *The IUCN Red List of Threatened Plants*.

Life form: Phanerophytes.

Pollination: By insects.

Fruit dispersal: The fleshy fruits are eaten and the ingested seeds are dispersed by birds.

Animal feeders: Caterpillars of a number of butterflies and moths feed on leaves and flowers of *Ehretia* species: those of family Ethmiidae, such as *Ethmia* species; Lycaenidae, for example Hairy Lineblue (*Erysichton lineata*); Nymphalidae, including common Aeroplane (*Phaedyma sherpherdi*); Sphingidae, such as Rustic Sphinx (*Manduca rustica*).

USES

GARDEN

Specimen trees and shrubs grown for ornamental foliage, fragrant flowers and sometimes fruits, mostly in warm climates. *Ehretia anacua* is especially suitable for bonsai, and *E. dicksonii* and *E. thyrsiflora* can also be cultivated in this way.

MEDICINAL

Leaves of *E. cymosa* are used in Ghana for treating bone fractures, and those of *E. philippensis* are said to be effective for dysentery and other intestinal disorders in the Philippines. Leaves and twigs of *E. acuminata* are used in Chinese medicine.

CULINARY

The fresh fruit of *E. anacua*, *E. acuminata* and *E. dicksonii* is sweet and edible; the unripe fruit is sometimes pickled.

ECONOMIC

The light and tough wood of *E. acuminata* and *E. dicksonii* is used for carrying poles. Heavy wood of *E. anacua* is made into handles of agricultural implements, yokes, axles, wheel spokes.

CULTIVATION

The majority of ehretias are unfortuately too tender to be grown in temperate regions. However, the hardiest species, such as *E. acuminata*, *E. dicksonii* and *E. anacua*, are tolerant of at least −10°C (14°F) when mature. They require a very sheltered position if grown outdoors. Full sun is essential in moist maritime climates, as in the UK, to ensure ripening of the wood, but in continental climates light shade is acceptable. A south-west-facing wall would probably provide optimum shelter – do not grow by an east-facing wall as rapid thawing in the morning after a frosty night will result in more severe damage to new growth. Young specimens of even the hardiest species are especially susceptible to frost damage and will need to be protected by such as polythene sheeting, if freezing temperatures are expected. If the worst happens, at least some ehretias regrow from the base. In autumn, give a deep mulch. Loamy soil of medium fertility is suitable for most ehretias. *Ehretia acuminata* and *E. dicksonii* are tolerant of moderate alkalinity; very good drainage is essential for all species. Soils that are too rich will encourage production of sappy growth that will be liable to be damaged by frosts.

Species that are not hardy enough to be grown outdoors, such as *E. hottentotica* and *E. laevis*, can be accommodated in containers in a cool glasshouse. They will need sharply drained compost and should be watered moderately while in growth and little in winter. Those ehretias that are grown as bonsai, *E. anacua* in particular, have similar requirements to those of the Fukien tea, *Carmona retusa*, but are more tolerant of variable growing conditions, and are less susceptible to pest attack. They should be repotted every year or two.

PROPAGATION

Seed Ideally should be sown as soon as it is ripe or in early spring, under glass. Sow stored seed in late winter or early spring. When they are large enough to handle, prick the seedlings out into individual pots and grow them on in the greenhouse for at least their first winter. Plant the young plants out in late spring or early summer, after the last expected frosts.

Vegetative Propagate by softwood cuttings in early summer, or semi-ripe cuttings in late summer, in a propagator or frame. Some species, like *E. anacua*, may produce suckers and can be propagated by division.

PROBLEMS

Marginal hardiness of even the toughest species prevents wider cultivation of ehretias in temperate climates.

REFERENCES
BEAN, 1981; FACCIOLA, 1990; GENDERS, 1994; HUXLEY, 1992; LEWIS & AVIOLI, 1991; OHWI, 1965; PFAF, 1992–02; WALTERS, 2000; WU & RAVEN, 1995.

Ehretia acuminata R. Brown

Synonyms: *Cordia thyrsiflora* Siebold & Zuccarini; *Ehretia acuminata* var. *grandifolia* Pampanini; *E. a.* var. *obovata* (Lindley) I.M. Johnston; *E. serrata* var. *obovata* Lindley; *E. thyrsiflora* (Siebold & Zuccarini) Nakai; *E. ovalifolia* Hasskarl.

Common names: *English:* Kodo wood; *Chinese:* hou ke shu; *Japanese:* Chisha-no-ki.

Description: A deciduous tree to 25m (80ft), usually 5–10m (16–33ft) when grown in temperate climates, with black-grey bark, and oblong or elliptical, pointed leaves, rounded or heart-shaped at base, to 17cm (6$^1/_2$in) long and 6cm (2$^1/_2$in) wide, mostly hairless. It bears dense paniculate clusters, 5–8 cm (2–3$^1/_4$in) wide, of fragrant white flowers, about 6mm across, followed by yellow or orange fruit, ripening to black, 3–4mm ($^1/_8$in) in diameter.

Distribution: China, Japan, Korea, the Philippines.

Habitat: Hills, open forests, thickets on slopes; 100–1700m (330–5570ft).

Hardiness: Zone 7.

Ehretia anacua (Teran & Berl.) I.M. Johnst.

Synonyms: *Ehretia ciliata* Miers.; *E. elliptica* DC, *E. exasperata* Miers.; *E. lancifolia* Sesse & Moc.; *E. scabra* Kunth & Bouche.

Common names: Sugarberry, Knock-away, Anaqua, Anacua, Manzanita, Sandpaper tree.

Description: An evergreen or semi-evergreen tree, 3–10m (10–33ft), usually with several trunks, with dark green, elliptical to oval leaves to 9cm (3$^1/_2$in) long and 3.8cm (1$^1/_2$in) wide, very rough, especially above, bearing terminal clusters of fragrant white flowers about 7mm ($^3/_8$in) long, from mid-autumn until spring, followed by yellow or orange fruit, to 8mm ($^3/_8$in) in diameter.

Distribution: South-western north America (Texas, Mexico).

Habitat: Forests, river valleys, dry ridges.

Associated plants: In Rio Grande River delta in subtropics of Texas forms the so-called 'Texas Ebony – Anacua Forest' with the co-dominant Texas ebony (*Pithecellobium ebano*), where other species may include *Leucaena pulverulenta*, *Phaulothamnus spinescens*, *Sideroxylon celastrinum*, *Havardia pallens*, *Zanthoxylum fagara*, *Condalia hookeri*, *Celtis pallida*, *Ziziphus obtusifolia* and mesquite (*Prosopis glandulosa*).

Hardiness: Zone 8.

Ehretia dicksonii Hance

'It makes a small sturdy tree with large glossy leaves and in June, on the ripened wood of the previous season, bears handsome panicles of creamy-white flowers which emit a powerful spicy fragrance.' (Roy Genders)

Synonyms: *Ehretia macrophylla* Shiras. not Wallich.; *E. m.* var. *tomentosa* Gagnepain & Courchet.

Common name: *Chinese:* cu kang shu.

Description: A deciduous tree to 12m (40ft) high, with grey-brown bark, and broadly oval, elliptical or obovate, glossy dark green toothed leaves, to 25cm (10in) long and 15cm (6in) wide, densely felty below. Terminal clusters, 6–9cm (2$^1/_2$–3$^1/_2$in) wide, of fragrant, tubular-bell-shaped white or pale yellow flowers, about 12mm ($^1/_2$in) long, are borne in spring or summer and followed by rounded, yellow fruits, 10–15mm ($^7/_{16}$–$^5/_8$in) in diameter. Introduced in 1897 by E.H. Wilson. Can be grown as a bonsai.

Distribution: China, Bhutan, Japan, Nepal, Vietnam.

Habitat: Open forests on slopes, shaded moist hillsides, rocky valleys, 100–2300m (330–7540ft).

Hardiness: Zone 7 or 8.

Ehretia hottentotica Burchell

Common names: *English:* Cape lilac, Puzzle bush; *Afrikaans:* Deurmekaarbos.

Description: A multi-stemmed shrub or small tree to 3.6m (12ft), multi-stemmed, with clustered or alternate, oval to lanceolate, rounded leaves to 5cm (2in) long and wide, minutely hairy on margins, and clusters of fragrant lilac flowers appearing before the leaves, followed by rounded orange fruits that turn blue-black.

Distribution: South Africa, Botswana.

Hardiness: Zone 9.

Ehretia laevis Roxburgh

Common name: *Chinese:* mao e hou ke shu.

Description: A tree to 10m (33ft) high with grey-brown bark, oval-elliptical or obovate leaves to 18cm (7in) long and 11mm ($^7/_{16}$in) wide, shiny above and softly hairy below, and branched clusters of tiny white flowers, almost wheel-shaped, each to 3.5mm ($^1/_8$in), followed by orange or yellow fruits, 3–4mm ($^3/_{16}$in) in diameter.

Distribution: India, Bhutan, Kashmir, China, Laos, Pakistan, Vietnam.

Habitat: Forests, roadsides.

Associated plants: In Uttar Pradesh (India), grows in

forests that include characteristic species such as *Mitragyna parviflora*, *Adina cordifolia*, sissoo (*Dalbergia sissoo*), Bengal quince (*Aegle marmelos*), *Kydia calycina*, emblic (*Phyllanthus emblica*), ber (*Ziziphus mauritana*), wedge-leaf fig (*Ficus semicordata*) and beggarweeds (*Desmodium triangulare* and *D. pulchellum*).
Hardiness: Zone 9.

Ehretia macrophylla Wallich not Shiras.

Description: A deciduous shrub or tree to 6m (20ft), with oval, irregularly toothed, bristly leaves, to 15cm (6in) long and 10cm (4in) wide, and rounded panicles of white flowers followed by fruit to 12mm ($^{7}/_{16}$in) in diameter. If killed back to the ground by frost, vigorous upright shoots are produced during the following summer.
Distribution: Himalaya to China.
Hardiness: Zone 9.

Ehretia rigida (Thunb.) Druce.

'Along the arching branches numerous short cymes of lilac flowers appear among the shining olive green leaves. In this early summer dress the shrub is most attractive. Later the small green berries turn yellow and are eagerly sought by birds.' (R. Dyer, I. Verdoon & L. Codd, 1962)

Common names: *English:* Puzzle-bush; *Afrikaans:* Deurmekaarbos.
Description: An irregularly branched, deciduous shrub or small tree 1.8–5m (6–16ft), with often drooping branches, and olive-green leaves with minutely hairy margins. Dense clusters of scented lilac or lavender flowers in spring are followed by edible yellow-orange fruit that turns red and eventually black.
Distribution: South Africa.
Habitat: Bushveld, on ridges, and in clumps in grassland.
Hardiness: Zone 8.

The genus *Hackelia* Opiz, 1838
Stickseed, Beggar's lice

NAMING
Etymology: *Hackelia* – for the Czech botanist Josef Hackel, who died in 1869.

Synonyms: Some species have been included in *Echinospermum* Sw. in Lehm., *Eritrichium* Schrad. and *Lappula* Moench.

GENUS CHARACTERISTICS AND BIOLOGY
Number of species: About 45.
Description: A genus of annual, biennial or perennial herbs. Leaves alternate. Inflorescence terminal or axillary, often a panicle, sometimes with bracts. *Pedicels* deflexed or recurved in fruit. *Calyx* deeply 5-lobed. *Corolla* funnel- or bell-shaped, cylindrical or almost wheel-shaped, throat appendages usually conspicuous. Stamens included in the corolla, style simple. Nutlets with a ventral keel, with a series of marginal bristles.
Distribution and habitats: North America, Europe, Asia; in mountains, dry places, scrub and forests. 11 species and forms are included on *The IUCN Red List of Threatened Plants*.
Life forms: Therophytes, hemicryptophytes.
Pollination: By bees.
Fruit dispersal: Bristly nutlets become attached to animal fur or clothing and are carried some distance away from the parent plant, hence the common name.
Animal feeders: *Hackelia californica* (Gray) Johnston is the main foodplant for the moth *Gnophaela latipennis* (family Arctiidae) in Sierra Nevada.

USES
GARDEN
North American species are sometimes grown in native plant collections. Some of the more compact species are worth trying in a rock garden, though the AGS *Encyclopedia of Alpines* states that they 'cannot be considered as choice plants for this purpose'.

CULTIVATION
Very little is known about the cultivation requirements of hackelias. Most species are hardy to Zone 7. Try providing conditions similar to those in plants' native habitats, sun and well-drained soil for those from open, dry places, light shade and moisture-retentive soil for species inhabiting forest and scrub. Small species from high elevations may need protection from winter wet and will probably fare best in the alpine house.

PROPAGATION
Hackelias are said to be easily raised from seed. Division may be possible for perennials.

REFERENCES
BECKETT, 1993–94; EDSON et al. 1996; GODFREY & CRABTREE, 1986(1987); HUXLEY, 1992; PHILLIPS & RIX, 1993; RYDBERG, 1954; SLABY, 2001; WEBBER, 1988; WEBBER & WITTMAN, 1996.

Hackelia cusickii (Piper) A. Brand
Common name: *English:* Cusick's stickseed.
Description: A woody-based perennial, 15–30cm (6–12in) high or more, with several branched stems, hairy, bluish-green, lanceolate leaves 2.5–6cm (1–2$^1/_2$in) long, and panicles of blue flowers to 10mm ($^7/_{16}$in) across.
Distribution: California and Oregon.
Habitat: Dry, rocky mountain slopes, 1650–2600m (5400–8500ft).

Hackelia diffusa (Lehmann) I.M. Johnston
Common name: *English:* Diffuse stickseed.
Description: A bristly perennial 20–80cm (8–32in) high, with several upright stems and linear leaves to 20cm (8in) long and 10mm ($^7/_{16}$in) wide, and white, yellow-eyed flowers 6–15mm ($^1/_4$–$^5/_8$in) across, borne from late spring into midsummer.
Distribution: Washington and British Columbia.
Habitat: Open or lightly wooded, dry slopes, often with sagebrush or ponderosa pine.

Hackelia floribunda (Lehmann)
Synonym: *Echinospermum floribundum* Lehmann.
Common names: *English:* False forget-me-not, Many-flowered stickseed.
Description: A biennial or perennial to 1,2m (4ft) tall, with few hairy stems, elliptical or lanceolate leaves to 21cm (8$^1/_4$in) long and 2.5cm (1in) wide, and cymes of blue or occasionally white flowers, to 7.5mm ($^3/_8$in) in diameter, with hairy calyces.
Distribution: Western North America (Minnesota to British Columbia and California).
Habitat: Scrub, woodland.
Associated plants: In Colorado Springs, can be found among pinyon pine (*Pinus edulis*) and peach-leaf willow (*Salix amygdaloides*), with such plants as narrow-leaf puccoon (*Lithospermum incisum*), lanceleaf bluebells (*Mertensia lanceolata*), scarlet gaura (*Ipomopsis aggregata*), cut-leaf evening primrose (*Oenothera coronopifolia*), *Yucca glauca*, *Penstemon lanceolata* and *Astragalus drummondii*.

Hackelia jessicae (McGreg.) Brand
Common name: *English:* Jessica's stickseed.
Description: A softly hairy, upright perennial, 30–40cm (12–16in) high or more, with a woody base, several stems, lanceolate leaves 7–15cm (2$^3/_4$–6in) long with winged stalk, and pale blue flowers 6mm across, in a panicle-like cluster, borne in late summer.
Distribution: California to British Columbia.

Habitat: Montane coniferous forests, sagebrush scrub, grassland on moist soils.

Hackelia micrantha (Eastwood) J.L. Gentry
Synonym: Sometimes considered synonymous with *H. Jessicae* (McGreg) Brand.
Common name: *English:* Blue stickseed.
Description: A hairy and bristly perennial, 30–110cm (12–44in) tall, with many upright stems, elliptical to lanceolate leaves to 28cm (11in) long and 3cm (1$^1/_4$in) wide, and blue or, sometimes, white flowers to 10mm ($^7/_{16}$in) across, with yellow scales in throat. Rather similar to *H. Floribunda*, but it is always perennial.
Distribution: Western North America (Cascades to Canada).
Habitat: Clearings, scrub.

Hackelia setosa (Piper) Jtn.
Description: A bristly, clump-forming perennial to 50cm (20in) high, with upright stems, narrowly oblanceolate leaves held more or less upright, 5–10cm (2–4in) long, and blue flowers 10–15mm ($^7/_{16}$–$^5/_8$in) across, borne in summer.
Distribution: California and Sierra Nevada to Oregon.
Habitat: Open, grassy places in the conifer forest zone, 300–1800m (1000–5900ft).

Hackelia sharsmithii I.M. Johnston
Synonym: *Hacklia ibapensis* L. M. Schultz & J.S. Schultz.
Description: A tufted perennial, 10–30cm (4–12in) high, with upright or ascending stems, elliptical-oval leaves 1.5–4cm ($^5/_8$–1$^1/_2$in) long, and raceme-like clusters of white or pale blue flowers, about 6mm ($^1/_4$in) across, with yellow scales in throat, borne from summer to autumn.
Distribution: California.
Habitat: Alpine to subalpine regions in mountains, often under overhanging rocks, 3600–4000m (11,800–13,100ft).
Associated plants: May be found growing among shrubs such as *Holodiscus dumosus* and *Philadelphus microphyllus*.

Hackelia uncinata (Royle in Bentham) C. Fisher
Description: A branched perennial 30–60cm (1–2ft) high, with upright stems, pointed oval leaves with a heart-shaped base, and pale blue flowers to 12mm ($^1/_2$in) across, borne in summer.

Distribution: The Himalayas, from Pakistan to south-west China.
Habitat: Open places in forest and in scrub, 2700–4200m (8850–13,780ft).

Hackelia venusta (St. John)
Common name: *English:* Showy stickseed.
Description: An upright perennial, 15–40cm (6–16in) high, with several branched stems, lanceolate leaves and panicle-like clusters of white flowers, 13–20mm ($^1/_2$–$^3/_4$in) across. Rare, endangered species which has been successfully micropropagated (Edson *et al.*, 1996). Rather similar to *H. cusickii*.
Distribution: Washington State (Chelan county).
Habitat: Ponderosa pine forest in mountains.

The genus *Lithospermum* Linnaeus, 1753
Gromwell, Puccoon

NAMING
Etymology: *Lithospermum* – from Greek *lithos*, stone, and *sperma*, seed.
Synonyms: *Batschia* Gmel. A large number of plants previously included within this genus have now been moved to *Lithodora* Grisebach, *Moltkia* Lehm., *Buglossoides* Meench., *Alkanna* Tausch., *Arnebia* Forsk., *Amsinckia* Lehm. or *Mertensia* Roth.

GENUS CHARACTERISTICS AND BIOLOGY
Number of species: About 60.
Description: A genus of annual or perennial herbs. Stems upright or spreading, bristly or shaggy-hairy. Leaves alternate, hairy, often unveined. Flowers borne in terminal 1-sided cymes or singly in leaf-axils; bracts usually numerous, leaflike. *Calyx* 5-lobed more or less to the base, enlarged in fruit. *Corolla* white, yellow, orange or blue, cylindrical to funnel-shaped, hairy on the outside; tube often with downy or glandular appendages at the throat. Stamens included within the corolla. Style also included with a 2-lobed stigma. Nutlets 4, ovoid or ellipsoid, stony, white or pale yellowish-brown, often smooth but can be rough or warty.
Distribution and habitats: Temperate regions throughout the world, except Australia, in a wide variety of habitats, mostly dry. 2 species are included on *The IUCN Red List of Threatened Plants*.
Life forms: Hemicryptophytes and therophytes.
Pollination: By insects.

Fruit dispersal: No specialized mechanism.
Animal feeders: Lithospermum species are foodplants for a number of moths of genera *Ethmia* and *Coleophora*.

USES
GARDEN
Herb garden (*L. officinale, L. erythrorhizon*), wild garden, mixed border (*L. multiflorum, L. carolinense*), rock garden or alpine house (*L. canescens, L. hancockianum*).

MEDICINAL
Many gromwells have been extensively used for medicinal purposes. *Lithospermum officinale* is said to have contraceptive, diuretic and sedative properties, with its seeds, leaves and roots used to treat a wide variety of ailments. All parts of the plant contain a substance that inhibits the secretion of the pituitary gonadotropic hormone.

Lithospermum ruderale and *L. incisum* have both been used as a contraceptive and for kidney problems; the former is also anti-rheumatic, astringent and diuretic, and the latter has been used in treating coughs and colds, as an eyewash and for stomach ache.

All parts of *L. erythrorhizon* have been extensively used to treat tumours, as salve for many irritant skin conditions, burns, cuts, wounds and so on and, like other *Lithospermum* species, it is reported to have contraceptive properties.

Lithospermum canescens, L. carolinense and *L. multiflorum* have also been used in herbal medicine.

CULINARY
The leaves of *L. officinale* are used as a tea substitute, and the root of *L. incisum* has also been made into tea, or eaten boiled or roasted. The seeds of *L. ruderale* and *L. multiflorum* have been used for food.

ECONOMIC
Lithospermum erythrorhizon is an ingredient in commercial skincare creams.

OTHER
A purple dye is obtained from the roots of *L. erythrothizon, L. officinale* and *L. multiflorum*, a red dye from the roots of *L. canescens* and *L. carolinense* and a blue dye from the roots of *L. incisum*. *Lithospermum ruderale* has been used as a dye and body paint by Native Americans, and its shiny white seeds made into beads, as were the seeds of *L. incisum*. The dried parts of the latter have been burned as incense.

CULTIVATION
Most *Lithospermum* are fully hardy – some can grow

in areas as cold as Zone 3 (*L. incisum, L. canescens, L. multiflorum*). The majority of species need a sunny situation, although *L. officinale* and *L. erythrorhizon* will tolerate a little shade. American lithospermums especially need a warm position in full sunlight, and are difficult to cultivate in the UK. All species require a well-drained but ideally moisture-retentive soil of moderate fertility, preferably acid for *L. carolinense* and *L. canescens* and neutral to alkaline for *L. erythrorhizon, L. multiflorum* and *L. incisum*. One of the most desirable of American species is *L. canescens,* which is tricky to grow and is perhaps best accommodated in the alpine house, in pots of gritty compost composed of equal parts of lime-free loam, leaf mould and coarse sand, or in tufa. It should be watered carefully with care using clean rainwater.

PROPAGATION

Seed Seed can be sown in spring or as soon as ripe, in a cold frame. There is evidence that cold stratification at 0–4°C (32–39°F) for three months may help improve germination.
Vegetative *Lithospermum erythrorhizon* and probably some other species can also be propagated by division in autumn.

AVAILABILITY

Rarely available, *Lithospermum* species may be obtainable as seed from various exchange schemes. *Lithospermum erythrorhizon* and *L. officinale* may occasionally be offered by specialist herb nurseries.

REFERENCES

BOWN, 1995; DUKE & AYENSU, 1985; HUXLEY, 1992; INGWERSEN, 1991; MOERMAN, 1998; OHWI, 1965; PFAF, 1992–02; SLABY, 2001; USHER, 1974, WU & RAVEN, 1995.

Lithospermum canescens (Michaux) Lehmann

'I have never persuaded it to do more than produce rather weak, straggling shoots, ending in clusters of orange-yellow flowers which promise more than they ultimately give. Lime-free gritty soil and a sunny position seem to be required for even a small measure of success' (Will Ingwersen)

Synonym: *Batschia canescens* Michaux.
Common names: *English:* Puccoon, Hoary puccoon, Paint Indian.
Description: A grey-hairy perennial 10–40cm (4–16in) high, with linear- or oval-oblong leaves to

4cm (1½in) long and 1cm (7/16in) wide, and, in spring and early summer, bearing orange-yellow, funnel-shaped flowers, to 18mm (3/4in) long and 15mm (5/8in) across, with glandular appendages in throat.
Distribution: Eastern north America (Ontario to Georgia, west to Saskatchewan and Texas).
Habitat: Dry or sandy open woodland, prairies.
Hardiness: Zone 3.

Lithospermum carolinense (Walt. in J.F. Gmel.) MacM.

Synonyms: *Batschia caroliniensis* Walt. in J.F. Gmelin, *Lithospermum carolinianum* Lam., *L. hirtum* (Muhlenb.) Lehm.
Common names: *English:* Hairy puccoon, Carolina gromwell.
Description: A bristly perennial to 1m (3ft) high, with stalkless, pointed, lanceolate or narrowly lanceolate stem leaves to 4cm (1½in) long and 1cm (7/16in) wide, and, in summer, loose clusters of orange-yellow, salver-shaped flowers to 25mm (1in) across, slightly hairy on the outside.
Distribution: Eastern North America (New York to Florida, Minnesota, Montana and New Mexico).
Habitat: Sandhills and dry sandy woods.
Hardiness: Zone 6.
Related forms and cultivars: *Lithospermum carolinense* var. *croceum* (Fern.) Cronq. (syn. *L. caroliniense* subsp. *croceum* (Fern.) Cusick, *L. croceum* Fern.) is a very similar plant, native to Oklahoma.

Lithospermum erythrorhizon Siebold & Zuccarini

Synonyms: *Lithospermum murasaki* Sieb., *L. officinale* var. *japonica* Miq., *L. o.* var. *erythrorhizon* (Sieb. & Zucc.) Maxim.
Common names: *English*: Red-rooted gromwell; *Chinese:* Zi Cao; *Japanese:* Murasaki; *Russian:* vorobeinik krasnokornevishnyi.
Description: A bristly perennial 40–90cm (16–36in) high, arising from thick roots that stain purple when dried, with few stems clothed in stalkless lanceolate leaves to 8cm (3¼in) long and 1.7cm (5/8in) wide, and terminal clusters of inconspicuous white flowers to 9mm (3/8) long and 4mm (3/16in) across, borne in summer. Used as a medicinal plant by Chinese herbalists since ancient times.
Distribution: East Asia – eastern China, Japan, Korea, far eastern Russia.
Habitat: Grassy slopes of hills and mountains.
Hardiness: Zone 6.

Lithospermum hancockianum Oliver

Synonyms: *Lithospermum mairei* H.Léveillé; *Lithodora hancockianum* (Oliver) Handel-Mazzetti.
Common name: *Chinese*: Shi sheng zi cao.
Description: A tufted, upright perennial 5–15cm (2–6in) tall, with crowded, stalkless, narrowly lanceolate, white-hairy leaves to 15cm (6in) long, and, in spring and summer, dense clusters of purple-red flowers 10–22mm ($^7/_{16}$–$^7/_8$in) long, subtended by leaflike bracts.
Distribution: Endemic to China.
Habitat: Rocks, at 2200–2400m (7200–7870ft).

Lithospermum incisum Lehmann

Synonyms: *Batschia linearifolia* (Goldie) Small; *Lithospermum angustifolium* Michaux; *L. linearifolium* Goldie; *L. mandanense* Spreng.
Common names: *English:* Fringed gromwell, Narrow-leaf puccoon, Narrow-leaf gromwell.
Description: An upright perennial 30–60cm (1–2ft) high, with solitary to numerous, bristly stems, oblanceolate, stalked basal leaves to 12cm (4$^1/_2$in) long and 1cm ($^7/_{16}$in) wide, and smaller, bristly, linear to narrowly lanceolate, stalkless stem leaves, bearing bright yellow flowers to 35mm (1$^3/_8$in) long and 20mm ($^3/_4$in) across in spring and summer.
Distribution: Eastern and Central North America (British Columbia to Manitoba, south to Illinois, Texas and Arizona).
Habitat: Dry soils of plains, foothills and ridges in mountains, to 2100 m (6550ft).
Hardiness: Zone 3.

Lithospermum multiflorum Torr.

Common names: *English:* Many-flowered pucoon, Many-flowered gromwell.
Description: An upright, grey-hairy perennial to 50cm (20in) high, with stems clothed in numerous, stalkless lanceolate leaves to 7cm (2$^3/_4$in) long and 1cm ($^7/_{16}$in) wide and, in summer, bearing terminal clusters of yellow-orange, funnel-shaped flowers, with a cylindrical tube to 10mm ($^7/_{16}$in) long and limb to 9mm ($^3/_8$in).
Distribution: Western North America.
Habitat: Gravelly soils, mainly in the juniper and pine belts, at 1800–3600m (5900–11,800ft) in Texas.
Hardiness: Zone 3.

Lithospermum officinale Linnaeus

Common names: *English*: Common gromwell, Pearl gromwell; *Chinese*: Ti Hsueh, Tzu Tan, Tzu Ts'Ao, Ya Hsien Ts'Ao; *Dutch*: Glad Parelzaad; *Finnish*:

Rohtorusojuuri; *French*: Grémil officinal; *German*: Gebraäuchlicher steinsame, Echter steinsame; *Italian*: Erba perla maggiore, Migliarino; *Norwegian*: Legesteinfro; *Russian*: vorobeinik lekarstvenny.
Description: A bristly upright perennial, 20–100cm (8in–3$^1/_2$ft) high, one to many-stemmed, with bristly, pointed lanceolate leaves 10cm (4in) long and 2cm ($^3/_4$in) wide, middle and upper stalkless, in summer bearing inconspicuous green, yellow or yellowish-white flowers 4–6mm ($^1/_4$in), followed by white smooth nutlets.
Distribution: Europe, Asia; naturalized in the USA.
Habitat: Hedges, bushy places and woodland borders, usually on basic soil.
Hardiness: Zone 6.

Lithospermum ruderale Dougl. in Lehmann

Synonym: *Lithospermum pilosum* Nutt.
Common names: *English:* Western gromwell, Western stoneseed, Columbian puccoon, Wayside gromwell, Whiteweed.
Description: An upright or decumbent perennial 20–60cm (8–24in) high, with many grey-green, hairy stems clothed in narrowly lanceolate leaves to 10cm (4in) long and 2cm ($^3/_4$in) wide, clustered at tops of stems and, in summer, bearing pale yellow to yellow-green, funnel- to bell-shaped flowers to 13mm ($^7/_{16}$in) across.
Distribution: Western North America (British Columbia to California).
Habitat: Open, fairly dry places from the foothills to moderate elevations.
Hardiness: Zone 4.

The genus Lobostemon Lehmann, 1830

NAMING
Etymology: *Lobostemon* – from the Greek *lobos*, a lobe, and *stemon*, thread.
Synonyms: *Echiopsis* Rchb., *Echiostachys* Levyns, *Lobostema* Spreng., *Traxara* Raf.

GENUS CHARACTERISTICS AND BIOLOGY
Number of species: About 28 species.
Description: A genus of shrubs or subshrubs, hairy or almost hairless. Leaves alternate, stalkless, sometimes armed with spiny hairs. Flowers borne in a one-sided cyme with bracts, or, occasionally, may be axillary, solitary or paired. *Calyx* usually divided almost to the base. *Corolla* blue, pink, red, white or pale yellow, tubular or funnel-shaped, regular or zygomorphic (in the latter case, 2 lobes are larger

than the rest), with hairy scales or a tuft of hairs in the throat, just below the point where the stamens are attached in the tube. Stamens have small, almost globose anthers and long filaments; they may vary in length. Style minutely lobed into two. Nutlets 4, attached at the base to the top of the flower stalk, erect, finely warty or rarely almost smooth, occasionally with glass-like spikes.

Distribution and habitats: Endemic to South Africa; in dry places. 5 *Lobostemon* species are included on *The IUCN Red List of Threatened Plants*.

Life forms: Phanerophytes.

Pollination: By bees.

Fruit dispersal: No specialized mechanism.

USES

GARDEN
Suitable only for glasshouses in temperate climates.

MEDICINAL
Lobostemon fruticosus is known as the 'eight-day-healing-bush' for its wound-healing properties.

CULTIVATION
They should be grown in a sandy loam soil with a slight admixture of peat, and require protection from frost.

PROPAGATION
Seed germinates without special treatment. Also by semi-ripe cuttings.

REFERENCES
ADAMSON & SALTER, 1950; PHILLIPS & RIX, 1997; WALTERS, 2000.

Lobostemon fruticosus (Linnaeus) Buek.

'Particularly pretty when young, it has young stems that show their usual rose colour through the foliage. The blooms are exquisite. The buds are the palest shell-pink and open into funnel-shaped flowers of pastel shades of powder blue, blush pink or pale rose.' (Adamson and Salter)

Common names: *English:* Eight-day-healing bush; Agdae-geneesbos, Luibossie, Douwurmbossie.

Description: An evergreen shrub to 1m (3ft) high, with stems pinkish when young, clothed in oblanceolate, obovate or linear, hairy leaves to 6cm (2½in) long and 1.2cm (½in) wide, and bearing cymes of blue, pink, or rarely white, funnel-shaped flowers to 25mm (1in) long, shortly hairy outside, flowering in August to November in the wild.

Distribution: South Africa.

Habitat: Flats and slopes.

Related species: Other attractive *Lobostemon* species include **L. argenteus** (Lehmann) Buek, a much-branched shrub growing to 1m (3ft) tall, with densely hairy oblong-lanceolate leaves and deep blue funnel-shaped flowers borne singly in the axils of bracts. **L. glaucophyllus** (Persoon) Buek is a shrub of similar size, with mostly hairless, lanceolate leaves and blue or occasionally pink flowers. Both species are native to South Africa.

The genus *Macromeria* D. Don, 1832
Giant-Trumpets

NAMING
Etymology: *Macromeria* – probably referring to the large dimensions of the plant's flowers.

Synonyms: Some species previously included within *Onosmodium* Michaux.

GENUS CHARACTERISTICS AND BIOLOGY
Number of species: About 10.

Description: A genus of perennial herbs. Stems usually upright and unbranched. Leaves alternate, lanceolate, stalkless. Flowers borne in few-flowered racemes, with bracts. Corolla usually whitish, funnel-shaped. Calyx divided into 5 sharply pointed, erect lobes. Stamens 5, a little unequal, with threadlike filaments and narrowly oblong anthers. Style thread-like, hairless. Nutlets smooth, shiny and usually white.

Distribution and habitats: Mexico to South America. One species, *M. alba* Nesom, is included on *The IUCN Red List of Threatened Plants*.

Life form: Hemicryptophytes.

Pollination: By hummingbirds.

Fruit dispersal: No specialized mechanism.

USES

GARDEN
If obtainable, *M. viridiflora* could be grown in a native wildflower garden (in the USA), herb garden, herbaceous and mixed border.

OTHER USES
The dried *M. viridiflora* has been smoked by Native Americans with mullein (*Verbascum*), to treat fits. The plant was dried by Hopi Indians, mixed with *Nicotiana* (wild tobacco) and used in rain-making ceremonies.

CULTIVATION

Very little information is available in the literature about cultivation of these plants. *Macromeria viridiflora* is hardy to Zone 7 (-15°C/5°F), and should be grown in a sunny situation on well-drained soil. In warm climates it will probably tolerate light shade.

PROPAGATION

The seed should probably be sown as soon as ripe or in spring under glass.

REFERENCES

DON, 1832; MOERMAN, 1998; JOHNSTON, 1954; PFAF, 1992–02; PHILLIPS & RIX, 1991.

Macromeria viridiflora De Candolle

Synonyms: *Onosmodium thurberi,* Gray.
Common names: *English:* Giant-trumpets, Green puccoon, Hopi Smoke, Silver Trumpets.
Description: A bristly perennial, to 90cm (3ft) high, with several stems and broadly lanceolate leaves 5–12cm (2–4^1/$_2$in) long, bearing, from midsummer until autumn, clusters of long trumpet-shaped flowers to 4cm (1^1/$_2$in), with shortly protruding stamens.
Distribution: New Mexico and eastern Arizona to northern Mexico.
Habitat: Rocky slopes and valleys, in pine forest and in scrub, at 1000–2750m (3280–9020ft).

The genus *Plagiobothrys* Fischer & C.A. Meyer, 1836

Popcorn-flowers

NAMING

Etymology: *Plagiobothrys* – from Greek *plagio,* oblique, and *bothros,* pit, referring to the nutlet scar.
Synonyms: *Allocarya* E. Greene, *Allocaryastrum* Brand, *Glyptocaryopsis* Brand, *Echinoglochin* (A. Gray) Brand. Previously included in the genus *Eritrichium.*

GENUS CHARACTERISTICS AND BIOLOGY

Number of species: About 50.
Description: A genus of annual or perennial herbs, usually hairy. Basal leaves clustered, opposite or in a rosette, oblong to linear. Flowers borne in a slender spike or a raceme, often with bracts. Calyx mostly persistent, deeply lobed, with oblong or lanceolate lobes. Corolla white, with a short tube, and overlapping, rounded, spreading lobes. Style short

and slender. Fruit of 1–4 ovoid, smooth or angled, incurved or erect nutlets.
Distribution and habitats: Western America and Australia, mostly dry places. 18 species are included on *The IUCN Red List of Threatened Plants.*
Life forms: Therophytes and hemicryptophytes.
Pollination: Bees.
Fruit dispersal: No specialized mechanism.

USES

GARDEN
For a native plant collection in their countries of origin, or scented garden. No known culinary or medical uses.

CULTIVATION

Very little information exists in the literature on cultivation of these plants, but the only species that is likely occasionally to appear in cultivation, *P. nothofulvum,* is a hardy annual, and requires a sunny position and a well-drained soil.

PROPAGATION

From seed only; should be possible to sow *in situ.*

AVAILABILITY

Unavailable, but seed may crop up in exchange lists.

REFERENCES

GENDERS, 1994; GILLETT & WALTER, 1998; HUXLEY, 1992.

Plagiobothrys nothofulvum (A. Gray) A. Gray

'A pretty annual, very similar to the forget-me-not, but pure white and emitting a powerful honeysuckle perfume. Treated as a biennial and massed as a carpet for summer flowering annuals, it will cover the ground with its racemes of purest white and when so used, the perfume is outstanding.' (Roy Genders)

Synonym: *Eritrichium nothofulvum* A. Gray.
Common name: *English:* Popcorn flower.
Description: A hairy, upright annual to 50cm (20in) high, branched above, with a rosette of slightly shaggy-hairy oblanceolate leaves, and a few narrowly lanceolate stem-leaves, and slender forked spikes of sweetly scented, white flowers to 8mm (^{3}/$_8$in) across.
Distribution: North America (Washington to California).
Habitat: Sandy places.

The genus *Rindera*
Pallas, 1771

NAMING
Etymology: Not known.
Synonyms: *Cyphomattia* Boiss.; *Mattia* Schult; *Bilegnum* Brand. Some species were previously included in the genus *Cynoglossum* L.

GENUS CHARACTERISTICS AND BIOLOGY
Number of species: 27 species.
Description: A genus of perennial herbs. Stems usually upright, unbranched and overtopping the basal leaves, usually softly hairy. Basal leaves in a rosette, long-stalked; stem-leaves gradually becoming stalkless upwards. Flowers borne in corymb-like scorpioid cymes. *Calyx* lobed, but not quite to the base; lobes oblong, oval or lanceolate, enlarged in fruit, woolly. *Corolla* cylindrical to bell-shaped, 1.5–2 times longer than the calyx; tube shorter or longer than the limb; throat-scales present. Stamens and style usually protruding from corolla. Nutlets flat, almost circular, each with a broad membranous wing, the central area of the nutlet's dorsal surface with or without barbed appendages.
Distribution and habitat: Mostly central Asia, also Balkans and south-western Asia eastwards to China (Xinjiang). Two species, including *R. graeca*, are on *The IUCN Red List of Threatened Plants*.
Life form: Hemicryptophytes.
Pollination: By bees.
Fruit dispersal: The nutlet's wing facilitates dispersal by wind.

USES
No known medicinal or culinary uses. Contains toxic alkaloids.

GARDEN
Suitable for larger rock gardens and front of the border. Smaller choice species can be admitted into the alpine house.

CULTIVATION AND PROPAGATION
These plants are very rarely grown, and little is known about their requirements, but it is suggested by the *European Garden Flora* (Walters, 2000) that they should be cultivated and propagated as for *Cynoglossum*. The densely hairy species are likely to need protection from winter wet. The seed should germinate without any special treatment.

Rindera species. (Rosemary Steel)

AVAILABILITY
Generally unavailable, but may crop up in some seed catalogues or exchange schemes.

REFERENCES
POLUNIN, 1980; SLABY, 2001; TUTIN *et al.*, 1972; VVEDENSKY, 1962; WALTERS, 2000.

Rindera caespitosa Bunge
Synonyms: *Cynoglossum umbellatum* Aucher; *Mattia caespitosa* A.DC.
Description: A tufted, grey-hairy perennial, with leafy stems 10–30cm (4–12in) tall, narrowly lanceolate basal leaves 7.5–12cm (3–4½in)in long and linear, pointed stems leaves, and clusters of red-purple flowers, 8–12mm (⅝–½in)across, in late spring and summer.
Distribution: Asia Minor (eastern Anatolia).
Habitat: Rocky slopes, at 1500–3100m (4900–10,170ft).

Rindera graeca (A. De Candolle) Boissier & Heldr. in Boissier
Description: A tufted, grey-hairy perennial 6–25cm (2½–10in) high, with linear to oblanceolate, woolly leaves 3–12cm (1¼–4½in) long, and an umbel-like cluster of purple flowers 10–12mm (⁷⁄₁₆–½in) long.
Distribution: Endemic to Greece.

Habitat: Mountain rocks.
Associated plants: In the Peloponnese in community with *Aster alpinus*, *Convolvulus cochlearis*, *Linum flavum*, *Minuartia stellata* and *Globularia stygia*.

Rindera holochiton M. Popov

Description: An upright perennial 15–25cm (6–10in) tall, with oblong basal leaves 5–8cm (2–3in), with stalks up to twice as long, and lanceolate stem leaves 2–4cm ($^3/_4$–$1^1/_2$in), and corymb-like cluster, to 5cm (2in) long, of violet flowers, 8–10mm ($^5/_8$–$^7/_{16}$in) long, in summer.
Distribution: Tien Shan only (Kyrgyzstan).
Habitat: Stony slopes, in damp places, at about 3000m (9850ft).

Rindera lanata (Lamarck) Bunge

Synonyms: *Mattia umbellata* K. Koch; *Rindera pubescens* K. Koch; *Cyphomattia lanata* (Lamarck) Boissier.
Description: An upright perennial 15–55cm (6–22in) high, with hairy or hairless stems, long-stalked oval to linear, usually hairy basal leaves to 15cm (6in) long, and broadly oval, clasping upper leaves and, in summer, pink flowers 7–12mm ($^5/_8$–$^1/_2$in), with a white-woolly calyx.
Distribution: Turkey, Armenia, Iraq, Iran.
Habitat: Mountains.
Associated plants: In Armenia it may be found with *Arenaria graminea*, *Stachys lavanulifolia*, *Lotus gebelia*.

Rindera tetraspis Pallas

Synonyms: *Rindera cyclodonta* Bunge; *Rindera laevigata* Roem. & Schult., *Cynoglossum laevigatum* L.f.
Common name: *Chinese*: Chi guo cao.
Description: A tufted perennial growing 20–35cm (8–14in) high, with hairless, long-stalked, oblong to lanceolate leaves 4–8cm ($1^1/_2$–3in) long, smaller and stalkless stem leaves, and blue-purple flowers 13–15mm ($^1/_2$–$^5/_8$in), in spring.
Distribution: Central Asia, southern Russia, western China.
Habitat: Rocky deserts in China at 500–600m (1650–1970ft).
Associated plants: In the deserts of Kazakhstan it may be found among *Artemisia* spp, *Salsola arbusculiformis*, with *Rheum tataricum*.

The genus
Solenanthus Ledebour, 1829

NAMING
Etymology: *Solenanthus* – from Greek *solen*, tube, and *anthos*, flower, referring to the shape of the corolla.
Synonyms: *Kuschakewiczia* Regel & Smirn. Previously some species were included in *Cynoglossum* L.

GENUS CHARACTERISTICS AND BIOLOGY
Number of species: From 10 to about 17.
Description: A genus of hairy biennial or short-lived perennial herbs. Leaves alternate, flowers in dense cymes that often become paniculate, with or without bracts. *Calyx* lobed to base, with linear, lanceolate to oblong lobes, scarcely enlarged in fruit. *Corolla* blue or purplish-red, tubular, cylindrical or rarely bell-shaped, small; with poorly defined, vertical or somewhat spreading limb, and also oblong sac-like folds/scales within the corolla, inserted at below middle of tube. Stamens inserted above the folds protruding from the corolla; anthers oblong to broadly ellipticical, with a blunt tip. Style usually protruding, with minute stigma. Nutlets compressed, oval to nearly round, 5–10 mm ($^1/_4$–$^3/_8$in), flat or slightly concave, with dense hooked bristles.
Distribution and habitat: Central and western Asia, south-east Europe; two species in China; mostly in grassy places and woodland. Seven *Solenanthus* species are included on *The IUCN Red List of Threatened Plants*.
Life form: Hemicryptophytes.
Pollination: Bees.
Fruit dispersal: Nutlets are armed with hooked, anchor-like bristles (glochiols) that attach to animal fur and human clothing.

USES
ECONOMIC
In Central Asia, *S. circinnatus* is valued as a nectar source in bee-keeping. The plants contain alkaloids.

GARDEN
Could work in a mixed border, or bee garden.

CULTIVATION
Little is known about cultivation of these plants. They should be hardy to Zone 7 and can presumably be grown quite easily in sun or light shade on well-drained, reasonably fertile soil. Tall plants may need staking.

PROPAGATION
Large nutlets should probably be sown as soon as

ripe, or possibly in spring, covering lightly with compost, and germinated in a cold frame.

AVAILABILITY
Generally unavailable, but may appear on seed-exchange lists.

REFERENCES
HUXLEY, 1992; PHILLIPS & RIX, 1991; POLUNIN, 1980; TUTIN et al., 1972; VVEDENSKY, 1962; WU & RAVEN, 1995.

Solenanthus apenninus (Linnaeus) Fisch. & C.A. Mey
Synonym: *Cynoglossum apenninum* L.
Common name: *Italian*: Lingua di cane appenninica.
Description: A biennial 60–120cm (2–4ft) tall, with elliptical to broadly lanceolate basal leaves 30–50cm (12–20in) long and smaller stem-leaves, and panicles of purple or blue flowers 7–9mm ($^{3}/_{8}$in)long.
Distribution: Italy, Sicily.
Habitat: Mountain woods and pastures.

Solenanthus circinantus Ledebour
Synonyms: *Solenanthus amplifolium* Boissier; *S. coronatus* Regel; *S. petiolaris* De Candolle; *S. rumicifolium* Boissier.
Common names: *Chinese*: Chang riu liu li cao; *Russian*: Trubkotzvet zavitkovy.
Description: An upright, usually unbranched perennial, to 80–100cm (32–40in), arising from stout rootstock to 2cm ($^{3}/_{4}$in) in diameter, with sparsely bristly, long-stalked oval-oblong basal leaves, heart-shaped at base, to 30cm (12in) long and 20cm (8in) wide, stalkless stem-leaves and many cymes of broadly tubular, purplish-red or bluish-purple flowers, 5–8mm long, borne in late spring and early summer.
Distribution: China, Afghanistan, Pakistan, central Asia, Kazakhstan, eastern Turkey, Siberia.
Habitat: Meadows, scrub, at 2100–3000m (6890–9850ft).

Solenanthus scardicus Bornm.
Description: A branched perennial or biennial, 50–100cm (20–40in) tall, with oval or broadly lanceolate basal leaves 25–35cm (10–14in) long, and smaller stem leaves, and panicles of funnel-shaped, purple flowers about 7mm ($^{3}/_{8}$in)long.
Distribution: Eastern Albania and western Macedonia.
Habitat: Beech (*Fagus*) woods.

Solenanthus circinantus in fruit, Kazakhstan. (Rosemary Steel)

The genus *Trichodesma* R. Brown, 1810

NAMING
Etymology: *Trichodesma* – literally 'hairy bonds', referring to the hairs which bind the stamens together.
Synonyms: Some species previously assigned to this genus are now included in *Cynoglossum* and *Lacaitaea* Brand, while others have been included in the genus *Borago* L. in the past.

GENUS CHARACTERISTICS AND BIOLOGY
Number of species: 40–45 species.
Description: A genus of annual and perennial herbs. Leaves entire, opposite, or upper sometimes alternate. Flowers in loose inflorescences, with bracts. *Calyx* often divided nearly to the base, enlarging in fruit, with pointed lobes. *Corolla* more or less bell-shaped or wheel-shaped, with pointed, spreading or reflexed lobes. Filaments very short, with large, narrowly lanceolate, hairy anthers that protrude far from the corolla, joining

together to form a cone, ending in long, awn-like appendages twisted together. Style thread-like, with a small, head-like stigma. Nutlets 4, flattened, more or less oval, smooth or wrinkly, often with toothed margin.
Distribution and habitat: Mostly tropics and subtropics of Africa, Asia and Australia, primarily in dry places. One species, *T. Scotti* Balf.f., is included in *The IUCN Red List of Threatened Plants.*
Life forms: Hemicryptophytes and some chamaephytes.
Pollination: Probably by bumblebees.
Fruit dispersal: No specialized mechanism.

CULTIVATION

Trichodesma species are very rarely cultivated, but probably require an open, sunny, well-drained position. *Trichodesma boissieri* appreciates plenty of moisture in combination with perfect drainage. Though many species are tender, some are probably marginally hardy, about Zone 8.

PROPAGATION

From seed, that should germinate without any special treatment.

REFERENCES

ALON, 1993; DYER *et al.*, 1962; FEINBRUN-DOTHAN, 1978; VAN WYK & MALAN, 1997.

Trichodesma boissieri Post

Description: A grey-hairy annual or perennial, covered in silky or somewhat felty hairs, 30–50cm (12–20in) high, with opposite, elliptical to oval, pointed leaves, upper ones stalkless and clasping the stem, and loose clusters of borage-like pale blue flowers 15–25mm ($^5/_8$–1in) across, with spreading or reflexed lobes, in spring and early summer.
Distribution: Endemic to the Middle East.
Habitat: Rocky and grassy slopes, especially by springs; sometimes on roadsides.

Trichodesma physaloides (Fenzl.) A. De Candolle

> 'The charming bell-like white flowers beautify stretches of our veld in the early spring. The flowers turn brown when touched and as they fade, and this, together with the chocolate-coloured calyx, gave rise to the local name Chocolate-creams.'
> (R.A. Dyer, I.C. Verdoon, L.E. Codd)

Common names: *English:* Chocolate-creams, Chocolate bells.

Trichodesma boissieri. (Masha Bennett)

Description: A bushy, woody-based perennial, covered with numerous white tubercles, with upright, purplish-brown stems, stalkless lanceolate leaves, and branched clusters of nodding, bell-shaped flowers, with dark purplish-brown calyx that enlarges in fruit, and a white corolla with the margin fading to brown; borne in spring.
Distribution: South Africa.
Habitat: Grassland.

The genus *Trigonotis* Steven, 1851

NAMING

Etymology: *Trigonotis* – three-angled, referring to nutlets.
Synonyms: Some species previously included within *Myosotis* L., *Omphalodes,* Miller, and *Eritrichium* Schrad.

GENUS CHARACTERISTICS AND BIOLOGY

Number of species: About 50 species.
Description: A genus of annual, biennial or slender perennial herbs, hairy to nearly hairless. Leaves alternate, entire. Flowers in simple or branched racemes, sometimes with bracts at the base, rarely with the flowers axillary in the leaf-like bracts. Flowers stalked or nearly stalkless, blue, pink or white. *Calyx* deeply 5-lobed, slightly enlarging after flowering. *Corolla* wheel-shaped, with 5 scales in the throat, and

5 lobes imbricate in bud. Stamens included in the corolla. Nutlets 4, erect, smooth, sometimes downy, 3–angled and slightly flattened dorsally.

Distribution and habitats: From Europe (one species), through Central Asia and Siberia, to New Guinea. Two species are included on *The IUCN Red List of Threatened Plants*.

Life forms: Therophytes and hemicryptophytes.

Pollination: Probably by bees.

Fruit dispersal: No specialized mechanism.

USES

CULINARY
Young leaves of *Trigonotis peduncularis* and *T. radicans* can be eaten when cooked as an emergency food.

GARDEN
Rock garden, some species suitable for light woodland or streamside.

MEDICINAL
T. peduncularis has diuretic and emollient properties, and is also used in the treatment of diarrhoea and dysentery.

CULTIVATION
These smallish plants are superficially similar to some woodland species of *Omphalodes* or *Myosotis,* and, judging by their natural habitats, *Trigonotis brevipes* and *T. radicans* can probably be grown in similar moist, lightly shaded woodland conditions. *Trigonotis rotundifolia*, on the other hand, should do well in a open position in a rock garden. They are probably hardy to at least Zone 6 or 7.

PROPAGATION
Very little information exists in literature, but try sowing the seed in spring or as soon as ripe in well-drained compost in a cold frame. Division should also be possible in autumn or spring.

AVAILABILITY
May occasionally be sold by some alpine nurseries, or the seed could appear on exchange lists.

REFERENCES
BECKETT, 1993–94; HUXLEY, 1992; MALYSHEV, 1997; OHWI, 1965; PFAF Database; WU & RAVEN, 1995.

Trigonotis brevipes Maxim.

Synonym: *Eritrichium brevipes* Maxim.

Common name: *Japanese:* Mizu-tabirako.

Description: A tufted or mat-forming perennial, spreading by rhizomes and stolons, with upright stems 10–40cm (4–16in), oval or oblong, stalked leaves, and light blue flowers about 3mm across, borne in dense racemes up to 10cm (4in) long in summer.

Distribution: Japan (Honshu, Shikoku, Kyushu, rather common).

Habitat: Wet places along streams and ravines in mountains.

Trigonotis radicans (Turcz.) Steven

Synonyms: *Omphalodes aquatica* Brand; *O. sericea* Maxim.; *Trigonotis sericea* Ohwi; *Myosotis radicans* Turcz.

Common names: *Chinese:* Bei fu di cai; *Japanese:* Ke-ruri-so.

Description: A somewhat tufted, hairy perennial, to about 30cm (12in) high, with elongated decumbent stems, oval leaves with rounded or heart-shaped base, 2.5–5cm (1–2in) long, the basal ones long-stalked, and blue or white flowers about 10mm across, on 1–2cm ($^7/_{16}$–$^3/_4$in) *pedicels* from the upper leaf axils in summer.

Distribution: Japan (Kyushu – Higo Province, rare), Korea, eastern Siberia.

Habitat: Mountains, damp rocks, woods, by streams.

Associated plants: In the Russian Far East, it sometimes grows with the famous ginseng (*Panax ginseng*), also wood sorrel (*Oxalis acetosella*), meadow-rue *Thalictrum filamentosum*, bedstraw *Galium dahurica* and shield fern *Dryopteris amurensis*.

Trigonotis rotundifolia (Wallich in Bentham) Bentham in C.B. Clarke

Description: A small tufted or mat-forming perennial, about 8cm (3in) or more, spreading, with hairy, elliptical or rounded leaves with a heart-shaped base, 1–2cm ($^7/_{16}$–$^3/_4$in) long, and terminal clusters of bright blue, yellow-eyed flowers, 6–8mm ($^1/_4$–$^3/_8$in) across, borne in summer.

Distribution: Himachal Pradesh to south-eastern Tibet.

Habitat: Rock ledges, screes, 3600–5600m (11,800–18,350ft).

APPENDIX I
BORAGINACEAE: FULL LIST OF GENERA AND SYNONYMS

KEY:

Bold Italics – currently recognized genera

Italics – synonyms

= – synonym for…

** – genus described in Chapters 1 and 2

* – genus described in Chapter 3

– genus not in cultivation, but of potential garden value

☠ – genus includes species known to contain alkaloids.

⊕ – genus includes species listed on *The 1997 IUNC Red List of Threatened Plants*.

Actinocarya Benth. 1 sp., *A.tibetica* Benth. NW India, Tibet, Pakistan.

Adelocaryum Brand = *Lindelofia*** Lehm.

Aegonychon Gray = *Lithospermum** L.

Afrotysonia Rauschert *Tysonia*) 3 spp. S&E Africa.

Aipyanthus Steven = *Arnebia*** Forssk.

Alkanna Tausch 25–30spp, S Europe, Mediterranean to Iran. ☠⊕

Allocarya E.Greene = *Plagiobothrys** Fisch & C.A.Mey

Allocaryastrum Brand = *Plagiobothrys** Fisch & C.A.Mey

Amblynotopsis J.F.Macbr. = *Antiphytum* DC. in Meisn.

Amblynotus (A.DC.) I.M.Johnst. 1 sp., *A.rupestris* (Pall.) M. Popov. W Siberia to Mongolia.

Amphibologyne A. Brand. 1 sp., *A.mexicana* Brand. Mexico.

Amsinckia Lehm. 15 spp. W USA, W temperate South America. ☠⊕

Anchusa L. ~35 spp. Europe, N & S Africa, W Asia. ☠⊕

Ancistrocarya # Maxim. 1 sp., *A.japonica* Maxim. Japan

Anisanthera Raf. = *Caccinia** Savi

Anoplocaryum # Ledeb. 1–3 spp. Siberia, Mongolia, China.

Antiotrema Hand.–Mazz. 1 sp., *A.dunnianum* Diels. SW China

Antiphytum DC. in Meissner. 10 spp. Mexico and tropical America.

Antrophora I.M.Johnst. = *Lepidocordia* Ducke

Argusia # Boehm. 4 spp. Romania to Japan. N

Arnebia ** Forssk. 25 spp. Mediterranean, tropical Africa, Himalayas. ☠⊕

Arnebiola Chiov. = *Arnebia*** Forssk.

Asperugo L. 1 sp., *A.procumbens* L. ("madwort"). Europe. ☠

Austrocynoglossum M.Popov in R.Mill. 1 sp., *A. latifolium* (R.Br.) R.Mil. E Australia.

Auxemma Miers. (Ehretiaceae) 2 spp. Brazil

Baphirhiza Link = *Alkanna** Tausch

Batschia Gmel. = *Lithospermum*

Beruniella Zakirov & Nabiev = *Heliotropium*** L.

Beurreria Jacq. (SUO) = *Bourreria* P.Browne

Bilegnum Brand = *Rindera** Pall.

Boraginella Kuntze = *Trichodesma** R.Br.

Borago ** L. 3 spp. Mediterranean Europe, Asia. ☠⊕

Borrachinea Lavy = *Borago*** L.

Borraginoides Boehm. = *Trichodesma** R.Br.

Bothriospermum Bunge. 5 spp. Tropical and NE Asia.

Bourjotia Pomel = *Heliotropium*** L.

Bourreria * P.Browne. ~ 30 spp. East Africa, Madagascar, Mascarenes, warm zones of America. ⊕

Brachybotrys # Maxim. in Oliv. 1 sp., *B.paridiformis* Maxim. in Oliv. Manchuria, E Siberia, Korea, China.

Brandella R.R.Mill. 1 sp., *B.erythraea* R.R.Mill. NE Africa, Saudi Arabia

Brunnera ** Steven. 3 spp. E Mediterranean to W Siberia. ⊕

Bucanion Steven = *Heliotropium*** L.

Buglossoides ** Moench. = *Lithospermum* L. (in this book described separately) ⊕

Caccinia * Savi. 6 spp. W & C Asia N

Camptocarpus K.Koch = *Alkanna** Tausch

Campylocaryum DC. in Meisn. = *Alkanna** Tausch

Carmona * Cav. 1 sp., *C.retusa* (Vahl) Masam. E Asia.

Caryolopha Fisch in Trautv. = *Pentaglottis*** Tausch

Ceballosia G.Kunkel = *Tournefortia* L.

Cerinthe ** L. 10 spp., Europe, especially the Mediterranean, Turkey, Siberia.

Cervia Rodr. in Lag. = *Rochelia* Rchb.

Chamissoniophila Brand 2 spp. S Brazil.

Chionocharis * I.M.Johnst. 1 sp., *C.hookeri* I.M.Johnst. Himalaya.

Choriantha Riedl. 1 sp., *C.popoviana* Riedl. Iraq

Cochranea Miers = *Heliotropium*** L.

Coldenia L. – 1 sp., *C.procumbens* L. Old World, tropics & warm temperate regions. ⊕

Colsmannia Lehm. = *Onosma*** L.

Cordia * L. 320 spp. Tropics, mostly of N & S America. ⊕

Cordiada Vell. = *Cordia** L.

Cordiopsis Desv. = *Cordia** L.

Cortesia Cav. (Ehretiaceae). 2 spp. Temperate S
America

Craniospermum # Lehm. 4–5 spp. C Asia, China.

Crucicaryum O. Brand 1 sp., *C.papuanum* O.Brand.
New Guinea

Cryptantha* G.Don. ~100 spp. Western N America. ⊛

Cynoglossopsis Brand. 2 spp., Somalia & Ethiopia.

Cynoglossum** L. ~75 spp. Temperate & warm
regions, especially Old World. ⚥⊛

Cynoglottis (Gusul.) Vural & Kit Tan. 2 spp. SE
Europe, SW Asia. (In this book included under
*Anchusa*** L).

Cyphomattia Boiss. = *Rindera* Pall.

Cystistemon Post & Kuntze (SUO) = *Cystostemon*
Balf.f.

Cystostemon Balf.f. 13 spp. Tropical Africa to Saudi
Arabia.

Dasynotus # I.M.Johnst. 1 sp., *D.daubenmirei*
I.M.Johnst. North-west USA. ⊛

Decalepidanthus Riedl. 1 sp., *D.sericophyllus* Riedl.
– for 70 years known only from type specimen
(rediscovered 1971). Pakistan.

Echinoglochin (A.Gray) Brand = *Plagiobothrys*** Fisch
& C.A.Mey.

Echinospermum Sw. in Lehm. = *Lappula* Moench

Echiochilon Desf. 17 spp. Arid NE Africa & Arabia
to Iran and Baluchistan.

Echiochilopsis Caball. = *Echiochilon* Desf.

Echioides Ortega (SUH) = *Arnebia*** Forssk.

Echiopsis Rchb. = *Lobostemon*** Lehm.

Echiostachys Levyns = *Lobostemon*** Lehm

Echium** L. 60 spp., Macronesia (27 endemic)
Europe, W Asia, N & S Africa. ⚥⊛

Eddya Torr. & A.Gray = *Tiquilia* Pers.

Ehretia* P.Browne 75 spp. Tropical and warm
regions. ⚥⊛

Elizaldia Willk. 5 spp., W Mediterranean.

Embadium J.M.Black. 3 spp. S Australia.

Endogonia Turcz. = *Trigonotis*** Steven

Eremocarya Greene = *Cryptantha*** G.Don

Eritrichium** Schrad. in Gaudin. ~30 spp. N
temperate regions. ⊛

Euploca Nutt. = *Heliotropium*** L.

Friedrichsthalia Fenzl = *Trichodesma*** R.Br.

Galapagoa Hook.f. = *Tiquilia* Pers.

Gastrocotyle Bunge. 2 spp. E Mediterranean to C
Asia and NW India.

Gerascanthus P.Browne = *Cordia*** L.

Glyptocaryopsis Brand = *Plagiobothrys*** Fisch. &
C.A.Mey.

Greeneocharis Gurke & Harms = *Cryptantha*** G.Don

Gymnoleima Decne. = *Moltkia*** Lehm.

Gyrocaryum Vald–s. 1 sp., *G.oppositifolium* Vald–s.
Spain. ⊛

Hackelia* Opiz. 45 spp., N temperate regions, C &
S America. ⚥⊛

Halacsya # Dorefl. 1 sp., *H.sendtneri* Doerfl. W
Balkans. ⊛

Halgania Gaudich. 18 spp., Australia.

Harpagonella A.Gray. 1 sp., *H.palmeri* A.Gray. SW
North America. ⊛

Havilandia Stapf = *Trigonotis*** Steven

Heliocarya Bunge = *Caccinia*** Savi

Heliotropium** L. ~250 spp. Tropical & temperate
regions. ⚥⊛

Henryettana Brand = *Antiotrema* Hand.–Mazz.

Heterocaryum A.DC.= *Lappula* Moench

Hormuzakia Gusul. = *Anchusa*** L.

Huynhia Greuter = *Arnebia*** Forssk.

Isorium Raf. = *Lobostemon*** Lehm.

Ivanjohnstonia Kazmi. 1 sp., *I.jaunsariensis* Kazmi.
NW Himalaya

Johnstonella Brand = *Cryptantha*** G.Don

Kuschakewiczia Regel & Smirn. = *Solenanthus***
Ledeb.

Lacaitaea Brand 1 sp., *L.calycosa* Brand. E
Himalaya.

Lappula Moench. ~40 spp. ("bur forget-me-not").
Temperate Eurasia, N America. ⚥⊛

Lasiarrhenum I.M.Johnst. 2 spp. Mexico.

Lasiocaryum I.M.Johnst. 4–7 spp. C Asia,
Himalayas, China.

Lepechiniella # Popov. 9 spp. C Asia.

Lepidocordia Ducke (Ehretiaceae). 2spp. Tropical S
America. ⊛

Leptanthe Klotzsch = *Arnebia*** Forssk.

Leurocline S.Moore = *Echiochilon* Desf.

Lindelofia** Lehm. 11 spp. C Asia to Himalayas. ⚥⊛

Lithocardium Kuntze = *Cordia*** L.

Lithococca Small ex Rydb. = *Heliotropium*** L.

Lithodora** Griseb. 7 spp. NW France & W
Mediterranean to Asia Minor. ⊛

Lithospermum* L. (incl. *Buglossoides*) ~ 45 spp.
Temperate regions of the world, excluding Australia.
⚥⊛

Lobophyllum F.Mueller = *Coldenia* L.

Lobostema Spreng. = *Lobostemon*** Lehm.

Lobostemon* Lehm. 28 spp. S Africa. ⊛

Lycopsis L. = *Anchusa*** L.

Maccoya F.Muell. = *Rochelia* Rchb.

Macromeria* D.Don. 10 spp. Mexico to S America. ⊛

Macrotomia DC. = *Arnebia*** Forssk.

Maharanga A.DC. 10 spp. E Himalayas to SW
China, Thailand.

Mairetis I.M.Johnst. 1 sp., *M.microsperma* I.M.Johnst. Canary Islands, Morocco.

Mallotonia (Griseb.) Britton. 1 sp., *M.gnaphalodes* (L.) Britton ("iodine bush") Florida, Mexico, West Indies.

Massartina Maire = *Elizaldia* Willk.

Mattia Schult. = *Rindera* Pall.

Mattiastrum (Boiss.) Brand = *Paracaryum* (A.DC.) Boiss.

Megacaryon Boiss. = *Echium*** L.

Megastoma Coss. & Durieu = *Ogastemma* Brummitt

Mertensia** Roth. ~ 45 spp. N temperate regions of the world, especially N America.

Messerschmidia Hebenstr. = *Argusia* Boehm. ⊛

Messerschmidia Roem. & Schult. = *Tournefortia* L.

Messersmidia L. = *Argusia* Boehm.

Metaeritrichium W.T.Wang = *Eritrichium*** Schrad. in Gaudin.

Microcaryum I.M.Johnst. 4 spp. C Asia to W China.

Microparacaruym (Popov in Riedl) Hilger & Podlech = *Paracaryum* (A.DC.) Boiss.

Microula Benth. 30 spp. Himalayas to W China.

Mimophytum Greenm. 1 sp., *M.omphaloides* Greenm. Mexico.

Moltkia** Lehm. 6 spp., 3 in N Italy to N Greece, 3 in SW Asia. ⊛

Moltkiopsis I.M.Johnst.1 sp., *M.ciliata* I.M.Johnst. N Africa to Iran.

Monomesia Raf. = *Tiquilia* Pers.

Moritizia DC. in Meissn. 5 spp., tropical America

Munbya Boiss. = *Arnebia*** Forssk.

Myosotidium** Hook. 1 sp., *M.hortensia* Hook. New Zealand ⊛

Myosotis** L. ~100 spp. Temperate regions of the world and at high altitudes in tropics. ⚘⊛

Neatostema I.M.Johnst. 1 sp., *N.apulum* (L.) I.M.Johnst. Macronesia to the Mediterranean & Iraq.

Nephrocarya Candargy = *Nonea*** Medik.

Nesocaryum I.M.Johnst.1 sp., *N.stylosum* I.M.Johnst. Desventuradas Is., Chile. ⊛

Nogalia Verdc. 1 sp., *N.drepanophylla* (Baker) Verdc. Somalia & Saudi Arabia

Nomosa I.M.Johnst. 1 sp., *N.rosei* I.M.Johnst. Mexico. ⊛

Nonea** Medikus. 35 spp. Mediterranean, SW Asia, to Siberia. ⊛

Octosomatium Gagnep. = *Trichodesma** R.Br.

Ogastemma Brummitt. 1 sp., *O.pusillum* (Bonnet & Barratte) Brummitt; Canary Is. & N Africa.

Omphalodes** Mill. 30 spp. Temperate Eurasia, Mexico. ⊛

Omphalolappula Brand – 1 sp., *O. concava* (F.Muell.) Brand; temperate Australia.

Omphalotrigonotis W.T.Wang = *Trigonotis** Steven

Onochiles Bubani = *Alkanna** Tausch

Onosma** L. ~150 spp. Mediterranean, especially Turkey, to the Himalayas & China. ⊛

Onosmodium Michaux. 5 spp. ("marbleseed", "false gromwell"). N America. ⊛

Oplexion Raf. = *Lobostemon* Lehm.

Oreocarya Greene = *Cryptantha** G.Don

Oreocharis (Decne.) Lindl. = *Pseudomertensia**** Riedl.

Oreogenia I.M.Johnst. = *Lasiocaryum* I.M.Johnst.

Oxyosmyles Speg. 1 sp., *O.viscosissima* Speg. Argentina

Paracaryopsis (H.Riedl) R.R.Mill. 3 spp. India

Paracaryum (A.DC.) Boiss. 9 spp. Meditteranean to C Asia. ⚘⊛

Paracynoglossum Popov = *Cynoglossum*** L.

Paramoltkia** Greuter. 1 sp., *P.doerfleri* (Wettst.) Greuter & Burdet. (in this book included under *Moltkia*). Albania. ⊛

Paraskevia W.Sauer & G.Sauer. 1 sp., *P.cesatiana* (Fenzl & Friedr.) W.Sauer & G.Sauer. Greece. ⊛

Pardoglossum E.Barbier & Mathez. 6 spp. West Mediterranean. (In this book *P.cheirifolium* included under *Cynoglossum* L.)

Patagonula L. (Ehretiaceae). 2 spp. Brazil, Argentina.

Pectocarya DC. in Meisn. 10 spp., N America.

Pedinogyne Brand = *Trigonotis** Steven

Pentaglottis** Tausch. 1 sp., *P.sempervirens* (L) L.Bailey. SW Europe.

Penthysa Raf. = *Lobostemon** Lehm.

Perittostema I.M.Johnst. 1 sp., *P.pinetorum* I.M.Johnst. Mexico.

Phyllocara Gusul. = *Anchusa** L.

Plagiobothrys* Fisch. & C.A.Mey. ~ 70 spp. W America, E Asia (1 sp.), Australia (4 spp.). ⊛

Podonosma Boiss. = *Onosma*** L.

Procopiana Gusul = *Symphytum*** L.

Procopiania Gusul. = *Symphytum*** L.

Pseudomertensia Riedl. 8 spp. Iran to the Himalayas. (In this book included under *Mertensia*** Roth).

Psilolaemus I.M.Johnst.1 sp., *P.revolutus* I.M.Johnston. Mexico.

Pteleocarpa Oliv. (Ehretiaceae). 1 sp., *P.lamponga* (Miq.) Heyne. W Malesia.

Ptilocalyx Torr. & A.Gray = *Tiquilia* Pers.

Pulmonaria** L. 10–18 spp. Europe. ⚘⊛

Raclathris Raf. = *Rochelia* Rchb.

Rhabdia Mart. = *Rotula* Lour.

Rhabdocalyx Lindl. = *Cordia* L.

Rhytispermum Link = *Alkanna** Tausch

Rindera Pall. ~25 spp. Mediterranean to C Asia. ⚥⊛

Rochefortia Sw. (Ehretiaceae). 12 spp. West Indies, 1 sp. C America to Colombia. ⊛

Rochelia Rchb. 20 spp. Eurasia.

Rotula Lour. (Ehretiaceae). 1(–3) spp., *R.aquatica* Lour. E Brazil, Old World tropics.

Saccellium Bonpl.(Ehretiaceae). 3 spp., tropical S America.

Scapicephalus Ovcz. & Czukav. 1 sp., *S.rosulatus* Ovcz. & Czukav. C Asia.

Schistocaryum Franch. = *Microula* Benth.

Sclerocaryopsis Brand = *Lappula* Moench.

Selkirkia Hemsl. 1 sp., *S.berteroi* Hemsl. Juan Fernandez. ⊛

Sericostoma Stocks. 1 sp., tropical E Africa to NW India.

Sinojohnstonia Hu. 3 spp. W China. (*S.moupinensis* W.T.Wang is described under *Omphalodes* Miller in this book).

Solenanthus* Ledeb. ~17 spp. Mediterranean to C Asia & Afghanistan. ⚥⊛

Spilocarpus Lem. = *Tournefortia* L.

Stegnocarpus Torr. & A.Gray = *Tequilia* Pers.

Stenosolenium # Turcz. 1 sp., *S.saxatile* (Pallas) Turcz. C Asia, China, Siberia, Mongolia.

Stephanocaryum # Popov. 2 spp., C Asia.

Strophiostoma Turcz. = *Myosotis** L.

Suchtelenia Karel. in Meissn. 1(–3) spp., *S.cerinthifolia* Karel. in Meissn. Caucasus to C Asia.

Symphytum** L. 35 spp., Europe, Mediterranean to Caucasus. ⚥⊛

Tetraedrocarpus O.Schwarz = *Echiochilon* Desf.

Thaumatocaryon Baillon. 4 spp., Brazil.

Thyrocarpus Hance. 3 spp., China, Vietnam.

Tianschaniella # B.Fedtsch. in Popov. 1 sp., *T.umbellulifera* B.Fedtsch. in Popov. C Asia. ⊛

Tiquilia Pers. (Ehretiaceae). 27 spp. American deserts. ⊛

Tiquiliopsis A.Heller = *Tiquilia* Pers.

Tournefortia L. ~100 spp. Tropics and warm temperate regions. ⚥⊛

Toxostigma A.Rich. = *Arnebia*** Forssk.

Trachelanthus Kunze. 3 spp., Iran to C Asia. N

Trachystemon** D.Don. 1–2 spp., Mediterranean.

Traxara Raf. = *Lobostemon** Lehm.

Traxilum Raf. = *Ehretia** P.Browne

Trichodesma* R.Br. 45 spp. Tropical & warm temperate regions of the Old World. ⚥⊛

Trigonocaryum Trautv. 1 sp., *T.prostratum* Trautv. Caucasus.

Trigonotis* Steven. 55 spp. E Europe, C Asia to New Guinea. ⊛

Tysonia Bolus = *Afrotysonia* Rauschert

Ulugbeckia Zakirov = *Arnebia*** Forssk.

Valentina Speg. = *Heliotropium*** L.

Valentiniella Speg. = *Heliotropium*** L.

Varronia P.Browne = *Cordia** L.

Vaupelia Brand. = *Cystostemon* Balf.f.

Wellstedia Balf.f. (Wellstediaceae). 3 spp. Somalia, Socotra, Ethiopia. ⊛

Wheelerella G.B.Grant = *Cryptantha** G.Don

Zwackhia Sendtn. in Rchb. = *Halacsya* Doerfl.

REFERENCES

MABBERLEY, 1997; MALYSHEV, 1997; THE PLANT NAMES PROJECT, 1999; WALTER & GILLETT, 1998; VVEDENSKY, 1962; WU & RAVEN, 1995. www.rbgkew.org.uk/cgi-bin/web.dbs/genlist.pl?BORAGINACEAE

APPENDIX II
USEFUL ADDRESSES

PLANT SUPPLIERS

Monksilver Nursery
Oakington Road, Cottenham, Cambridgeshire, UK CB4 8TW
Tel: 01954 251555; fax: 01223 502887
E-mail: plants@monksilver.com
Website: www.monksilver.com

The Beth Chatto Gardens Ltd
Elmstead Market, Colchester, Essex, UK CO7 7DB
Tel: 01206 822007; fax: 01206 825933
E-mail: info@bethchatto.fsnet.co.uk
Website: www.bethchatto.co.uk

Beeches Nursery
Village Centre, Ashdon, Saffron Walden, Essex, UK CB10 2HB
Tel: 01799 584362; fax: 01799 584362
E-mail: sales@beechesnursery.co.uk
Website: www.beechesnursery.co.uk

Stillingfleet Lodge Nurseries
Stillingfleet, Yorkshire, UK YO19 6HP
Tel: 01904 728506; fax: 01904 728 506
E-mail: vanessa.cook@still-lodge.freeserve.co.uk
Website: www.stillingfleetlodgenurseries.co.uk

Crûg Farm Plants
Griffith's Crossing, Caernarfon, Gwynedd, UK LL55 1TU
Tel: 01248 670232; fax: 01248 670232:
E-mail: bleddyn&sue@crug-farm.co.uk
Website: www.crug-farm.co.uk

Cally Gardens
Gatehouse of Fleet, Castle Douglas, UK DG7 2DJ
Fax: 01557 815029

Cotswold Garden Flowers
Sands Lane, Badsey, Evesham, Worcestershire, UK WR11 5EZ
Tel/fax: 01386 47337
E-mail: cgf@star.co.uk; Website: www.cgf.net

Elworthy Cottage Plants
Elworthy Cottage, Elworthy, Lydeard St Lawrence, Taunton, Somerset, UK TA4 3PX
Tel: 01984 656 427; fax: 01984 656427;
E-mail: elworthycottage@care4free.net.uk

Hans Simon
StandenWeg 2, 97828 Marktheidenfeld, Germany

Hans Kramer
Hessenweg 41, 6718 TC EDE, The Netherlands
Tel: 0031 3186 7334

Coen Hansen
Ankummer ES15, 7722 RD Dalfsen, The Netherlands. Tel: 0031 529 434086

Ernst Pagels
Deichstrasse 4, 26789 Leer, Germany
Tel: 0049 491 3218

Un Jardin de Beuchigranges
88640 Granges sur Vologne, Vosges, France
Tel: 0033 329 514719

Terra Nova Nurseries (wholesale only)
P.O. Box 23938, Tigard, OR 97281-3938, USA
Nursery site: 10051 South Macksburg Road, Canby, OR 97013, USA
E-mail: info@terranovanurseries.com
Website: www.terranovanurseries.com

Heronswood Nursery Ltd.
7530 NE 288th Street, Kingston, WA 98346, USA
Tel: 001 360 297 4172; fax: 001 360 297 8321
E-mail info@heronswood.com
Website: www.heronswood.com

JDS Gardens
RR#4 2277 County Road, Harrow, Ontario, Canada N0R 1G0
Tel: 001 519 738 9513; fax: 001 519 738 3539
Website: www.jdsgardens.com

SEED SUPPLIERS

Chiltern Seeds
Bortree Style, Ulverston, Cumbria, UK LA12 7PB
Tel: 01229 581137; fax: 01229 584549
E-mail: info@chilternseeds.co.uk
Website: www.edirectory.co.uk/chilternseeds

PlantWorld Botanic Gardens
St Marychurch Road, Newton Abbot, Devon, UK TQ12 4SE
Tel: 01803 972939; fax: 01803 875018.

Thompson & Morgan
Poplar Lane, Ipswich, Suffolk, UK IP8 3BU
United Kingdom
Tel: 01473 688588; fax: 01473 680199
Website: www.thompson-morgan.com

B & T World Seeds
Rue des Marchandes, Paguignan, 34210 Olonzac, France
Tel: 00 33 468 912 963; fax: 00 33 468 913039
E-mail: ralph@b-and-t-world-seeds.com
Website: www.b-and-t-world-seeds.com

Jelitto Perennial Seeds Ltd
PO Box 1264, D-29685 Schwarmstedt, Germany
Tel: 00 49 5071 9829-0; fax: 00 49 5071 9829-27;
Website: www.jelitto.com

Rareplants
Bjørn Malkmus, Am Parkfeld 14 E, D-65203 Wiesbaden, Germany
Website: www.rareplants.de

Alplains
PO Box 489, Kiowa, CO 80117-0489, USA
Tel: 303 621 2247 (not for orders); fax: 303 621 2864

Silverhill Seeds
PO Box 53108, Kenilworth 7745, South Africa
Tel: 00 27 21 762 4245; fax: 00 27 21 797 6609
E-mail: rachel@silverhillseeds.co.za

Southern Seeds
The Vicarage, Sheffield, Canterbury, NZ 8173
Tel/fax: 00 64 3318 3814

PLANT SOCIETIES

The Alpine Garden Society
Avon Bank, Pershore, Worcestershire, UK WR10 3JP
Tel: 01386 554790; fax: 01386 554801
E-mail: webinfo@alpinegardensoc.demon.co.uk
Website: www.alpinegardensociety.org

The Cottage Garden Society
'Brandon', Ravenshall, Betley, Cheshire,
UK CW3 9BH
Website: www.alfresco.demon.co.uk/cgs

Hardy Plant Society (including Pulmonaria Group)
Mrs Pam Adams, Little Orchard, Great Comberton,
Pershore, Worcestershire, UK WR10 3DP
Tel: 01386 710317; fax: 01386 710117
E-mail: admin@hardy-plant.org.uk
Website: www.hardy-plant.org.uk

The Herb Society
Sulgrave Manor, Sulgrave, Banbury, UK OX17 2SD
Tel: 01295 768899
Fax: 01295 768069
E-mail: email@herbsociety.co.uk
Website: www.herbsociety.co.uk

National Council for Conservation of Plants & Gardens
The Stable Courtyard, RHS Garden, Surrey, UK
GU3 6QP
Tel: 01483 211465; fax: 01483 212404
Website: www.nccpg.org.uk

The Royal Horticultural Society
80 Vincent Square, London, UK SW1P 2PE
Tel: 020 7834 4333
E-mail: info@rhs.org.uk
Website: www.rhs.org.uk
RHS Garden Wisley, Woking, Surrey GU23 6QB
Tel: 01483 224234; fax: 01483 211750

Scottish Rock Garden Club
Membership Secretary SRGC
Mr A.D. McKelvie
43 Rubislaw Park Crescent, Aberdeen, UK AB15 8BT
E-mail: enquiries@srgc.org.uk
Website: www.srgc.org.uk

Klub skalnickaru Praha, Ceska republika (KSP)
[The Rock Garden Club Prague, Czech Republic
(RGCP)]
Marikova 5, 162 00 Praha 2, Czech Republic.
E-mail: holubec@vurv.cz
Website: www.backyardgardener.com/cz.html

Nederlandse Rotsplanten Werkgroep (Dutch Alpine
Garden Society)
Mr. Rob Koolbergen, Stelvio 15, 1186 EE
Amstelveen, The Netherlands.
Email: NRWKoolbergen@hetnet.nl
Website: home.wish.net/~nrwned

The Société des Amateurs de Jardins Alpins
(French Alpine Garden Society)
SAJA BP 432, 75233 Paris, Cedex 05 France
Website: saja.free.fr

Alpine Garden Club of British Columbia
c/o Dana Cromie Alpine Garden Club, 2208 Alder
Street, Vancouver, B.C., Canada V6H 2R9
Tel: 001 604 733 7566
E-mail: dana@vancouverbc.net
Membership secretary: Joy Curran RR1, B38 Bowen
Island B.C., Canada V0N 1G0
Website: www.hedgerows.com/Canada/
clubbrochures/alpinegardenclub.htm

The Canadian Herb Society
5251 Oak Street, Vancouver, B.C., Canada
V6M 4H1
E-mail: info@herbsociety.ca
Website: www.herbsociety.ca

The Herb Society of America
9019 Kirtland Chardon Road
Kirtland, OH 44094, USA
Tel: 00 1 440 256 0514; fax: 00 1 440 256 0541
E-mail: herbs@herbsociety.org (general information)
Website: www.herbsociety.org

The North American Rock Garden Society
PO Box 67, Millwood, NY 10546, USA
E-mail: nargs@advinc.com
Website: www.nargs.org

Australian Herb Society
The Secretary, Australian Herb Society, PO Box 110,
Mapleton, QLD 4560, Australia
E-mail: aherbsoc@hotmail.com
Website:
www.cybersayer.com/sunweb/groups/aherbsoc.html

The New Zealand Alpine Garden Society
NZAGS (Inc.), PO Box 2984, Christchurch, New
Zealand
Website: www.backyardgardener.com/nz.html

APPENDIX III
GLOSSARY OF BOTANICAL AND HORTICULTURAL TERMS

Acid in horticulture, usually referring to low pH of the soil.

Actinomorphic rasially symmetrical, i.e. having more than one plane of symmetry, referring to the shape of flower (in most Boraginaceae).

Alkaline in horticulture, usually referring to high pH of the soil.

Alkaloids nitrogen-containing compounds produced by many members of Boraginaceae; they have toxic effects if consumed in large quantities.

Alpine a term normally applied to plants that are found at high altitudes in mountains, above the tree line; in gardening, often applied to any plants that are compact enough for growing in a rock garden.

Alpine house a specially designed glasshouse, normally used to accommodate the more difficult and special alpines and other small plants; alpine houses are normally unheated and have extra ventilation.

Alternate usually referring to the arrangement of leaves on stem; strictly speaking it means that they are arranged in two rows on the stem and are not opposite, but also commonly refers to the spiral arrangement of leaves.

Annual plant that completes its life cycle in less than 12 months, dying after flowering and setting seed, e.g. *Omphalodes linifolia*, *Myosotis ramosissima*, *Borago officinalis*.

Anther the part of the stamen containing the pollen grains.

Ascending sloping or curving upwards, often referring to plant stems.

Biennial plant that completes its life cycle within two years (but not within one year), and dies after flowering and setting seed, e.g. many *Cynoglossum* and *Echium*.

Bract a modified leaf, found in the inflorescence; it may subtend flowers or inflorescence branches; normally much smaller in size than the normal leaves.

Calyx part of a flower, usually green in colour, that protects the flower bud, and is composed of a number of lobes (normally 5 in Boraginaceae), also called sepals.

Capitate headlike, usually referring to the shape of stigma.

Caudex a swollen stem base, for example, in some *Mertensia* and *Eritrichium* species.

Chamaephyte a Raunkiaer life form, plants with perennating buds above the soil level but below 25cm (10in).

Cincinnus a one-sided, coiled inflorescence, uncoiling as flowers open so that newly opened flowers face in the same direction.

Corolla part of flower, often brightly coloured (frequently blue in Boraginaceae, also yellow, white, pink), that protects the reproductive organs and attracts pollinators; when in bud it is enclosed in the calyx. In Boraginaceae, the petals of the corolla are joined, and form a bell-, funnel-, salver- or wheel-shaped, or tubular corolla.

Corolla limb the upper part of the corolla which in Boraginaceae is usually lobed into 5 rounded or pointed lobes, and is often expanded or spreading.

Corolla tube the bottom, narrow part of the corolla, where the stamens are usually attached.

Cryptophyte a Raunkiaer life form, plants with their perennating buds below the soil (or water) level, can be subdivided into geophytes, hydrophytes and helophytes.

Cultivar a cultivated variety, selected for certain desirable characteristics, and maintained by man. Cultivar names should be written in single quotation marks, e.g. *Omphalodes linifolia* 'Starry Eyes'.

Cutting a portion of stem or root that is used in vegetative propagation with the aim of producing a new plant (or plants).

Division a vegetative method of propagation whereby a plant's rootstock is split into two or more parts, each containing some shoots or buds, to produce new plants.

Drupe a fleshy fruit with one or more seeds, each surrounded by a stony layer; in Boraginaceae found in *Ehretia* and *Cordia*.

Endemic native only in one country or a particular restricted area.

Entire not toothed or cut (usually about leaves).

Epigeal a type of germination where the seed cotyledons emerge above the soil.

Family a taxonomic rank above the level of genus and below the order. Sometimes subdivided into subfamilies and further into tribes.

Filament a part of stamen that acts as a stalk for the anther.

Fruit dispersal a variety of methods that plants employ to spread their seeds, including anemochory (by wind), hydrochory (by water), or zoochory (by animals).

Genera plural of genus.

Generic adjective pertaining to genus.

Genus a taxonomic rank above the level of species.

Geophyte a Raunkiaer life form, plants with their perennating buds below the soil level, e.g. *Symphytum tuberosum*.

Germination a sequence of complex changes within

the seed, beginning with the uptake of water, and culminating in the seedling appearing above the soil.

Glaucous bluish, usually referring to leaves of some plants, e.g. *Mertensia maritima*.

Glochid a hooked bristle on seeds of some Boraginaceae, such as *Cynoglossum*.

Helophyte a Raunkiaer life form, a marsh plant, such as *Myosotis scorpioides*.

Hemicryptophyte a Raunkiaer life form, a plant with its perennating buds at soil level.

Herb botanically, any non-woody plant, but in horticulture is commonly used to refer to culinary and medicinal herbs.

Herbaceous non-woody.

Hydrophyte a Raunkiaer life form, a water plant, with its perennating buds below soil level; hardly any Boraginaceae are hydrophytes, apart from perhaps *Myosotis scorpioides*.

Inflorescence portion of stem bearing flowers, often with bracts, and frequently branched.

Lanceolate usually referring to leaves, shaped like a 'lance'.

Leafmould organic material remaining after leaves have decayed, excellent ingredient for composts or mulching woodland plants.

Life form see Raunkiaer life forms.

Mulch material applied to soil surface for conserving moisture and for weed control, e.g. bark chippings, leafmould, gravel, plastic sheeting.

Nutlet the fruit in Boraginaceae, a smallish dry structure which does not dehisce (split), containing only one seed.

Oblanceolate with the shape reverse to that of lanceolate, i.e. with the tip wider than the base.

Obovate with the shape reverse to that of oval, i.e. with the tip wider than the base.

Order taxonomic rank above the level of family.

Panicle a term generally applied to a branched inflorescence.

Pedicel the stalk of a single flower.

Pollination the transfer of pollen from an anther to a stigma.

Pollinator any animal (most commonly insect) that facilitates pollination.

Propagation the increase of plants either from seed, or vegetatively.

Raunkiaer life forms classification of plants according to the position of their perennating buds in relation to soil level.

Receptacle the upper part of the flowering stem from which the parts of the flower arise; it can be flat, concave or convex.

Rhizome an underground stem which is often horizontal, and may be fleshy or thin.

Scales (fornices) appendages found in the corolla throat of many Boraginaceae.

Scorpoid cyme see Cincinnus.

Slow-release fertilizer referring to fertilizers that release their nutrients in the soil over a long period of time.

Species a taxonomic rank below the level of family; this is the rank that defines particular, 'real' plants, not just taxonomic categories. Definition of species is a complex one, but basically it is a group of plants (or other organisms) that resemble each other closely and can interbreed between each other but not with other species.

Stamen the male organ of a flower, consisting of a filament and anther.

Stellate star-shaped, referring to mostly to the shape of hairs in some *Onosma* species.

Stratification treatment of seed that may involve sowing or storing them in a moist medium that is then subjected to a period of, usually, low temperature – commonly around 4°C (39°F) (or sometimes alternating warm and cold periods), to improve germination.

Subfamily a taxonomic rank below the level of family, but above the level of tribe.

Subspecies a taxonomic rank immediately below species, often geographically isolated from other populations of a species, but still able to interbreed with them.

Succulent applied to plants with thick, fleshy leaves or stems that are adapted for water storage; a few species of Boraginaceae are more or less succulent, e.g. *Myosotidium hortensia*, *Heliotropium curassavicum*.

Taproot a long fleshy root found in some Boraginaceae, which often makes transplanting difficult.

Therophyte a Raunkiaer plant form, referring to summer annuals – plants that spend the unfavourable season only as seeds.

Tribe an intermediate taxonomic rank above the level of genus, but below the level of subfamily.

Variety a colloquial term used for any plant variant, including cultivars and naturally occurring forms. Cultivated varieties should correctly be called cultivars.

APPENDIX IV

BIBLIOGRAPHY

ADAMSON, R.S. & SALTER, T.M. (eds). (1950). *Flora of the Cape Peninsula*. Juta & Co Ltd., Cape Town.

ALEKSEICHIK, N.I. & SANKO, V.A. (1994). *Gifts of forests, fields and meadows*. Fizkultura i sport, Moscow. pp.157–161. (in Russian).

ALLAN, H.H. (1961). *Flora of New Zealand: Volume 1*. R.E.Owen.

ALLARDICE, P. (1993). *A-Z of Companion Planting*. Cassell Publishers Ltd.

ALON, A. (1993). *300 wild flowers of Israel*. Steimatzky/Society for the Protection of Nature in Israel.

BEAN, W. (1981). *Trees and shrubs hardy in Great Britain*. John Murray.

BECKETT, K.A. (1993–94). *Encyclopedia of alpines. Vols.1&2*. Alpine Garden Society.

BIRD, R. 1993. *The propagation of hardy perennials*. Batsford.

BIRD, R. & KELLY, J. (1992). *The complete book of alpine gardening*. Ward Lock.

BLAMEY, M. & GREY-WILSON, C. (1989). *The illustrated flora of Britain and Northern Europe*. Hodder & Stoughton, pp.320–328.

BLAMEY, M. & GREY-WILSON, C. (1993). *Mediterranean wild flowers*. HarperCollins, pp.186–191, 385–387.

BOWN, D. (1995). *The RHS Encyclopedia of herbs & their uses*. Dorling Kindersley.

BOYD, A. E. (accessed 2002). *Geographic variation in morphology and pollinator taxa in Macromeria viridiflora* (University of Arizona, Tucson) at www.ou.edu/cas/botany-micro/botany2000/section13/abstracts/35.shtml

BRAMWELL, D. & BRAMWELL, Z. (1974), *Wild flowers of the Canary Islands*. Stanley Thornes Publishers Ltd, Cheltenham, pp.181–186.

BRICKELL, C. (ed.) (1996). *The RHS A-Z encyclopedia of garden plants*. Dorling Kindersley.

BRUMMIT, R.K. (1992). *Vascular plant families and genera*. Royal Botanic Gardens, Kew.

BUCZACKI, S.T. & HARRIS, K.M. (1991). *Collins guide to the pests, diseases and disorders of garden plants*. Collins.

CHOPRA, R. N., NAYAR S, L. & CHOPRA, I. C. (1986). *Glossary of Indian Medicinal Plants*. Council of Scientific and Industrial Research, New Delhi.

CLAPHAM, A.R., TUTIN, T.G. & WARBURG, E.F. (1952). *Flora of the British Isles*. Cambridge University Press, pp.829–846.

CLARKE, G. & TOOGOOD, A. (1992). *The complete book of plant propagation*. Ward Lock.

COLORADO NATIVE PLANT SOCIETY (1997). *Rare plants of Colorado. 2nd Edition*. Falcon Press/The Rocky Mountain Nature Association in cooperation with the Colorado Native Plant Society.

COOMBES, A.J. (1990). *Dictionary of plant names*. Collingridge.

CORREL, D.S. & JOHNSTON, M.C. (1970). *Manual of the vascular plants of Texas*. Texas Research Foundation.

DAVIS, P.H. (ed.) (1978). *Flora of Turkey: vol.6*. Edinburgh University Press.

DEAN, J. (1999). *Wild Colour*. Mitchell Beazley, p.71.

DON, D. (1832). *Characters and affinities of certain genera in the Flora Peruviana*. Edinb.N.Phil.Journ., Jul–Oct, p.239.

DUKE, J.A. & AYENSHU, E.S. (1985). *Medicinal Plants of China*. Reference Publications, Inc.

DYER, R.A., VERDOORN, I.C. & CODD, L.E. (1962) *Wild flowers of the Transvaal*. Trustees, Wild Flowers of the Transvaal Book Fund.

EDSON, J.L., LEEGE-BRUSVEN, A.D., EVERETT, R.L. & WENNY, D.L. (1996). Minimizing growth regulators in shoot culture of an endangered plant, Hackelia venusta (Boraginaceae). In Vitro Cellular and Developmental Biology, 32(4), pp.267–271.

ELLIOT, J. (1996). Recent Arrivals. The Hardy Plant, Vol.18, no 2, pp.19–25.

FACCIOLA, S. (1990). *Cornucopia – A Source Book of Edible Plants*. Kampong Publications.

FEINBRUN-DOTHAN, N. (1978). *Flora Palaestina. Part three: Ericaceae to Compositae*. The Israel Academy of Sciences and Humanities.

FISH, M. (1961). *Cottage garden flowers*. W.H. & L. Collingridge Ltd. (Reissued by B T Batsford, 2001)

FLORA OF CHINA CHECKLIST (2001). Hosted by Harvard University Herbaria at hua.huh.harvard.edu/china/

GENDERS, R. (1994). *Scented flora of the world*. Robert Hale.

GILLETT, H.J. & WALTER, K.S. (1998). *1997 Red List of Threatened Plants*. IUCN.

GODFREY, G.L. & CRABTREE, L. 1986(1987). Natural history of Gnophaela latipennis Boisduval Arctiidae, Pericopinae in northern California USA. J Lepid. Soc., 40(3): 206–213.

GREY-WILSON, C. (ed.) (1989). *A manual of alpine and rock garden plants*. Christopher Helm/Timber Press.

GRIFFITH, A.N. (1964). *Collins guide to alpines*. Collins.

HALLIWELL, B. (1992). *The propagation of alpine plants and dwarf bulbs*. Batsford.

HARKNESS, M.G. (1993). *The Bernard E. Harkness seedlist handbook.* Batsford.

HAY, R. (1955). *Annuals.* The Garden Book Club.

HEATH, R. (1964). *Collector's alpines.* Collingridge.

HEIMS, D. (1996). *Pulmonarias in America.* Garden Web at www.gardenweb.com/cyberplt/people/pulmonar.html

HEWITT, J. (1995). Pulmonaria 'Dawn Star'. The Hardy Plant, vol.17, No.1., p.10

HEWITT, J. (1994). Pulmonarias. The Hardy Plant Society.

HICKEY, M. & KING, C. (1988). *100 families of flowering plants.* Cambridge University Press.

*** (1981) *Hillier's manual of trees and shrubs.* David & Charles.

HITCHCOCK, C.L., CRONQUIST, A., OWNBEY, M. & THOMPSON, J.W. (1955). *Vascular plants of the Pacific North-west: Part 4.* University of Washington Press.

HUXLEY, A. (ed.) (1992). *The New Royal Horticultural Society Dictionary of Gardening.* Macmillan.

INGWERSEN, W. (1991). *Manual of alpine plants.* Cassell.

JOHNSTON, I.M. (1954). Studies in the Boraginaceae, XXVI. Journal of the Arnold Arboretum, 35(1), pp.1–12

KOSHEYEV, A.K. & KOSHEYEV, A.A. (1994). *Edible wild plants.* Kolos. (in Russian)

KRAEMER, M. & SCHMITT, U. (1997). Nectar production patterns and pollination of the Canarian endemic *Echium wildpretii* Pearson ex Hool. fil. Flora, 192: pp.217–221.

LESLIE, A. (1997). Plant introductions from Monksilver Nursery. The Hardy Plant, vol.19, no.2, pp.2–8.

LEWIS, W.H. & AVIOLI, L.V. (1991). "Leaves of Ehretia cymosa (Boraginaceae) used to heal fractures in Ghana increase bone remodeling." Economic Botany 45: pp.281–282.

LIPSCOMBE Vincett, B.A. (1977). *Wild flowers of central Saudi Arabia.* SILPI S.R.L. – Milano.

LORD, T. (ed.) (2002). *The RHS Plant Finder 2002–2003.* Dorling Kindersley.

LUSBY, P. & WRIGHT, J. (1996). *Scottish wild plants: Their history, ecology and conservation.* RBGE/The Stationery Office Ltd.

MABBERLEY, D.J. (1997). *The Plant-Book. 2nd edition.* Cambridge University Press.

Index Kewensis.

MABEY, R. (1997). *Flora Britannica.* Chatto & Windus, pp.306–311.

MALKMUS, B. (2002). Endemic plants of the Canary Islands at www.rareplants.de/plants/seeds/canary_islands_endemics.htm

MALYSHEV, L.I. (ed.) (1997). *Flora of Siberia.* Vol.11: Pyrolaceae-Lamiaceae. Nauka, Novosibirsk, pp.99–156 (in Russian)

MANSFIELD, T.C. (1949). *Annuals in colour and cultivation.* Collins.

MATHEW, B. Pulmonaria in Gardens, The Plantsman 4(2): pp.100–111 (1982).

MAUSETH, J.D. (1988). *Plant Anatomy.* The Benjamin/Cummings Publishing Company.

MOERMAN, D. (1998). *Native American ethnobotany.* Timber Press.

NABIYEV, M.M. (ed.) (1986). *Flora of Central Asia: Vol. VIII.* Pp.85–167. Fan, Tashkent.

OBERRATH, R., ZANKE, C. & BOHNINGGAESE, K.(1995) *Triggering and ecological significance of floral colour-change in lungwort* (Pulmonaria *spec.)* Flora, 1995, 190(2), pp.155–159

OHWI, J. (1965). *Flora of Japan.* Smithsonian Institution.

OLSEN, O.P. (1999). Some Sino-Himalayan plants in Arctic Norway. Rock Garden, vol. 26, part. 2, no.103, p.134.

PHILLIPS, R. & RIX, M. (1991a). *Perennials Vol.1: Early Perennials.* Pan Books Ltd.

PHILLIPS, R. & RIX, M. (1991b). *Perennials: Vol. 2: Late Perennials.* Pan Books Ltd.

PHILLIPS, R. & RIX, M. (1997). *Conservatory and indoor plants. Vol.2.* Macmillan.

PHILLIPS, R. & RIX, M. (1999). *Annuals and biennials.* Macmillan.

PHILIPSON, W.R. & HEARN, D. (1965). *Rock garden plants of the Southern Alps.* Merlin Press, pp.41–47.

PLANTS FOR A FUTURE DATABASE (PFAF) (1992–2002). at www.scs.leeds.ac.uk/pfaf

THE PLANT NAMES PROJECT (1999). International Plant Names Index at www.ipni.org.

POLUNIN, O. (1987). *Flowers of Greece and the Balkans: a field guide.* Oxford University Press, pp. 376–386.

POLUNIN, O. & STAINTON, A. (1997). *Flowers of the Himalaya.* Oxford University Press, pp.276–283.

POPOV, A.P. (1992). *Healing plants of the woodland.* Ecologia. (in Russian)

PRESTON, C.D. & CROFT, J.M. (1997). *Aquatic plants in Britain and Ireland.* Harley Books, pp.126–127.

POLUNIN, O. (1980). *Flowers of Greece and the Balkans.* Oxford University Press.

POLUNIN, O. & STAINTON, A. (1984). *Flowers of the Himalaya.* Oxford University Press.

PULMONARIA GROUP Newsletters.

RANDUSHKA, D., SHOMSHAK, L. & GABERO-VA, I. (1990). *Colour atlas of plants.* Obzor. (In Russian)

RECHINGER, K.H. (1964). *Flora of lowland Iraq.* Stuttgart, pp.492–513.

RICE, G. (1999). *Discovering annuals.* Frances Lincoln.

RICE, G. & STRANGMAN, E. (1993). *The Gardener's Guide to growing hellebores.* David & Charles.

SHARMAN, J. (1998). *A breath of fresh air.* Gardens Illustrated, Feb–Mar, pp. 60–67.

SLABY, P. 2001 Rock Garden Plants Database. at web.kadel.cz/flora/kvInfo.html

ROBINSON, W. (1898). *The English Flower Garden. 6th edi.* John Murray.

ROYAL HORTICULTURAL SOCIETY (2000). *Award of Garden Merit Plants 2000.* RHS.

RYDBERG, P.A. (1954) *Flora of the Rocky Mountains and adjacent plains.* Hafner Publishing Co.

SHISHKIN, B.K. & BOBROV, E.G. (eds.) (1967–77). *Flora of the USSR (Flora SSSR)* (English Translation). Israel Program for Scientific Translations, Jerusalem.

SHPILENYA, S.E. & IVANOV, S.I. (1989). *Nature's Alphabet: Medicinal Plants.* Znaniye. (in Russian).

SINGH, Dr. G. & KACHROO, Prof. Dr. P. (1976). *Forest Flora of Srinagar.* Bishen Singh Mahendra Pal Singh.

STACE, C. (1991). *New Flora of the British Isles.* Cambridge University Press, pp.640–656.

STEARN, W.T. (1993). The gender of the generic name *Onosma (Boraginaceae).* Taxon, 42(3), pp.679–681.

SWINDELLS, P. & MASON, D. (1992). *The complete book of the water garden.* Ward Lock.

THOMAS, G.S. (1992). *Ornamental shrubs, climbers and bamboos.* John Murray.

THOMAS, G.S. (1993). *Perennial garden plants, or, the modern florilegium.* 3rd ed. J.M.Dent & Sons with The Royal Horticultural Society.

THOMAS, G.S.(1990). *Plants for ground-cover.* J.M.Dent & Sons.

THOMPSON, P. (1992). *Creative propagation: A grower's guide.* B.T.Batsford.

TREHANE, R.P. et al. (1995). *International Code of Nomenclature for Cultivated Plants – 1995.* Quarterjack Publishing.

TUTIN, T.G., HEYWOOD, V.H., BURGES, N.A., MOORE, D.M., VALENTINE, D.H.,

USDA, NRGS (1999). The Plants Database. National Plant Data Centre, Baton Rouge, LA 70874-4490, US at plants.usda.gov/

USHER, G. (1974). *A dictionary of plants used by man.* Constable.

VAN WYK, B. & MALAN, S. (1997). *Field guide to the wild flowers of the Highveld.* Struik.

VVEDENSKY, A.I. (ed.) (1962). Flora of Kyrghyz SSR. Academy of Sciences of Kyrghyz SSR, Frunze. pp.20–139.

WALTERS, S.M. (ed.) *et al.* (2000). *The European Garden Flora. Vol.6.* Cambridge University Press.

WALTERS, S.M., & WEBB, D.A. (1972.) *Flora Europaea. Vol.3: Diapensiaceae to Myoporaceae.* Cambridge University Press, pp. 83–122.

WEBBER, W.A. (1988). *Colorado Flora: Western Slope.* Colorado Associated University Press, pp.157–166.

WEBBER, W.A. & WITTMANN, R.C. (1996). *Colorado Flora: Eastern Slope.* University Press of Colorado, pp.117–122.

WILLIAMS, L.O. (1937). *A monograph of the genus Mertensia in North America.* Annals of the Missouri Botanical Garden 14: 17159.

YEUNG. Him-Che. (1985). *Handbook of Chinese Herbs and Formulas.* Institute of Chinese Medicine, LA.

Catalogues

BETH CHATTO LTD catalogue

CHILTERN SEED catalogue

COTSWOLD GARDEN FLOWERS catalogue

MONKSILVER NURSERY catalogue

STILLINGFLEET LODGE NURSERIES catalogue

TERRA NOVA NURSERIES catalogue (online)

THOMPSON & MORGAN catalogue

APPENDIX V

HARDINESS ZONES

One of the first considerations when choosing a plant is whether it is hardy in your area. The United States Department of Agriculture (USDA) has developed a system of temperature zones as a basis for assessing which plants can be grown in different areas. The zones are based on the annual average minimum temperature in an area, and are illustrated on the maps on the next few pages of Europe, North America, Australia, New Zealand and South Africa. The maps have been divided into the USDA climatic zones, numbered from Zone 1, the coldest, with a winter minimum of -50°F (-44°C), up to Zone 11, the warmest, with a winter minimum of +40°F (+5°C). Every species entry in Chapters 1 and 2 of this book cites the plant's hardiness zone.

To establish whether a perennial will be hardy in your garden, refer to the map of hardiness zones and find the rating for your area. A plant with a zonal rating equal to or lower than the rating for your area should be hardy in your garden. If the species you want to grow is not completely hardy in your area, however, not all is lost. If, for example, the plant of your choice is hardy to Zone 8 and you live in Zone 7, it may still be possible to cutivate it outdoors, provided it is given some sort of protection. This could be a deep mulch, a cover of straw, bracken or woven polypropylene, or perhaps a cloche.

Another consideration is that every garden has a number of microclimates – that is, some parts of the garden are warmer than others. It may be that the zonal rating for your area does not apply to every part of your garden. So if your garden is rated Zone 7, the warmest corner, for example at the foot of a south-facing wall, may well be Zone 8. The only way to find out is to experiment by growing Zone 8-rated plants in that site.

Plants originating in cold climates, such as those of boreal and cold temperate regions, will not grow well in areas warmer than Zone 8, especially in dry conditions. The chances of success with such plants in warm areas can be increased by choosing the coolest, semi-shaded spot in the garden and not allowing the soil to dry out.

Average minimum winter temperatures
ZONE 1: below -50°F (below -45°C)
ZONE 2: -50 to -35°F (-45 to -37°C)
ZONE 3: -35 to -20°F (-37 to 29°C)
ZONE 4: -20 to -10°F (-29 to 23°C)
ZONE 5: -10 to -5°F (-23 to -21°C)
ZONE 6: -5 to -5°F (-21 to -15°C)
ZONE 7: 5 to 10°F (-15 to -12°C)
ZONE 8: 10 to 20°F (-12 to -7°C)
ZONE 9: 20 to 30°F (-7 to -1°C)
ZONE 10: 30 to 40°F (-1 to -4°C)
ZONE 11: above +40°C (above +5°C)

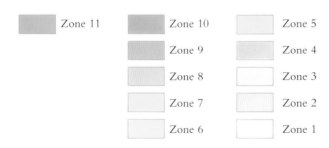

Zone 11	Zone 10	Zone 5
	Zone 9	Zone 4
	Zone 8	Zone 3
	Zone 7	Zone 2
	Zone 6	Zone 1

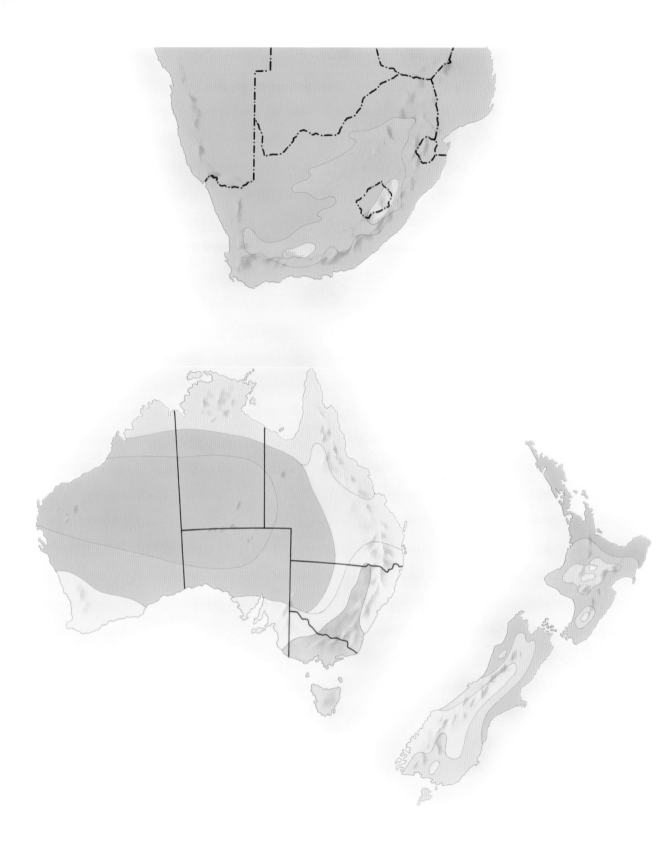

APPENDIX VI

COMMON NAMES IN ENGLISH

Alkanet see *Anchusa officinalis*
Alkanet, Dyer's see *Alkanna tinctoria*
Alkanet, Green see *Pentaglottis sempervirens*
Alkanet, Large Blue see *Anchusa azurea*
Alkanet, Oriental see *Alkanna orientalis*
Alkanet, True see *Anchusa officinalis*
Alkanet, Undulate see *Anchusa undulata*
Anacua see *Ehretia anacua*
Anaqua see *Ehretia anacua*
Arabian Primrose see *Arnebia cornuta*
Azores Forget-me-not see *Myosotis azorica*
Beggar's lice see *Hackelia* spp.
Bethlehem lungwort see *Pulmonaria saccharata*
Bethlehem Sage see *Pulmonaria rubra*
Bindweed Heliotrope see *Heliotropium
    convolvulaceum*
Bluebells see *Mertensia* spp.
Bluebells, Virginian see *Mertensia pulmonarioides*
Blue-eyed Betty see *Omphalodes cappadocica*
Blue-eyed Mary see *Omphalodes verna*
Blue Hound's-tongue see *Cynoglossum creticum*
Blue-weed see *Echium vulgare*
Borage see *Borago officinalis*
Borage, Eastern see *Trachystemon orientalis*
Bugloss see *Anchusa arvensis (Lycopsis arvensis)*
Bugloss, Pale see *Echium italicum*
Bugloss, Purple Viper's see *Echium plantagineum*
Bugloss, Sea see *Mertensia maritima*
Bugloss, Viper's see *Echium vulgare*
Bugloss, White-leaved see *Echium albicans*
Bulbous Comfrey see *Symphytum bulbosum*
Bur Forget-me-not see *Lappula squarrosa*
Camel bush see *Cynoglossum zeylanicum*
Cherry Pie see *Heliotropium x hybridum*
Chiming bells see *Mertensia ciliata*
Chinese Forget-me-not see *Cynoglossum amabile*
Chinese Hound's-tongue see *Cynoglossum amabile*
Chocolate-creams see *Trichodesma physaloides*
Christmas Cowslip see *Pulmonaria rubra*
Comfrey see *Symphytum* spp.
Comfrey, Bulbous see *Symphytum bulbosa*
Comfrey, Common see *Symphytum officinale*
Comfrey, Tuberous see *Symphytum tuberosum*
Comfrey, Turkish see *Symphytum ottomanum*
Corn Gromwell see *Lithospermum arvensis*
Cowslip, Christmas see *Pulmonaria rubra*
Cowslip, Jerusalem see *Pulmonaria officinalis*

Cowslip, Virginian see *Mertensia pulmonarioides*
Cryptantha, Paradox Valley see *Cryptantha paradoxa*
Eastern Borage see *Trachystemon orientalis*
Eight-day-healing-bush see *Lobostemon fruticosum*
Fiddleneck see *Amsinkia* spp.
Forget-me-not, Alpine see *Myosotis alpestris or
    Eritrichium* spp.
Forget-me-not, Bur see *Lappula squarrosa*
Forget-me-not, Chatham Island see *Myosotidium
    hortensia*
Forget-me-not, Changing see *Myosotis discolor*
Forget-me-not, Chinese see *Cynoglossum amabile*
Forget-me-not, Creeping see *Omphalodes verna*
Forget-me-not, Early see *Myosotis ramosissima*
Forget-me-not, Field see *Myosotis arvensis*
Forget-me-not, Giant see *Myosotidium hortensia*
Forget-me-not, Water see *Myosotis scorpioides*
Forget-me-not, White see *Cryptantha* spp.
Forget-me-not, Yellow and Blue see *Myosotis discolor*
Fukien Tea see *Ehretia buxifolia*
Giant Forget-me-not see *Myosotidium hortensia*
Giant Trumpets see *Macromeria viridiflora*
Giant Viper's-bugloss see *Echium piniana*
Golden Drop see *Onosma* spp.
Green Alkanet see *Pentaglottis sempervirens*
Green Mertensia see *Mertensia viridis*
Green Puccoon see *Macromeria viridiflora*
Gromwell see *Lithospermum officinale*
Gromwell, Blue see *Buglossoides purpurocaerulea*
Gromwell, Corn see *Lithospermum arvensis*
Gromwell, Greek see *Lithodora zahnii*
Gromwell, Purple see *Buglossoides purpurocaerulea*
Gromwell, Rosemary-leaved see *Lithodora
    rosmarinifolia*
Gromwell, Scrambling see *Lithodora diffusa*
Gromwell, Shrubby see *Lithodora fruticosa*
Heliotrope see *Heliotropium europaeum, Heliotropium*
    spp.
Heliotrope, Bindweed see *Heliotropium
    convolvulaceum*
Honeywort see *Cerinthe major*
Honeywort, Lesser see *Cerinthe minor*
Honeywort, Smooth see *Cerinthe glabra*
Honeywort, Violet see *Cerinthe retorta*
Hopi Smoke see *Macromeria viridiflora*
Hound's-tongue, Blue see *Cynoglossym creticum*
Hound's-tongue, Common see *Cynoglossum officinale*
Jerusalem Sage see *Pulmonaria officinalis or
    P. saccharata*
King of the Alps see *Eritrichium nanum*
Knock-away see *Ehretia anacua*
Kodo Wood see *Ehretia acuminata*
Languid-ladies see *Mertensia* spp.
Lungwort, Common see *Pulmonaria officinalis*

Lungwort, Narrow-leaved see *Pulmonaria longifolia*
Madwort see *Asperugo procumbens*
Manzanita see *Ehretia anacua*
Marbleseed, Soft-hair see *Onosmodium molle*
Mertensia, Alpine see *Mertensia alpina*
Mertensia, Foothill see *Mertensia lanceolata*
Mertensia, Green see *Mertensia viridis*
Mertensia, Lanceleaf see *Mertensia lanceolata*
Miner's candle see *Oreocarya virgata*
Monk's-wort, Yellow see *Nonea lutea*
Monk's-wort, Rose see *Nonea rosea*
Monk's-wort, Red see *Nonea vesicaria*
Nonea, Brown see *Nonea pulla*
Nomeolvides alpina (Spanish) see *Myosotis alpestris*
Oysterleaf see *Mertensia maritima*
Oysterplant see *Mertensia maritima*
Paterson's Curse see *Echium plantagineum*
Philippine Tea see *Ehretia buxifolia*
Popcorn Flowers see *Plagiobothrys nothofulvus*
Pride of Madeira see *Echium candicans*
Pride of Tenerife see *Echium simplex*
Primrose, Arabian see *Arnebia cornuta*
Prophet Flower see *Arnebia pulchra*
Puccoon see *Lithospermum* spp.
Puccoon, Green see *Macromeria viridiflora*
Puccoon, Many-flowered see *Lithospermum multiflorum*
Puccoon, Narrow-leaf see *Lithospermum incisum*
Puzzle Bush see *Ehretia rigida*
Sandpaper tree see *Ehretia anacua*
Soldiers and Sailors see *Pulmonaria saccharata*
Stickseed see *Hackelia* spp.
Stickseed, Blue see *Hackelia micrantha*
Stickseed, Many-flowered see *Hackelia floribunda*
Sugarberry see *Ehretia anacua*
Tall fringe bluebells see *Mertensia ciliata*
Tower of Jewels see *Echium wildprettii*
True Alkanet see *Anchusa officinalis*
Turkish Comfrey see *Symphytum ottomanum*

APPENDIX VII

EXISTING *PULMONARIA* CULTIVARS NOT DESCRIBED IN THIS BOOK

Below is the list of pulmonarias that, for various reasons, have not been described in this book. They may be new cultivars for which I could find little or no information, and some are varieties that have been introduced in the past but are no longer available.

'Ballyrogan Blue'
'Blue Haze'
'Blueberry Muffin'
'Bofar Red'
'Botanic Hybrid'
'Buckland'
'Eleanor'
'Clent Skysilver'
'Glauei'
'High Contrast'
'Howard Eggins'
'Lady Lou's Pink'
'Lime Close'
'Marjory Lawley'
'Magenta'
'Melancholia'
'Middleton Red'
'Patrick Bates'
'Raspberry Ice'
'Pink Haze'
'Plas Meredyn'
'Silver Maid' (syn. 'Margaret Owen')
'Silver Shimmers'
'Silver Spring'
'Silver Surprise'
'Silverado'
'Skylight'
'Snow Queen'
'Snowy Owl'
'South Hayes'

INDEX

Boraginaceae